RECENT ADVANCES IN
ANIMAL NUTRITION — 2001

Recent Advances in Animal Nutrition

2001

P.C. Garnsworthy, PhD

J. Wiseman, PhD

University of Nottingham

NOTTINGHAM
University Press

Nottingham University Press
Manor Farm, Main Street, Thrumpton
Nottingham, NG11 0AX, United Kingdom

NOTTINGHAM

First published 2001
© The several contributors names in the list of contents 2001

British Library Cataloguing in Publication Data
Recent Advances in Animal Nutrition — 2001:
University of Nottingham Feed Manufacturers
Conference (35th, 2001, Nottingham)
I. Garnsworthy, Philip C. II. Wiseman, J.

ISBN   1-897676-08-5

Typeset by Nottingham University Press, Nottingham
Printed and bound by Redwood Books, Trowbridge, Wiltshire

PREFACE

The 35th University of Nottingham Feed Conference was held at the University of Nottingham Sutton Bonington Campus 3-5th January 2001. As in previous years, the Organising Committee (made up of members of the Industry and University personnel) had detailed meetings to arrange a programme that was informative and topical. One major innovation this year was to increase the number of presentations.

A regular feature in recent years has been a paper on legislative developments; it is of course essential that the Industry is kept fully informed of this increasingly complex area and B Cooke (with considerable assistance from J Nelson, UKASTA) was able to provide a comprehensive description of recent developments. The wider, and again increasingly important, issues of meeting consumer demands with regulatory requirements had been addressed in the first paper of the conference by P Thomas who, with a depth and breadth of experience in this area, was well-qualified to provide a stimulating paper. There was much discussion at the end of these two papers, indicating that the subjects are at the centre of the Industry's activities.

Subsequent papers addressed specific topics of interest, either within a species or general format. The nutrition of companion species is continuing to occupy the Industry and the first paper in the non-ruminant session, presented by A Rodiek, considered equine nutrition in the context of practical diet formulation. Scientific understanding of the dietary needs of horses is still very much in its infancy and any presentation that attempts to offer sound advice is to be welcomed. The adult sow, somewhat similar to the horse, is able to derive much nutritional benefit from high fibre diets. The conference has, in previous years, had papers with a more scientific basis and the presentation by R Duran on the more applied aspects of sow nutrition was a logical development, responding as it did to requests from the industry that discussion of practical developments would now be welcome.

Poultry meat consumption is increasing worldwide and, in many countries, is by far the most popular meat item. Strategies to improve performance and carcass quality were presented by J Wiseman and are of considerable value in improving the overall efficiency of poultry systems. With current moves towards complete removal of growth promoters from animal diets, two papers considered likely consequences. The first by K Hillman discussed likely effects on gut micro flora and the second by C Kamel considered the development of understanding of various naturally occurring products as alternatives. It is probable that this overall theme will be extended in future years.

A general session, based on the science into practice approach that has characterised the conference, discussed developments in organic farming in terms of the implications for the feed industry (N Lampkin) and practicalities for diet formulation (S. Wilson). This is another area receiving greater interest and will no doubt be considered in subsequent conferences. A theme that has characterised animal science recently is the integration of different disciplines. Nutrition and immunity are two such subjects; the basic scientific principles were discussed thoroughly by K Klasing and J Meijer was able to take the subject forward into practical application by considering pig performance.

The ruminant session covered a wide range of subjects. It started with a presentation covering various aspects of dairy cow feeding systems by integrating concentrate provision, grazing management and grass quality (J-L Peyraud). Developing a theme that had been considered at the 34th Conference, D Bauman gave a comprehensive overview of conjugated linoleic acid (a component of milk with wide-ranging benefits in terms of human health). The final part of the ruminant programme considered various aspects of rumen fermentation from the scientific (J Newbold) to the more applied (L Kung) and specific means of altering its pattern through use of exogenous enzymes (K Beauchemin).

The overall conference was a lively event; there were heated discussions following papers / sessions and workshops (a new initiative in which speakers and delegates could meet outside the more formal presentations). Debate was aided considerably by the session chairmen, to whom we are most grateful (W Knock, M Bedford, J. Twigge, S. Papasolomontos). Finally, it always a pleasure to thank those people whose 'behind the scenes' activities ensured the success of the conference administration (S Golds, Nottingham postgraduates, Catering staff). Trouw nutrition provided generous financial support.

J. Wiseman
P.C. Garnsworthy

CONTENTS

1

MEETING REGULATORY REQUIREMENTS AND CONSUMER DEMANDS

P.C.THOMAS
Artilus, Edinburgh EH14 5AJ

Introduction

In this chapter I want to consider factors influencing the development of regulatory requirements and consumer demands on the animal feed industry in the UK. This will touch on EU issues as well as on some specific areas of legislation. However, the intention is not to provide a review of recent legislative developments - that is the subject of the next chapter. Rather the purpose here is to give an overview of the legislative and consumer environment, to consider its implications and to draw out some pointers for the future.

The topics considered and the sources of material referred to are selected to highlight some underlying themes rather than to be comprehensive. The views expressed are my own and do not necessarily reflect those of any organisation or body with which I am associated.

The Food Chain

BACKGROUND TRENDS

Food is important. Not only does it provide people with nutritional sustenance it is also a defining feature in national and regional cultures and societies; it is part of our social and cultural heritage.

Food is also highly political, or more precisely food supply and food policy are highly political matters. Historically, religious differences and food shortages have vied with each other as the main causes of public ferment. Very few countries have maintained stable governments in the face of major disruptions in food supply, and even quite small perturbations can produce substantial public reaction.

In many developing countries security of food supply is still a major preoccupation, and simply ensuring that the population has a varied and nutritionally satisfying diet is a substantial challenge. But in advanced developed countries, like the UK, food security is no longer an issue. For all but the poorest in society, a vast range and choice of food is available at affordable prices. Indeed food expenditure in the UK has consistently declined as a proportion of household budgets for over thirty years (Figure 1.1).

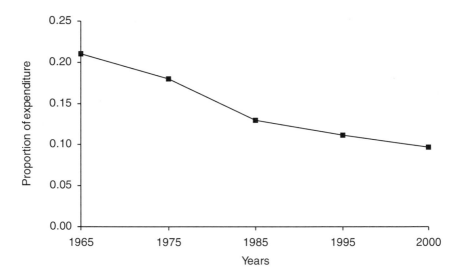

Figure 1.1 Expenditure on food as a proportion of total household expenditure in the UK. (Figures are extracted from MAFF statistical records).

Alongside the increased range and reduced cost of food, the UK food supply chain and consumers' purchasing and consumption patterns have grown infinitely more complex and sophisticated. Systems of local food production, local manufacturing and corner-shop retailing that would have been typical forty years ago have all but disappeared. Food supply chains, from production through processing, manufacturing and retailing to the consumer, have increased enormously in length, in many cases becoming global; manufacturing has become more and more dominated by international brands, and retailing by large multiple retailers.

The variety of packaged, partly-prepared and wholly prepared 'meal solutions' has also spiralled ever upwards, and eating habits and patterns have changed. The traditional emphasis on the family meal has declined; both within and outside the home, fast, convenient and individualised meals have become an accepted norm.

Most important of all, consumer choice has become the primary market driver, almost to the exclusion of all other considerations.

PERCEPTIONS OF SCALE

The evolution of the new food chain model is widely recognised not only amongst those who are professionally involved in the food business but also by the wider public. However, it is apparent from public debate of food and agricultural issues that there is a widespread lack of appreciation of the economic scale of the different parts of the food chain. Many people perceive agriculture, and by inference the feed manufacturing and agricultural supply industries, to be 'big businesses'; their size seems self-evident because the agricultural industry can be seen all around us.

However, a large part of the £15.5 billion annual economic activity of agriculture relates to trading within the industry itself. The net contribution of agriculture to the total national income is approximately £6.99 billion, and of that some 45% is accounted for by public funding from agricultural support schemes. In Table 1.1 these figures are shown alongside those for the livestock sector, the animal feed and pharmaceutical industries and the major UK food retailers. The figures illustrate that whilst agriculture and the supply industries may make an important contribution to UK food supply, they are very small in economic terms; the 'big businesses' are found very much further up the food supply chain.

Table1.1 Estimated economic size of UK agriculture industries and the main UK food retail companies.

Industry or company	Annual market turnover (£ billion)
Agriculture	6.99 (3.87)[1]
Livestock production	3.70 (2.05)[2]
Feed manufacturing	1.42
Non-manufactured feed supply	0.66
Livestock pharmaceuticals	0.16[3]
Food retail companies	
Tesco	18.80
Sainsbury	16.38
ASDA	8.20
Safeway	8.07
Somerfield	5.90
Morrison	2.97

1. Total agricultural turnover is *ca* £15.5 billion; net economic gross value is £6.99 billion; value in parenthesis is net economic gross value minus State funding, estimate in 1999 as £3.12 billion.
2. Total livestock turnover is *ca* £8.3 billion; values given are estimated net value and, in parenthesis, estimated net value minus State funding.
3. Pharmaceuticals refer to farm livestock sales. There are additional companion animal sales of £0.18 billion.

OPERATION OF THE FOOD CHAIN

There has been a recent focus of attention on the operation of the food chain and on its future development (MAFF, 1999a; Office of Science and Technology, 2000). This has placed emphasis on the different parts of the food chain working closer together and on finding better ways to communicate and respond to market signals. The Food Chain Group (MAFF, 1999a) identified what it considered to be needed under five main topic headings.

- **Enhance constructive dialogue between partners in the food chain.** This was seen as necessary to improve understanding of cost structures at each stage of the chain; and to better link production and marketing and drive down costs. The need to reconcile long-term food chain objectives with short-term business priorities was highlighted.

- **Better communicate with consumers, with both direct, targeted information and more informed media coverage.** The identified need was to improve consumer understanding of the economic and environmental realities of production, manufacturing and marketing; and to communicate the concepts of risk and the impact of technology on food, the environment and the countryside.

- **Strengthen consumer confidence through improved product labelling and dissemination of information at all points in the chain.** This recognised the requirement for simple and coherent labelling of foods, and for the information necessary to meet consumers' needs.

- **Better understanding of consumer concerns, going beyond product marketing to welfare and environmental concerns.** Here there was a perceived opportunity to add value to products through emphasising high welfare and environmental standards.

- **Improved Government Policies.** This was a call for 'joined up' Government and for public policies recognising the need to promote home food production.

These are objectives to which all parts of the food chain should be committed. But the inherent tensions between the different parts of the chain are indicated by the reference to the need to 'reconcile long-term food chain objectives with short-term business priorities'. This underscores the fact that each part of the food chain is subject to its own business pressures which, in the short term, may not always be in harmony.

Probably the most significant factor in this is that retailers have well-advertised objectives of providing consumers with ever increasing quality at ever reducing price. However, at the same time they must seek to maintain retail profit margins and shareholder dividends. Leaving aside whether these somewhat competing objectives are sustainable in the long term, it is obvious that the retailers' objectives will inevitably create business pressures further down the food supply chain.

For the feed manufacturers this is manifest in a variety of ways, the most common of which is the difficulty of responding to a combination of legislative requirements and divergent market signals (Figure 1.2). Typically, these situations involve increased feed specifications or manufacturing controls coupled with pressure for reduced prices i.e. the feed manufacturer experiences the market pressure of the retailers' commitment to the food consumer.

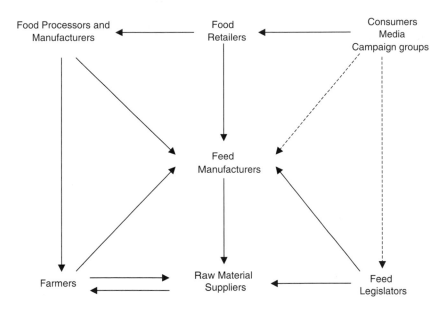

Figure 1.2 A schematic representation of market signals and legislative requirements affecting feed manufacturing businesses.

Legislation

DEVELOPMENT OF LEGISLATION

Feed legislation in the UK arises through the combined influences of international, EU and UK market considerations. However, because of the different time scales of the legislative processes, the linkages between the different tiers of activity are not always immediately apparent. However, EU legislation is increasingly taking account of World Trade Organisation (WTO) and Codex Alimentarius considerations, and is being based on 'centralised procedures', which remove the need for the legislation to be transposed into Member State law. Likewise, UK legislation is being modified to facilitate the direct transfer of new EU legislation (for example see, Feedingstuffs Regulations 2000 (MAFF 1999b)).

In the UK, recent years have also seen a general change in approach to policy development and legislation. Traditionally, administrations of all political persuasions were rather autocratic and developed policy and legislation through what has been described as a 'decide, announce and defend model'. However, modern government is much more consensual and now operates through what has been described as an 'ask, listen and seek public acceptance model'.

From a democratic standpoint the new approach has obvious advantages in more fully involving the wider community in decision-making. However, it makes the tacit assumption that point-in-time estimates of public acceptance, based on consultation and media indicators, are reliable measures of the 'view of the people', and that assumption remains to be substantiated.

There have also been important changes in the UK institutions with roles and responsibilities in feed legislation, in particular changes arising from the devolution of government and the establishment of the UK Food Standards Agency (FSA).

DEVOLUTION

When the New Labour Government came to power in 1997, it was committed to devolving certain powers from the House of Commons to a Parliament in Scotland and Assemblies in Wales and Northern Ireland. By 1999 the relevant constitutional changes had been made and the institutions established.

Devolution of itself is no extraordinary thing since many countries throughout the world operate federal government systems. However, most models of devolved government have been designed and established as symmetrically structured parliamentary systems, whereas this is not the case in the UK.

In a sense, the UK is still developing its devolution model, and what has been implemented can be seen as an interim stage. What exists now is a hybrid system. Varying degrees of power for certain areas of legislation have been devolved to Scotland, Wales and Northern Ireland, but the House of Commons has reserved powers for certain areas of UK government and for all areas of England's government.

The nuances of this arcane arrangement are potentially quite significant. Many areas of technical and secondary legislation, including those related to animal feed are now dealt with separately through the House of Commons, Scottish Parliament and Welsh and Irish Assemblies and, for a number of reasons, the outcome of these legislative processes may not necessarily be identical.

For instance, the business procedures and the Committee structures of the Scottish Parliament provide greater opportunity for public petition than is the case in the House of Commons. Conversely, the Scotland Act (1999), which established the Scottish Parliament, leaves less flexibility for Scottish legislation to vary from EU legislation than is allowed for in England under the House of Commons arrangements.

FOOD STANDARDS AGENCY

Background

The Food Standards Agency (FSA) has its origins in a discussion paper produced by Professor Philip James the former Director of the Rowett Research Institute in May 1997. Against a background of growing public unease about 'food scares' and about Bovine Spongiform Encephalopathy in particular, proposals were made for a new independent agency that would have a central role in food policy and enforcement covering the whole of the food chain *from farm to fork*.

However, it was not until after the White Paper *The Food Standards Agency: a Force for Change* published in 1998 and an unprecedented period of public consultation, that the overall responsibilities, status and structure of the FSA was agreed. It finally became operational on the 3 April 2000.

Aims and values

The FSA represents a new approach in UK public administration, deriving from a painstaking assessment of what would be needed to create a regulatory body operating at arm's length from Government but with sufficient power and authority to oversee the regulation of the food chain. The FSA was deliberately designed to be transparent in its approach and to create, or perhaps more precisely to restore, public confidence in the UK food supply.

From its first day of operation the FSA has unequivocally expressed its aim *to protect public health from risks which may arise in connection with the consumption of food, and otherwise to protect the interests of consumers in relation to food*. Likewise it has published and promoted its core values:

- to put the consumer first;
- to be open and accessible;
- to be an independent voice.

It has gone to great lengths to establish its openness and accessibility. Board meetings, and related regional Advisory Committee meetings in Scotland, Wales and Northern Ireland are all held in public, and there are opportunities at each meeting for the public to question Board and Committee members. Likewise, there is an extensive use of the Internet to provide open access to the Minutes and working papers of the Board, Regional Committees and Scientific Advisory Committees. As an innovation it is piloting public consultation via e-mailed comments, which are placed on a bulletin board for general public consideration.

In short, the FSA is making every effort to consult widely and to provide consumers with opportunities to comment fully on matters about which they have concerns.

Scope of FSA

The Headquarters of the FSA is in London, but there are executive branches in Scotland, Wales and Northern Ireland. These ensure that FSA policies and approaches are applicable throughout the country and provide expert resources locally for the Scottish Parliament and the Welsh and Northern Irish Assemblies.

The structure of the Headquarters Science Groups (Table 1.2) provides an indication of the scope of the FSA's responsibilities, which include support for both policy development and regulatory enforcement. In the latter functions, the FSA works closely with the Local Authorities (Environmental Health and Trading Standards) and with specialist Agencies such as the Meat Hygiene Service.

Table 1.2 Structure of the Science Divisions of the FSA Headquarters.[1,2]

Food safety policy divisions and units	Enforcement and food standards divisions and units
Additives and Novel Foods	Food Labelling Standards and Consumer Protection
Animal Feed	Local Enforcement (Policy)
Chemical Safety and Toxicology	Local Enforcement (Support)
Contaminants	Meat Hygiene
Food Chain Strategy	Veterinary Public Health Unit
Microbiological Safety	Food Emergencies Unit
Nutrition	
Research Co-ordination Unit	

1. Additionally, there is a Corporate Strategy Group, incorporating: Finance, Procurement and IT Division; Personnel and Establishment Division; Communications Division; and Legal Division.
2. The Meat Hygiene Service (MHS) is an Executive Agency of the Food Standards Agency

Many of the FSA's policy and statutory functions were formerly the domains of the Department of Health or Ministry of Agriculture, Fisheries and Food (MAFF). In some areas MAFF or agencies, such as the Veterinary Medicines Directorate (VMD) and Pesticides Safety Directorate (PSD), have continuing roles. However, in areas where food safety, food quality or consumer assurance are important considerations, the FSA is the 'competent authority'. Additionally, because of the way that policy specialists in animal nutrition have been clustered, the FSA has responsibility not only for livestock nutrition but also for the nutrition of horses and companion animals.

In its scientific operation, the FSA is able to call on the specialist capabilities of other Government Departments and Agencies, as well as drawing on the expertise of the Scientific Committees, which exist to provide advice (Table 1.3). Therefore, it is in an exceptionally good position to act in the co-ordination of science and legislative policy.

Table 1.3. Independent Advisory Committees.[1]

Advisory Committees	
Food Advisory Committee (FAC)	Expert Group on Vitamins and Minerals (EVM)
Advisory Committee on Novel Foods and Processes (ACNFP)	Committee on the Medical Aspects of Food and Nutrition (COMA)
Advisory Committee on Animal Feedingstuffs (ACAF)	Spongiform Encephalopathy Advisory Committee (SEAC)
Advisory Committee on the Microbiological Safety of Food (ACMSF)	Committee on the Carcinogenicity of Chemicals in Food, Consumer Products and the Environment (CoC)
Committee on the Toxicity of Chemicals in Food, Consumer Products and the Environment (CoT)	Committee on the Mutagenicity of Chemicals in Food, Consumer Products and the Environment (CoM)

[1] There are also Working Parties on: Chemical Contaminants in Food; Dietary Surveys; Nutrients in Food; Chemical Migration from Materials in Contact with Food; Food Additives; Food Authenticity; and Radionuclides in Food.

Functional Operation

In functional terms the main tasks of the FSA are to:

• provide advice and information to the public and to the Government on food safety from farm to fork, and on nutrition and diet;

- protect consumers through effective monitoring and enforcement; and
- support consumer choice through promoting accurate and meaningful labelling.

These tasks imply, and in some cases rely on, an underlying assessment of risk, and the FSA is committed to consulting widely across both the scientific and consumer communities, and to taking a precautionary approach where the information available is judged insufficient. Where there are consumer choices to be made, its policy is to ensure that the public has full and transparent information on which to base decisions. This approach has been widely welcomed, although it is recognised that it may create genuine challenges in public communication that have not been addressed before.

Wider FSA considerations

In many senses the FSA can be seen as part of a wider European development, which is reflected in the establishment of statutory food bodies in a number of EU Member States. Also the EU *White Paper on Food Safety* (Commission of the European Communities, 2000a) sets out proposals for an overarching European Food Authority, and there is the prospect of an EU network of food agencies linked with the central body.

At this stage it is difficult to know this regulatory model will operate. However, there is no question that it heralds an important political shift in food policy at both the EU and Member State level.

Consumer Demand

TERMINOLOGY

To the question, 'Who creates consumer demand?' the obvious answer is 'Consumers!'. However, in a modern society, that answer can only be regarded as simplistic. Advertising influences us all, although most of us resolutely insist we are immune to its effects. More significantly perhaps, we are constantly bombarded with information from the media - papers, magazines, books, television and radio, all of which influences our views about food. The media shapes our perceptions and understanding, whilst friends and family modulate the media messages, introducing their own perceptions on the topic of the day (see Figure 1.3).

By any measure, the factors influencing 'consumer demand' are complex, not least because the term is commonly used to describe two related but different

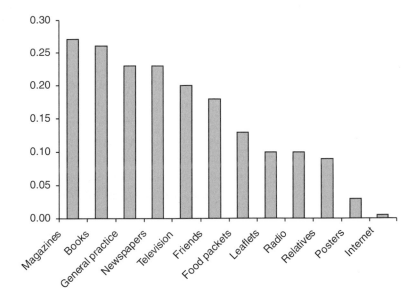

Figure 1.3 A summary of sources of information used for nutritional advice by consumers over 50-years old; proportion of survey group (n=205) who use the sources for nutritional advice. (After Nestle, 2000)

things. The first is the actual market demand for products, as measured by the sale of products at a given price. The second is the interpretation of consumers' views about products or about aspects of the food supply chain. Consumer demand, as reflected in sales, can be objectively measured; consumer demand, as reflected in interpretation of consumer views, is more subjective, and depends on the design and scale of market surveys and the correctness of their interpretation.

There are times when the two consumer demand concepts converge - when assessments of consumers' opinions can be demonstrably linked with measurable changes in market sales. But, there are also times when a 'demand', reflecting an interpretation of consumers' views, lacks supporting market evidence or the market evidence is open to a variety of interpretations.

In the remainder of this chapter 'consumer demand' is used to refer to estimated expressions of consumer views.

CONSUMER ATTITUDES

A visitor to Britain might well conclude that the British are very food conscious, with a great interest and concern for all aspects of food safety, food production and diet. Food issues have a high media profile and there is a very active range of food campaign organisations covering an almost complete spectrum of interests.

Many food campaign organisations are dedicated solely to food and consumer issues, and often have a strong interest in diet, nutrition and food quality. However, some of the most active food campaigners are environmental organisations, who are strong supporters of 'green' agriculture and organic farming and opposed to any practices or developments that can be construed as 'agricultural intensification'.

Notwithstanding the impression of a consuming British interest in food issues, market research indicates a rather different and more complex picture. In a recent study by the Institute of Grocery Distribution (IGD, 2000a) some 0.48 of the sample population expressed little interest in knowing more about the production of their food, whilst 0.42 had intentions of finding out more but rarely got around to it. Only about 0.11 of the sample group were keenly interested in knowing more about food production methods and were prepared to seek out information to improve their knowledge.

This result could be interpreted as providing evidence either of consumer confidence in the food supply or consumer apathy. The former interpretation is confirmed by other survey data which indicates that 0.76 of the population is 'relatively to very confident' about the safety of British food (IGD, 2000b) and also by 'focus group' studies. The latter show that when consumers go shopping their first considerations are convenience, price, value and quality, (meaning freshness and hygienic presentation); very few people have food safety issues on their minds (FSA, 2000a).

However, these results do not mean that consumers do not have views or concerns about food safety or production. Rather, it is that their views are elicited only when food safety or production issues are prompted through explicit questions. Consumers do not rank food concerns at the top of their list of worries (Figure 1.4) but where questions are specific and focused, consumer responses can be unequivocal.

This possibly explains such studies as the CWS 'Food Crimes' report (CWS, 2000) where, in contrast with the low consumer interest outlined above, a survey of Co-op shoppers showed high proportional responses to specific questions.

- 0.93 believed they had the right to know everything that had happened to their food;
- 0.84 were concerned about animals being treated badly and 0.61 wanted more information;
- 0.86 disapproved of legally approved animal products (blood, gelatine and tallow) being permitted in animal feeding;
- 0.87 disapproved of growth promoting antibiotics;
- 0.85 considered multinational companies had too much power over food supplies and 0.72 considered that global food production damaged the environment.

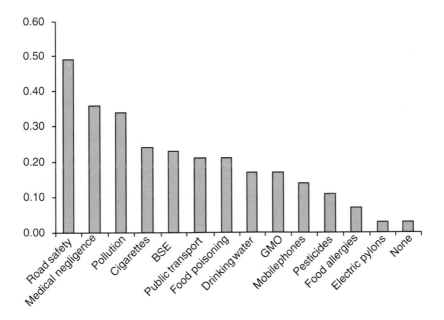

Figure 1.4 Consumer concerns about food safety and other issues; proportion of survey group (n=949) identifying concerns as in the top three concerns. (After IGD, 2000b).

IMPLICATIONS

The obvious message from this is that great care should be taken in interpreting surveys of 'consumer demand'. The indications are that the responses in survey studies may be heavily conditioned by the questions asked. Whilst surveys based on direct and specific questions may establish the consumers views on a particular food issue in isolation, they may give a misleading impression of the consumer's assessment of the importance of that specific issue in relation to a wider framework of concerns.

This highlights the dilemmas faced by the food retailers and legislators trying to respond to perceived 'consumer demand', and it may also help to explain the apparent polarisation of consumer views on various food and animal feed issues.

Food and Feed Issues

Whilst the public debate about food in the UK is wide ranging, it has focused particularly around a limited number of topic areas that have gained or retained a high media and consumer profile. Some of these, which have relevance to the feed industry, are considered in the following sections.

FOOD CONTAMINATION

Microbial contamination

UK consumers' initial concerns about food safety were caused by the dramatic rise in cases of food poisoning that began in the mid 1980s. At long last, this trend appears to have been contained (Figure 1.5). However, whilst there have been substantial reductions in the incidence of salmonella and listeria infections from their peak figures, there has been a substantial increase in reported isolations of Campylobacter species and verocytotoxic *E coli* O157 (Figure 1.6); these latter have implications for poultry producers and livestock farmers respectively.

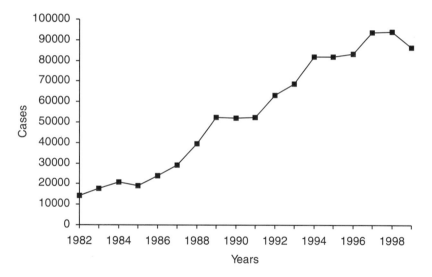

Figure 1.5 Total formally notified or otherwise ascertained cases of food poisoning in England and Wales (Data from Public Health Laboratory Service statistics).

The increase in *E coli* O157 is of particular concern for public health. Whilst the organism does not produce symptoms in livestock, its clinical manifestations in humans vary from mild diarrhoea to severe haemorrhagic colitis, haemolytic uraemia and death; severe symptoms occur most often in the young and the old. In 1996 it was responsible for Britain's most serious food poisoning outbreak when a single source of contamination led to 496 confirmed cases and 21 deaths in Central Scotland.

Although the organism is sometimes referred to as the 'burger bug' because of food poisoning incidences arising from ground beef, faecal contamination can affect vegetable crops and water supplies, so the organism can be transferred to humans on a wide range of foods. Recent evidence also shows that sporadic outbreaks of disease are probably due to environmental transfer rather than food consumption.

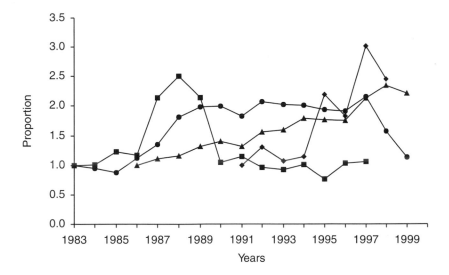

Figure 1.6 Trends in food poisoning bacteria: proportional increases in isolations of Salmonella (●), Listeria (■), Campylobacter (▲) and *E coli* 0157 (♦). (Data are taken from Public Health Laboratory Service statistics and are calculated as a proportion of the figures in a selected first year or in the starting, first year when reporting was begun. First year for Salmonella and Listeria is 1983; for Campylobacter is 1986; and for *E coli* 0157 is 1991).

The organism has a widespread occurrence, affecting an estimated 0.19 of cattle in England (Paiba, 2000) and 0.07- 0.10 of cattle (0.23 of herds) in Scotland (Synge, 2000). Other than a statistical association with barley diets there is no evidence that the occurrence of the organism is feed related. However, there is a greater faecal shedding of the organism in housed animals (Synge, 2000), which may lead to its indirect association with intensive livestock systems.

Chemical contamination

Chemical contamination of food or animal feed is not a frequent occurrence but where it does occur it requires emergency action, product recalls and a programme of public communication. In recent times faecal contamination of French animal feed, mineral oil contamination of Belgian feed and dioxin contamination of Belgian and German feeds have been reported. Whilst the animal feed impact of these problems was relatively localised, the food production impact was regional, and the media and consumer impact was EU-wide.

As a result of the Belgian oil incident, the EU has already acted to tighten regulatory controls on the use of recycled vegetable oil (RVO), and further controls on permitted feed ingredients are envisaged in the White Paper on Food Safety (Commission of the European Communities, 2000a). Revised regulations on the blending down of contaminants to below Maximum Permitted Limits (MPL) by approved manufacturers in authorised premises are also under consideration (see Advisory Committee on Animal Feedingstuffs, 2000).

ANTIBIOTICS

Over-use of antibiotics in human therapy, rather than the use of antibiotics in livestock production, is accepted as the main factor contributing to antibiotic resistance in human disease bacteria (see Advisory Committee on Microbiological Safety of Food, 1999; European Agency for Evaluation of Medicinal Products, 1999). However, there will be continued pressure to minimise antibiotic use in livestock both as a precaution against the development of antibiotic resistance and as a result of consumer concerns. Avoparcin, Tylosin, Spiramycin, Zinc Bacitracin, Virginiamycin, Carbadox and Olaquindox have already been removed from use as in-feed animal growth promoters, but other products and antibiotic uses will continue to be scrutinised.

Whilst antibiotic treatment of individual animals suffering from diagnosed disease is generally regarded as consumer-acceptable, metaphylaxis (where animals are treated on a group basis on first signs of disease) and prophylaxis are frequently-questioned practices. Whilst these treatments require veterinary supervision, the in-feed inclusion of medicines under a Medicated Feedingstuffs (MFS) prescription (Medicated Feedingstuffs Regulations, 1998) invites consumer comparisons with in-feed zootechnical products (Feedingstuffs (Zootechnical Products) Regulations, 1998). The feed and livestock industries should be aware of this and should seek to ensure that retailers and consumer groups understand the controls on antibiotic use that apply.

BSE

Any belief that the publication of the BSE Inquiry (Phillips et al, 2000) and the continued reduction of BSE disease in domestic cattle would bring to a close the worse public health crisis in UK history should be set aside. The continued rise in the UK incidence of human vCJD and the development of BSE as a recognised problem in Continental Europe will ensure that 'mad cow disease' remains a consumer concern for many years to come.

Many of the public health lessons from the BSE crisis, highlighted by Lord Phillips and his team, have already been incorporated into UK legislation. They are being implemented, but too slowly, in other European Member States, who until recently have tended to regard BSE as a 'British disease'. Given the linkage of the disease with Meat and Bone Meal (MBM), and the problems of cross contamination in feed manufacturing, the UK ban on MBM feeding to all livestock must be rigorously maintained. Moreover there are strong public health arguments for its extension to countries outside the EU. No one in the world should disregard the danger of sporadic BSE in cattle and the consequences of its inadvertent spread through the food chain.

On the other hand, taking account of all precautions, there needs to be a careful judgement about the use of animal-derived materials both in the food chain and for other purposes. Some consumer groups would like to see a blanket ban, but that would not reflect a logical assessment of risk. Moreover it would have significant consequences on the supply of tissue-extracted chemicals, which have important consumer applications and would be difficult to replace.

GM FEEDS

The application of recombinant DNA technology to produce genetically modified (GM) food and feed crops has created a new area of active consumer debate. On a global scale, consumer opinions for and against appear to be divided, but in the UK there is a strong anti-GM movement championed by environmental organisations, food-campaign groups, food journalists and some food retailers.

To date public discussion in the UK has been characterised by confusion and a good deal of acrimony. There has been a failure adequately to separate the environmental concerns about the introduction of GM crops to a small island like Britain from the legitimate, but separate, concerns about food safety and consumer choice. Also there has been insufficient recognition that it is not GM technology that should be under scrutiny but the individual products that are made through its application. It is as nonsensical to conclude that all GM crops are dangerous for human and animal consumption as it is to conclude that all GM crops will necessarily be safe.

The controversy, particularly over GM animal feeds, has been intensified as a result of the absence of specific EU legislation relating to the marketing of novel feed products. But there have also been shortcomings in the speed with which the biotechnology companies have begun to understand the UK consumer market. Add to this the direct action and 'no-holds barred' media campaign of the anti-GM groups and the competition amongst retailers to differentiate themselves on a basis of non-GM produce policies and it is easy to see why GM technology is facing difficulties of consumer acceptance.

In general, animal feed manufacturers are agnostic about GM products. However, the UK is a major importer of animal feed and will find it difficult to isolate itself from a world uptake of GM technology. Likewise there are national economic implications of UK agriculture adopting non-GM systems of livestock production when there is unrestricted UK importation of animal products and manufactured foods produced by GM systems elsewhere in the world.

A new EU Novel Feeds Directive is expected in early 2001, but the time scale of resolution of the broader GM issues is more difficult to predict. Evidence suggests that consumers wish to have more information about the use of GM feed in livestock production (Advisory Committee on Animal Feedingstuffs, 2000b; Greenpeace, 2000). However, the present situation in which each food retail company adopts its own GM-feed definitions and labelling policy is confusing, unsatisfactory and potentially open to abuse. This issue therefore needs be addressed alongside the implementation of the new feed legislation.

ORGANIC FARMING

One of the most interesting side effects of UK 'food scares' has been the re-emergence of organic farming and organic food products. Whilst remaining very small in relation to 'mainstream' food sales, organic food sales have been projected to increase to reach some 8% of the market by 2005 (Colman, 2000). Organic food has a high media and consumer profile, in part because of the activities of promotional bodies such as the Soil Association and the support of the environmental groups mentioned earlier. It also is much cherished by food writers with an enthusiasm for 'traditional food quality'.

In the UK, organic food is defined as that meeting the regulations for organic farming laid down by the United Kingdom Register of Organic Food Standards (UKROFS) i.e. it is defined by method of production rather than by any inherent attributes. Different regulations apply overseas and this can raise issues of consumer assurance since 60-70% of UK organic fruits and vegetables are imported. Organic animal feed is presently less of a problem because of the small organic market and the relative availability of home-produced organic feed ingredients.

Although promoted as environmentally friendly, there is little evidence that organic farming *per se* provides significant benefits in bio-diversity over non-organic farming of similar land-use intensity (Greenwood, 2000). Also, as many people have pointed out, organic farming leads to reductions in crop yield of 30-50% and is not a feasible global option for meeting the food requirements of the world's population (Avery, 1999). Less emphasised is that a substantial growth in UK organic farming, because of the yield reduction, would increase the land requirement of the UK population and have the dubious ecological benefit of converting a reduction in land-use intensity at home into an increase in land-use intensity somewhere else.

Organic food offers a significant market premium to both producers and retailers, and a proportion of consumers seem prepared to pay on the basis of perceived benefits. Around 0.36 of consumers believe organic food is healthier to eat, 0.36 believe it is safer to eat and 0.25 believe it tastes better (IGD, 2000c), although none of these perceptions has a firm basis in fact.

Whilst there is some evidence of compositional differences between organic and conventional crops, evaluation studies of the nutritional effects in animals and man or of benefits to human health are lacking or equivocal (Williams *et al*, 2000). Likewise, there is no evidence for organic foods being either safer or more dangerous than their conventional counterparts (Williams *et al*, 2000). Advocates of organic food point to the hazards of pesticide application to conventional crops, whilst sceptics draw attention to the hazards of animal excreta as a fertiliser (Williams *et al*, 2000) and to fungal toxins on crops untreated with fungicides (Williams *et al*, 2000). However, in none of these cases is there sufficient evidence to allow the hazard to be established as a quantifiable risk.

In a ruling on a Soil Association promotional leaflet the Advertising Standards Authority (2000) upheld complaints against four organic food claims: 'You can taste the difference'; 'Its healthy'; 'Its better for the environment'; and 'Organic means happy animals'. Also an FSA Position Paper (FSA, 2000) concluded there was not enough information to be able to say that organic foods differ significantly in their safety and nutritional value from those produced by conventional farming.

However, organic food remains a strong brand with very positive consumer associations, indicating that market image has a powerful effect on consumer demand, irrespective of other considerations.

Concluding remarks

EUROPEAN DEVELOPMENTS

In his speech, on the 8 November 2000, introducing the proposal for developing new European food law and establishing the European Food Authority, EC Health and Consumer Commissioner David Byrne made the following remarks.

'Safety is the most important ingredient in our food. Europe must have the capacity to ensure that we can deliver this to our consumers. This legislative package is designed to overcome the weaknesses of the past and put food firmly on the top of our agenda.'

'We have to regain public confidence in the capacity of the food industry and in public authorities to ensure that food is safe. The new food law provides the basic principles and requirements for marketing food and for the assurance of a safe food supply to consumers. It will address the safety of animal feeds particularly where these may have a direct or indirect effect on food safety.'- - -'A key element of the (Food) Authority is the close involvement of the food safety authorities in the Member States to facilitate the early identification of emerging risks and to avoid confusing and conflicting messages to consumers.'

He went on to say:

'Food Law shall pursue the protection of human life, taking account of the precautionary principle, the protection of consumers' interest, the traceability of food and feed and clearly established responsibilities for food and feed business operators and authorities.'

'Only safe food may be placed on the market and food shall be considered unsafe if is potentially injurious to health or unfit for human consumption or contaminated. Similarly no feed shall be placed on the market unless it satisfies the feed safety requirements. Food and feed business operators shall ensure that in all stages of production and distribution under their control this principle is respected.'

These passages set out the vision for the next phase of development of European food policy, and also reflect the tone of the approach that the Commission wishes to adopt. The resonance with recent developments in the UK, including the establishment of the FSA, is very clear.

EMERGING POLICIES

The details of legislative and regulatory developments that will emerge from the EU in due course will be subject to specific Commission proposals and European Parliament and Council of Ministers approval. Nonetheless a number of emerging policies can be discerned from the *White Paper on Food Safety* (Commission of the European Communities, 2000a) and other sources.

Firstly, the policies that are now being adopted are universally being designed to provide *farm to fork* coverage of the food chain. This can already be seen to be having legislative impact, for example in the proposed EU hygiene regulations for animal-derived foodstuffs (Commission of the European Communities, 2000b).

Secondly, there will be an emphasis on openness, transparency and consumer involvement at all levels of policy and legislative development. The emphasis on policies that can be publicly set out in scientific and ethical terms within a consensual framework will also have an impact on industry and commerce. Increasingly companies operating in the food chain will find a public expectation of openness and transparency.

Thirdly, whilst the precautionary approach is being stressed, there is also beginning to be recognition that regulatory controls must be proportionate to risk and must be uniformly applied both to home-produced and imported produce. Underlying this is a conviction that food safety must be the responsibility of businesses operating in the food chain, and that 'self-check' systems based on Hazard Analysis Critical Control Point (HACCP) are the way forward.

Finally, there is a growing appreciation that consumer safety must also embrace issues of consumer assurance. It is no longer sufficient just to provide safe food, consumers also wish to make informed choices and they require dependable information about food quality and methods of production.

LOOKING TO THE FUTURE

Education institutions

Graduate recruits to the feed industry have traditionally come mainly from animal nutrition or animal science courses with a strong emphasis on chemistry, biochemistry and nutritional science. However, the industry is now being repositioned within the framework of the food chain, and this will bring a greater emphasis on food policy, statutory regulation and legislation.

Against this background, the content of undergraduate courses requires reconsideration to ensure that graduates are equipped to meet the future needs of the feed industry. It is only a question of time before there are demands for statutory regulation of animal nutritionists and present-day graduates need to be trained with that development in mind.

Feed Industry

The feed industry provides the first link in the chain of food production from livestock agriculture and it is imperative that it develops its relationships with primary livestock producers, processors, manufacturers and retailers. However, it must also begin to embrace the understanding that its ultimate 'customers' are food consumers, and it must evolve strategies for developing better relationships with them.

It is no longer tenable to adopt the view that the food retailers will act as the sole consumer interface. Retailers are under enormous market pressures and, in creating market differentiation, they may provide consumers with messages and assurances that may be confusing or misleading. This should be a matter of concern for feed manufacturers since it risks the consumer-credibility of their own industry.

Feed manufacturers must also recognise the implications of the new consensual politics. Increasingly, companies will need to find time to engage with the consultation processes, to contribute their views to policy making and to plan their business strategies to deal with emerging legislation. Few feed manufacturers have traditionally allocated significant resources or priority to these activities, and the changed circumstances will have significant implications for company organisation.

Finally, individual companies and the feed industry as a whole must seek to address the challenge of creating a better consumer image.

References

Advisory Committee on Animal Feedingstuffs (2000a) Minutes of ACAF Meeting, 27 June. Food Standards Agency, London.

Advisory Committee on Animal Feedingstuffs (2000b) Review of Animal Feed Labelling (In Press).

Advisory Committee on Microbiological Safety of Food (1999) Synopsis on Microbial Antibiotic Resistance. HMSO, London.

Advertising Standards Authority (2000) Adjudication on the Soil Association leaflet '5 Reasons to Eat Organic'. Ed 9:3.1, 6.1, 7.1 July 2000. Advertising Standards Authority, London.

Avery, D. (1999) The fallacy of the organic Utopia. In Fearing Food, Ed. J. Morris and R. Bate, pp. 3-18. Butterworth-Heineman, Oxford.

Colman, D. R. (2000) Comparative economics of farming systems. In Shades of Green, Ed. P. B. Tinker, pp. 42-58. Royal Agricultural Society of England, Stoneleigh.

Commission of the European Communities (2000a) White Paper on Food Safety. COM (1999) 719, final. CEC, Brussels.

Commission of the European Communities (2000b) Draft regulations and directives. COM (2000) 438 final, incorporating 2000/0178 (COD), 2000/0179 (COD), 2000/0180 (COD), 2000/0181 (CNS) and 2000/0182 (COD), 14 July 2000. CEC Brussels.

CWS (2000) Food Crimes: A Consumer Perspective on the Ethics of Modern Food Production. CWS Ltd, Manchester, pp. 1-16.

European Agency for the Evaluation of Medicinal Products (1999) Antibiotic resistance in the European Union Associated with Therapeutic Use of Veterinary Medicines. EMEA/CVMP/342/99. EMEA, London.

Feedingstuffs (Zootechnical Products) Regulations (1998) Statutory Instrument 1998 No 1047. Stationery Office, London.

FSA (2000a) Consumer Attitudes to Food Safety. Qualitative Research to Explore public Attitudes to Food Safety, May 2000. Food Standards Agency, London, pp. 1-98.

FSA (2000b) Food Standards Agency View on Organic Foods. Position Paper, 25 August 2000. Food Standards Agency, London.

Greenpeace (2000) British public rejects meat, eggs and dairy products from animals fed on GM crops. Press Release, 25 September 2000. Greenpeace, London.

Greenwood, J.J.D. (2000) Biodiversity and environment. In Shades of Green, Ed. P. B. Tinker, pp. 59-73. Royal Agricultural Society of England, Stoneleigh.

IGD (2000a) Consumer Views on How the Food Industry Produces Food. Press Release, 5 September, Institute of Grocery Distribution, London.

IGD (2000b) Consumer Views on BSE and Food Safety. Press Release, 3 November, Institute of Grocery Distribution, London.

IGD (2000c) The Future For Organic Food. Press Release, 25 October, Institute of Grocery Distribution, London.

MAFF (1999a) Working Together for the Good of the Food Chain. Report of the Food Chain Group. MAFF, London, pp.1-61.

MAFF (1999b) The Feedingstuffs Regulations 2000. Draft Statutory Instrument, Ministry of Agriculture, Fisheries and Food, London.

Medicated Feedingstuffs Regulations (1998) Satutory Instrument 1998 No 1046. Stationery Office, London.

Nestle (2000) Nutrition and Lifestyle of the Over 50s. Nestle Family Monitor No 9, July 2000. Nestle, UK.

Pabia G. A. (2000) Prevalence of verotoxogenic *E coli* O157 in cattle, sheep and pigs at slaughter in GB and in cattle farms in England and Wales. In Zoonotic Infections in Livestock and the Risk to Public Health, pp.19-25, Scottish Centre for Infection and Environmental Health, Glasgow.

Phillips of Worth Matravers, Bridgeman, J. and Ferguson-Smith M. (2000) The BSE Inquiry. Vol.1, pp.1-308. Stationary Office, London

Office of Science and Technology (2000) Food Chain and Industrial Crops. Foresight Panel Consultation, Department of Trade and Industry, London.

Synge, B. A., Gunn, G. J., Ternent, H. E., Hopkins, G. F., Thomson-Carter, F, Foster, G and McKendrick, I. (2000) Preliminary results from epidemiological studies in cattle in Scotland. In Zoonotic Infections in Livestock and the Risk to Public Health, pp.10-21. Scottish Centre for Infection and Environmental Health, Glasgow.

Williams, C. M., Pennington, H., Bridges, O. and Bridges, J. W. (2000) Food quality and health. In Shades of Green, Ed. P. B. Tinker, pp.73-90. Royal Agricultural Society of England, Stoneleigh.

2

LEGISLATION AND THE FEED COMPOUNDER – LATEST DEVELOPMENTS

B.C. COOKE[1] AND J. NELSON[2]
[1]*1 Jenkins Orchard, Wick St Lawrence, Weston-super-Mare, BS22 7YP*
[2]*UKASTA Ltd, 3 Whitehall Court, London SW1A 2EQ*

Introduction

Virtually all the legislation relating to compound feed production in the UK now originates from the EU in the form of Directives, Regulations or Commission Decisions. Whilst Directives and some Commission decisions have to be implemented in the UK Regulations, EU Regulations and other Commission Decisions normally have immediate legal standing within the UK. The current legislation is reviewed and changes that have recently been introduced and proposals currently under debate within the EU are discussed. Most of these changes are as a result of the White Paper published by DG SANCO (Commission for Consumer Health) of the European Union, following various feed/food scares that have occurred in the last five years (EC 2000 a).

The current EU legislation and proposed changes currently under discussion

The legal basis for feed production applies directly to the 15 member states, but other countries applying for membership are in the process of adopting similar laws. Other non-EU countries of Europe, such as Norway, also have similar feed safety legislation. The EU legislation is in three forms:

(i)   Regulations, which have legal standing in all member states
(ii)   Directives, which introduce major new legislation or major modifications to existing legislation, must be incorporated into the national law of all member states

(iii)   Commission Decisions, which modify annexes and details in existing Regulations or Directives.

Many Regulations and Directives have been introduced that control what can go into animal feeds, how they should be manufactured and what information must be given to the farmer about those feeds. The following list is not exhaustive, but covers the main legislation of relevance to the manufacture of safe animal feeds.

THE ADDITIVES DIRECTIVE 70/524 (EEC 1970)

Concerning additives in feeding stuffs has now been amended some five times and the 6th amendment is currently under discussion. This directive controls the use of all additives in feed; it specifies what can be used and, where necessary, it lays down minimum and maximum inclusion rates for the additive. It also lays down details as to how each additive should be labelled on the compound feed. It covers vitamins, trace elements, antibiotics, coccidiostats and antiblackhead drugs, growth promoters, antioxidants, colourants, emulsifiers, stabilisers, thickeners and gelling agents, binders, aromatics and appetisers, preservatives, acidifiers, enzymes and micro-organisms.

All additives must be approved by entry into the annexes of this Directive before they can be used in animal feeds. This approval requires the presentation of a dossier covering the efficacy, quality and safety of the product to animals, man and the environment.

Proposed changes

In order to avoid potential threats to the environment, the Commission has proposed a reduction in the maximum permitted levels of trace elements in compound feeds (Table 2.1).

These new values are supposedly based upon nutritional requirements, but in some cases practical needs to avoid health and welfare problems are ignored. As a result, UKASTA has asked MAFF and FEFAC to appeal to the Commission to take account of on-farm experience when fixing the maxima. DG SANCO has now referred the proposal to the Scientific Committee for Animal Nutrition (SCAN) for its consideration. Possible revised proposals are expected early in 2001.

A working group for horses and pet animals has been established and UKASTA has requested a working group for fish. The early proposals from the working group in relation to maximum trace elements for horses are causing concern due to their restrictive nature.

The use of copper as a growth promoter in pigs will almost certainly be removed from the trace element sections of the Directive. It will largely depend on SCAN's opinion whether or not it is reinstated in the growth promoter section, but it must be emphasised that a high copper level needs to be accompanied by zinc levels above 100 mg/kg if the growth promoting effect is to be attained.

Further changes in conjunction with Directive 96/51 (EC 1996b) are proposed in order to ensure that zootechnical additives approved before 1988 are subject to re-evaluation.

Table 2.1 Current and proposed maximum permitted levels of trace elements in compound feed

Trace Element	Current maximum mg / kg	Proposed maximum mg / kg
Iron	1250	600 except ovines
		500 ovines
Iodine	4 equines	No change
	10 other species	
Cobalt	10	2
Copper[1]	35 pigs over 6 months	20 pigs over 2 months
	175 pigs to 16 weeks	30 pigs up to 2 months
	100 pigs from 16 weeks to 6 months	
	30 calf milk replacers	10 calf milk replacers
	50 other calf feeds	15 other calf feeds
	15 ovines	15 ovines and fish
	35 other species	30 bovines
		20 other species
Manganese		150 poultry
	250	55 fish
		130 other species
Zinc		100 pigs
	250	150 fur animals
		120 dairy cows
		100 other bovines
		200 fish
		120 other species
Molybdenum	2.5	No change
Selenium	0.5	No change

[1] Authorisation for the use of cupric oxide to be withdrawn

A consolidated version of the Directive and its Annexes is to be produced in order to ensure transparency and understanding.

Maximum residue limits for additives are to be proposed in 2001 for adoption by the end of 2002. These limits could prove to be problematical for the feed industry if they are not set at a realistic and practical limit. Tolerance limits for growth promoters, coccidiostats and antiblackhead medicines are also under review.

The European Parliament (EP) is suggesting that the coccidiostats, anti blackhead drugs and antibiotic growth promoters be removed from the Additives Directive, and be classified as medicines. This proposal, if pushed forward, could lead to these materials becoming available only on prescription. If this occurs it will clearly increase the compounders' work load and lead to greatly enhanced costs for the livestock, particularly the poultry, industry. In the case of growth promoters and antibiotics, since 1st September 1999 only 4 products are now authorised by the EU. These are: Salinamycin, Monensin Sodium, Flavomycin and Avilamycin. However, even these are banned in Norway, Sweden and Finland, whilst both Denmark and Holland are introducing a ban on their use. Pressure from the European Parliament for a complete ban continues to be applied on the Commission. With Sweden taking over the Presidency on 1st January 2001 we might well expect further moves toward banning these last four antibiotic growth promoters. The next major amendment of 70/524 may well be tabled during 2001.

UK position

Some of the recent changes to the Additives Directive have been incorporated into the Feeding Stuffs Regulations 2000 (MAFF 2000). These Regulations list many, but not all, of the authorised additives. For instance, whilst authorising the use of enzymes and micro-organisms, in order to find which specific materials are authorised, one has to refer to Commission documents published in the Official Journal (O.J.) and elsewhere. Of relevance are EN SANCO/3399/99 (Version of 28.2.2000), Commission Regulation (EC) No. 2437/2000 of 3rd November 2000 (EC 2000c) and Commission Regulation (EC) No.2697/2000 of 27 November 2000 (EC 2000d).

It should further be noted that in the Preservative section, whilst formic acid is not listed, it is in fact authorised for use.

THE UNDESIRABLE SUBSTANCES DIRECTIVE 99/29 (EC, 1999)

On the fixing of maximum permitted levels for undesirable substances and products in feedingstuffs, controls the levels of many contaminants in materials that might be used in animal feeds. Maximum levels in feed materials and compound feed are laid down for certain chemicals, including the heavy metals, e.g. arsenic, cadmium, fluorine, lead, mercury; dioxins, certain mycotoxins e.g. aflatoxin; natural anti-nutrients, e.g. gossipol in cotton, hydracyanic acid in linseed and volatile mustard oil in rape. Certain toxic plant materials, such as various mustard seeds and certain weed seeds, such as those containing alkaloids, are also controlled. It lays down certain maximum pesticide residues allowed in feed materials and in final compound feeds. Following the Brazilian citrus pulp and German clay dioxin scares, maximum levels for Dioxin have been established in these materials. These are 500 pg I – TEQ/kg (upper bound detection limit) for citrus pulp and 500 pg I-TEQ/kg for clays.

Proposed changes

The Commission has proposed maximum limits of Dioxin for fish oils at 6 ng/kg of fat, for fish meal at 1.5 ng/kg, for fish feed at 3 ng/kg, for animal fats at 2 ng/kg, for vegetable oil at 1 ng/kg and for compound feed at 0.75 ng/kg. The fish industry, in particular, has objected to these values and have been supported by the Standing Committee. As a result the proposal was referred to SCAN and the Scientific Committee for Food (SCF) for a risk assessment of Dioxin contamination in feed materials and compound feed.

These committees have now reported. The SCF recommends a temporary, tolerable weekly intake for humans of 7 picograms per kg bodyweight of Dioxin and Dioxin-like PCB's. SCAN has concluded that fish oil and fishmeal are the most heavily contaminated feed materials. Both have recommended an integrated approach to reduce Dioxin contamination along the food chain. As a result, SCAN recommends that European-produced fishmeal and fish oil that is heavily contaminated should be replaced by less contaminated sources of fish products, such as those from Chile and Peru. It further points out that animal fats are the next most contaminated feed materials. The Commission has, as a result of these opinions, stated that it will bring forward further legislative proposals early in 2001. The EP has passed amendments to limit the Dioxin limit on citrus pulp to 300 pg I-TEQ/kg and for the introduction of dioxin and PCB limits on all feedingstuffs within 6 months of September 2000.

These is clearly a danger that the Parliament's push for the application of the Precautionary Principle could lead to the maximum limits being set so low as to make some feed materials unusable in compound feed production. It must be remembered that dioxins are widespread contaminants, the levels of which are completely beyond the control of the feed-materials producer.

The Commission has proposed that the dilution option in the Directive be deleted. This would mean, for instance, that allowing authorised compounders to use some feed materials with levels of aflatoxin B_1 up to 0.2 mg/kg providing the finished feed contains no more than 0.005 to 0.05 mg/kg depending on its use would no longer be allowed. The majority of the Council supports this proposal, and the EP has endorsed it. Indeed, the Parliament has gone further and passed amendments calling for a ban on decontamination and re-export of materials high in undesirable substances. The Commission has rejected these amendments, but has proposed action limits set well below the maximum allowed, whereby any material containing levels of undesirables above the action limit would have to be reported to the authorities for investigation. Clearly a considerable tightening of the rules will come into operation in the not-too-distant future. This will have the effect of further restricting feed material availability to the compounder.

Suggestions have been made that mycotoxins other than aflatoxin should be brought under the control of this Directive. Both ochratoxin and fusarium toxins have been mentioned in this context, but no proposals for maximum limits have yet been made.

The scope of the Undesirable Substances Directive is to be extended to cover additives as well as feed materials. This is thought unlikely to have much effect on the industry.

The EP has called for a reduction in the maximum limit on mercury in feed materials of fish origin from 0.5 mg/kg to 0.1 mg/kg and the limit on cadmium in feed materials of animal origin from 2 mg/kg to 1 mg/kg. The Commission has agreed to ask SCAN to carry out a scientific review of these limits.

Adoption of a number of these proposals would automatically lead to the need for amendments to the Feed Materials Directive 96/25 (EC 1996) and the Establishments Directive 95/69 (EC 1995 b).

UK position

UKASTA is very concerned at the implications of the proposed amendments to the Undesirable Substances Directive. Whilst recognising that feed/food safety is paramount for public health, it considers that the Commission's proposals will not improve current safeguards for public health and could lead to greater use of undesirable substances in dairy imports from non-EU countries. UKASTA, in common with FEFAC, is calling for a proper risk assessment before any decision is made on new legislation.

THE CERTAIN PRODUCTS DIRECTIVE 82/471 (EEC, 1982)

Concerns certain products used in animal nutrition. It controls the use of proteins produced from fermentation processes, such as bacterial proteins and yeasts, non-protein nitrogen sources such as urea, by-products from amino acid production, amino acids and analogues of amino acids. It lays down not only labelling requirements for these constituents, but also labelling requirements for the final feed containing these constituents. The presentation of a dossier proving the product's efficacy, quality and safety is required for any product before it can be authorised under this Directive.

Proposed changes

Whilst it was suggested that this Directive would be substantially modified to introduce controls on GMO feed materials, this no longer appears to be the case. Thus no significant changes are expected in the near future.

THE FEED MATERIALS DIRECTIVE 96/25 (EC, 1996)

On the circulation of feed materials lists the most commonly used feed materials in animal nutrition. It controls these materials whether they are fed directly to animals or as constituents of compound feeds. It lays down requirements for certain declarations on these individual ingredients, and defines their source and the processing through which the ingredients have passed.

Proposed changes

Following the dioxin contamination of recycled fat, which occurred in Belgium, but also affected the Netherlands and France, and the sewage sludge problem in France, it is proposed that the definitions of oils and fats and animal products should be amended in order to avoid future abuse and ensure traceability and safety. These changes, which will be introduced shortly, will not have much impact on the UK industry because the major effects will come from changes to the prohibited list of ingredients (EEC, 1991).

Of much greater impact will be the White Paper which is expected to introduce an exclusive positive list of feed materials authorised for use in animal feeding. A proposal is expected to be published in 2002, with adoption by the end of 2003. However the EP is pressing for more immediate action during 2001. This will undoubtedly be a mammoth task, for it is suggested that there are some 15,000 different feed materials in use throughout the world. The danger from this proposal is that, due to oversight, some materials currently in use could be omitted and thus be prohibited from use. Further to this any new feed materials, such as a new by-product from the food industry, would have to pass through an authorisation process before it could be used in a compound feed. Under both the Compounds and Feed Materials Directives, feedingstuffs have to be wholesome, unadulterated and of merchantable quality. No-one in the industry can see any benefit to feed/food safety from the introduction of such a positive list of ingredients.

UK position

The Feed Materials Directive, which was adopted in 1996, is being implemented in UK legislation in the Feeding Stuffs Regulations 2000. The legislation for England was made on 22nd September and was brought into operation on 29th October; that for Scotland was signed on 28th December and will come into operation on 31st January 2001. At the time of writing this chapter, the Feeding Stuffs Regulations for Wales and Northern Ireland have not yet been published.

A major feature of these new Regulations is that all supplies of feed materials (formally known as 'Straight Feedingstuffs') will have to be labelled (if in a bag) or accompanied by a declaration (if delivered in bulk). This differs from previous Feedingstuffs Regulations whereby for bulk deliveries of straight feedingstuffs, the Statutory Statement may follow as soon as practical after delivery to the purchaser.

The Feedingstuffs Regulations 2000 also introduce new definitions and new compulsory declarations of nutritional parameters for some feed materials - these being provisions laid down in the Feed Materials Directive. The requirements to provide a Statutory Statement for feed materials becomes operative once the goods are 'put in circulation' within the EU. 'Putting into circulation' covers offering for sale, or any other form of transfer, whether free or not, to third parties. Those supplying feed materials to feed compounders and/or to feed merchants are responsible for providing a statutory statement, meeting the requirements of the legislation, with the deliveries of materials. This requirement covers imported and home-produced feed materials. UKASTA has been in discussion with the FSA on the provision of written guidance for the feed industry designed to assist successful adoption of this legislation

COMMISSION DECISION 91/516 (EEC, 1991)

Establishing a list of ingredients whose use is prohibited in compound animal feedingstuffs.

This Decision currently bans the following:

(a)   Faeces, urine and digestive tract contents
(b)   Hide treated with tanning substances
(c)   Treated seeds
(d)   Wood and sawdust treated with wood preservative
(e)   Sewage sludge including waste water from urban, domestic and industrial treatment processes
(f)   Solid urban waste
(g)   Untreated waste from eating places except vegetable material unsuitable for human consumption for reasons of freshness
(h)   Packaging and parts of packaging

This ban has recently been extended to ban the feeding of these materials in addition to the ban on their use in animal feeds.

Proposed changes

Following the various feed scares mentioned above, the Commission proposed a ban on "all fats and oils and their derivatives other than: those from animal origin manufactured by a plant approved in accordance with Council Dr. 90/667/EEC;

those from vegetable origin manufactured by an oil crusher and/or refiner and previously unused for any purpose;

those from animal origin not covered by Directive 90/667/EEC and of vegetable origin recovered from a food business, when the companies at each stage of the recovery process (collection, storage, blend process, transport, distribution, etc.) are monitored by a HACCP system in order to guarantee the nature, the origin, the full traceability and the safety as feed materials of the products; the recovering process shall be certified by a national authority or by an international or national recognised body."

This approach has been supported by UKASTA, FEFAC and the major fat blenders in the UK. However some members of APAG (representing the oleochemical industry) and UNEGA (representing the producers of animal fat for human consumption) are concerned that the definition may exclude certain products from animal feeding. The measure was nonetheless formally adopted in September 2000. It is further proposed that the scope of the Establishments/Approvals Directive 95/69 should be extended to cover RVO/UCO plants.

In the wake of the discussions at European level on measures to control BSE throughout the Community, the use of animal fats is being reviewed. Animal fats are currently excluded from the recently agreed definition of 'processed animal proteins'. FEFAC advised that the Commission Services will submit a proposal to the Standing Veterinary Committee introducing micro-filtration as a processing condition in order to guarantee a maximum impurity level of 0.15% based on the opinion of the SSC.

UK position

UKASTA and the fat blenders have launched the UK Used Cooking Oil (UCO) scheme. This involves independent audit of blenders, processors, collectors and brokers of used fats and oils by Product Authentication International Ltd (PAI). The audit aims to ensure compliance with the Commission Decision, traceability, implementation of HACCP and overall safety. A list of 'Certificated Suppliers' is now available.

Companies registered under UFAS – Code of Practice for the Manufacture of

Safe Compound Animal Feedingstuffs have been advised of these developments. At the same time they should check with their feed fat suppliers that no problems are anticipated in their achieving certification under the UCO scheme by the end of 2000.

THE PATHOGENS DIRECTIVE 90/667 (EEC, 1990 C)

Lays down veterinary rules for the disposal of animal waste, for its placing in the market and for the prevention of pathogens in feedingstuffs, of animal or fish origin. It is under this Directive that controls of salmonella, etc. are laid down. No comparable Directive, however, covers vegetable materials, although one is currently under consideration. No details are yet available.

THE MEDICATED FEEDINGSTUFFS DIRECTIVE 90/167

Laying down the conditions governing the preparation, placing on the market and the use of medicated feedingstuffs defines which medicines can be used and under what circumstances and at what levels. It also lays down the necessary labelling requirements for the compound feed containing medicines. These medicines are of course all under veterinary control and member states all have individual rules for the veterinary surgeon to write a prescription for the production of the animal feed containing the relevant medicine.

Proposed changes

Current discussions may lead to the coccidiostats and antiblackhead additives (D. coccidiostats and other medicinal substances) being placed under the control of this Directive. Should that happen, a veterinary prescription would be required for their use.

DIRECTIVE 90/220 ON THE DELIBERATE RELEASE INTO THE ENVIRONMENT OF GENETICALLY MODIFIED ORGANISMS

Directive 90/220 controls the authorisation for the growing of, import of, and use of GMO plants. Unfortunately its interpretation is not clear. In theory it only controls materials that contain genetically active material and thus any whole maize or soya from GMO varieties has to be authorised. The problem arises with raw materials such as extracted soya and maize gluten feed. Some national authorities, such as the UK and Holland, have stated that these materials do not contain genetically active material, because the processing fragments the DNA, and thus no authorisation is required. Other countries, such as France, take the contrary view and only allow material, even in by-product form, which has been specifically authorised.

However the Food Safety Information Bulletin, August 2000, produced by the FSA reported on discussions held in June at the Advisory Committee on Animal Feedingstuffs (ACAF). This included a report that ACAF members considered a paper summarising the results of research which indicate that DNA fragments large enough to contain potentially functional genes survived processing in many of the samples studied. It was considered that this could help inform the assessment of new GM crops. However, it was also reported that there was a greater degree of fragmentation when processing was more extensive, with the results varying considerable between different feed materials, and even between different samples of the same materials. Thus the position of what is and what is not a GMO under the Directive is by no means clear. Currently only four GMO maize, one GMO soya and three GMO rapeseed varieties are authorised in the EU, but many more varieties are authorised and grown in third countries, such as the US. Thus the European feed compounder is in a very awkward position with regard to knowing what can legally be used.

Proposed changes

A new Regulation on the circulation and use of genetically modified feed materials is currently under discussion and is likely to be proposed in 2001. A preliminary draft of the proposal requires specific authorisation and labelling of all GMO feed materials and those derived from GMO varieties of crops even if they do not contain intact genetically modified DNA or protein. All these materials will be known as 'Novel feed materials'. The authorisation will require the presentation of a dossier, including a full assessment of the risk to human and animal health, and to the environment, appropriate safety and emergency response, a monitoring plan where necessary, as well as precise instructions and conditions for use, labelling and packaging. It will lead to the phasing out by the end of 2004 of any novel feed materials containing antibiotic-resistant marker genes that have a current or potential use in medical or veterinary treatment.

UK position

If adopted as drafted, a major concern to the feed compounder is the definition of derived from GMO. This reads "feed material derived from GMO means any feed material which is either produced from or produced by, but does not contain, genetically modified organisms". This would not only include many by-products such as extracted soya and soya oil from GMO soya, but also products produced from fermentation, such as brewers' and distillers' by-products, some vitamins, amino acids and enzymes where the fermentation organism has been genetically modified. It would thus become virtually impossible to produce a "GMO free" feed for efficient animal production. Thus all animal products would have to be labelled as being fed on GMO derived feed.

COMMISSION DECISION 94/381 – BSE CONTROLS (EC, 1994)

Under a number of legislative approaches various controls existed on the use of Meat and Bone Meal or mammalian protein in compound feeds in the period up to the end of 2000.

i.   In the EU, mammalian Meat & Bone Meal was banned from use in ruminant feeds by Commission Decision 94/381, except where meat and bone is guaranteed to contain no ruminant protein.

ii.   In the UK and Portugal the use of meat and bone was banned in all farm feeds.

iii.   In Denmark and Ireland the use of Meat and Bone Meal was banned in all mills making ruminant feed.

iv.   In France the use of fallen animals in Meat & Bone Meal was banned.

v.   Various other bans were introduced during November and December 2000.

In June 2000 a further Commission Decision 2000/418 (EC, 2000 b) was introduced. This banned specified risk material from use in animal feed. The Annex to this Decision defines specified risk material as:

1.   (a)   The following tissues shall be designated as specified risk material
            (i)   the skull including the brains and eyes, the tonsils, the spinal cord and the ileum of bovine animals aged over 12 months.
            (ii)   the skull including the brains and eyes, the tonsils and the spinal cord of ovine and caprine animals aged over 12 months or that have a permanent incisor erupted through the gum, and the spleen of ovine and caprine animals of all ages.
      (b)   In addition to the specified risk material listed in point 1(a) the following tissues shall be designated as specified risk material in the United Kingdom of Great Britain and Northern Ireland and Portugal
            (i)   the entire head excluding the tongue, including the brains, eyes,

trigeminal ganglia and tonsils; the thymus; the spleen; the intestines from the duodenum to the rectum and spinal cord of bovine animals aged over six months;

(ii)   the vertebral column, including dorsal root ganglia, of bovine animals aged over 30 months.

Proposed changes

The Commission planned to bring forward a proposal early in 2001 to ban the incorporation of any fallen animal in feed materials for use in animal feed. However, following the French and German BSE problems this Decision has been tabled and will become effective on 1 March 2001

In late November/December 2000 the Commission tabled a draft Decision banning various processed animal proteins in animal feeds used for the production of meat, milk and eggs. This Decision was agreed after amendment and came into force on 1st January 2001.

The products banned temporarily were:

Meat & Bone meal, meat meal, bone meal, blood meal, dried plasma and other blood products, hydrolysed proteins, hoof meal, horn meal, dried grieves, feather meal, poultry offal meal, fish meal, dicalcium phosphate, gelatine and similar products, and feedingstuffs, feed additives and premixtures containing these products.

Exemptions allow the use of

1.   fish meal in non-ruminant diets made in mills which can separate ruminant feeds and where the feed is used on mixed farms that satisfy the authorities that controls prevent the feed going to ruminant animals or where used on farms with no ruminant livestock Further interpretation of this rule is still awaited.
2.   non-ruminant gelatine used for coating additives
3.   dicalcium phosphate and hydrolysed proteins produced under certain conditions yet to be defined.

All three products are subject to cleaning and disinfection of transport that carries them or feed containing them.

Milk products, animal fats and mineral dicalcium phosphate are not covered by the ban.

The ban includes storage and thus use of feed on farm.

The use of animal fats will be reconsidered in the second half of January 2001.

UK position

Whilst the FSA had proposed the banning of various animal products from animal feed, these materials are covered in the main by the EU ban. Thus the major emphasis at the moment is to comply with the EU ban. UKASTA has been, and continues to be, in constant touch with MAFF and the FSA with regard to interpretation and implementation of the EU decisions.

THE ESTABLISHMENTS DIRECTIVE 95/69 (EC, 1995 b)

Lays down the conditions and arrangements for approving and registering certain establishments and intermediaries in the animal feed sector. It has recently been extended to cover the approval/registration of manufacturers of additives, premixtures, compound feeds containing them and certain feed materials. This requires non-EU establishments to have a representative within the EU and ultimately will involve inspection and approval of establishments in third countries. It covers the production and use of additives, certain constituents and some feed materials high in undesirable substances. It lays down some aspects of Good Manufacturing Practice.

Proposed changes

A formal proposal to implement changes to the Directive is expected. This may propose:

1.   The deletion of registration
2.   The approval of all manufacturers of compound feedingstuffs, including farm mixers, and feed additives
3.   Approval of manufacturers of recycled products such as used cooking oil. This may also include food premises producing by-products for animal feed use, alternatively these may come under a feed hygiene directive
4.   Improved traceability of feed materials and identification of critical control points
5.   Setting up of specifications for feed materials
6.   The Establishment of a Code of Good Manufacturing Practice for animal feeding providing traceability of feed materials and additives. FEFAC has been requested by DG SANCO to help in the preparation of this document.
7.   Guidelines for implementation of HACCP systems.

The aim is to bring these proposals into effect by the end of 2001.

UK position

UKASTA FEED ASSURANCE SCHEME: The Establishments Directive provided foundations for the Code of Practice for the Manufacture of Safe Compound Animal Feedingstuffs. This Code of Practice is one of three elements of the UKASTA Feed Assurance Scheme (UFAS). It was introduced in 1998 and is a codification of legislation relating to the manufacture and supply of compound feedingstuffs. It also includes best industry practices, ranging from the sourcing and delivery of feed ingredients, both micro and macro, the production of finished feeds and their delivery to farms.

The other two elements of UFAS are:

Code of Safe Practice for the Supply, Storage and Packaging of Animal Feed materials which are destined for farm use.

Summary of Requirements applying to agricultural merchants/distributors supplying compound feedingstuffs to farm, but who do not manufacture feedingstuffs.

UFAS is subject to independent audit and certification, the administration of which is undertaken by Assured British Meat and Food Certification International (ABM/FCI), which operates as the certifying body for UFAS. The United Kingdom Accreditation Service (UKAS) recently granted ABM/FCI an extension to its scope for EN45011 to cover UFAS.

EN45011 is a European Standard for bodies offering product certification. In the UK, UKAS is responsible for ensuring that certification bodies wishing to offer product certification in accordance with EN45011 meet all the necessary criteria. Advantages of EN45011 accreditation include:

- demonstration to customers, retailers, government and others that UFAS is operated in accordance with accepted procedures, hence increasing integrity and acceptability

- improved uniformity of auditing standards and procedures by use of detailed check lists and instructions

- gives a standard of certification which is recognised throughout the EU with implications for international trade in livestock products.

RPSGB FEES

In Great Britain, manufacturers of compound feedingstuffs containing zootechnical additives and therefore seeking approval under the Establishments Directive, register with the Royal Pharmaceutical Society of Great Britain (RPSGB). Discussions are currently being undertaken between UKASTA, the Veterinary Medicine Directorate (VMD) and the RPSGB concerning the fees payable by an approved establishment for inspection under this Directive. In the UK the aspects of the Directive relating to zootechnical additives are implemented in the Feedingstuffs (Zootechnical Products) Regulations 1999. The RPSGB is also responsible for enforcing the Medicated Feedingstuffs Regulations 1998.

The VMD has decided that the RPSGB should also undertake responsibilities for sampling and analysis of medicated and zootechnical feedingstuffs. This role was formerly undertaken by Trading Standards Officers. Concern has been expressed by UKASTA and the British Association of Feed Supplement and Additive Manufacturers (BAFSAM), together with the National Farmers Union, at the level of fees being proposed by the RPSGB. Not only could the industry not afford the proposed increases, but it had already invested in UFAS, which included annual audits. At the time of writing this chapter, the level of fees for the RPSGB Inspectorate for the twelve-month period commencing April 2001 has not yet been finalised.

THE PESTICIDES RESIDUES DIRECTIVE 86/362 (EEC, 1986)

Fixes maximum levels for pesticide residues in and on cereals. It lays down maximum levels of various pesticides which may not be exceeded in cereals to be used in animal feeds.

Proposed changes

No specific changes are currently proposed, but a tightening of maximum pesticide residues is likely to be imposed via the regulation on the hygiene of foodstuffs and comparable legislation in relation to hygiene of animal feeds.

THE COMPOUNDS DIRECTIVE 79/373 (EEC, 1979)

On the marketing of compound feeding stuffs as amended, lays down general rules for the manufacture of all compound feeds for farm animals and pet animals. It lays down a labelling requirement for compound feeds, which not only includes nutritional parameters, but also safety information, such as the species of animal to be fed, the age and type of animal within the species, together with feeding instructions. All the feed materials included in the compound feed have to be declared either by category or by individual ingredients in descending order. Further, it states that no compound feed shall be harmful to the health of animal or man.

Proposed changes

The Commission has proposed changes to the method of declaration of ingredients present in compound feed. The European Parliament and Council have agreed to the withdrawal of the option to declare categories of feed materials. The Parliament is pushing for the % inclusion of each raw material in descending order to be declared. The Council has agreed a compromise whereby individual ingredients would be declared by ranges of % (less than 2, 2-5, 5-15, 15-30, 30 plus). The number and details of the ranges may be subject to further negotiation, as is the principle of ranges under the 'Conciliation Procedure' with the EP. This compromise would be accompanied by the requirement that compounders give specific percentage information to customers on request. Implementation of the changes is expected during 2001. Pressure still exists for full percentage declaration.

UK position

In briefing MPs and MEPs on the Commission's proposal, UKASTA advised that it welcomed the opening up of the debate on food safety and supported the clear labelling of individual ingredients in animal feed. However, it did not support the Commission's percentage proposal as this would hinder rather than help the industry to meet the nutritional requirements of farm animals. It was considered that existing legislation had already delivered safe feed for safe food in Europe, and that by discouraging research and development and imposing additional costs caused by new legislation, there was a real danger of Europe losing its livestock production.

THE DIETETIC FEEDS DIRECTIVE 93/74 (EEC, 1993)

On feedingstuffs intended for particular nutritional purposes lays down specific requirements for specific purposes in animal feeding, e.g. when animals have special nutritional requirements because of stress or some other cause. It specifies what materials can be used in that particular type of feed and at what level they can be included. It also lays down specific labelling requirements for the finished feed.

Proposed changes

The Commission has proposed that "nutritional supplements" should be brought under the control of this Directive. Nutritional supplements cover the uses of additives outside of feeds e.g. in drinking water. The proposal was recently taking up under the French Presidency, the measure envisaged that particular nutritional uses of supplements be listed together with conditions and periods of use. There was some doubt about the coverage of the Proposal and it was thought that some non-feed uses e.g. boluses would not be included. An alternative approach to control non-feed uses of additives was to amend the Additives Directive 70/524.

UK position

UKASTA believes that in order to ensure transparency and safety, all methods of supplying nutrients to animals should be controlled. However, any new controls should not preclude the safe use of any existing products or methods of supply to the animal.

THE ZOONOSIS DIRECTIVE 92/117 (EEC, 1992)

Concerns measures for protection against specified zoonoses and specified zoonotic agents, in animals and products of animal origin, in order to prevent outbreaks of food-borne infections and intoxications. In relation to compound animal feeds, this Directive at the moment only covers salmonella in poultry feeds. It calls for the authorities to sample compound feed at the final production stage, if that feed is suspected of involvement in a salmonella outbreak on a poultry farm.

Proposed changes

Whilst no firm proposals are currently on the table, the Directive does include Campylobacter Listeriosis, Taxoplasmosis and Yersinia, and some of these within feed could be brought under control.

UK position

At the beginning of December a voluntary Code of Practice for the prevention and control of salmonella on pig farms was issued by MAFF and SERAD. It contains a section on feed which reads as follows:

> "Purchased feed should be supplied from a feed mill which operates in accordance with the relevant MAFF and UKASTA codes of practice and using ingredients which have been obtained from sources with a consistently satisfactory bacteriological record.

> "If feed is supplied by a mill operating to other codes then your veterinary or technical adviser should be able to confirm with the

manufacturer that the processes being used are effective in the control of salmonella.

"It is equally important that home mixers follow advice given in the relevant codes. Farmers who feed their pigs catering waste ("swill") as defined by the Animal By-Products Order 1999 or equivalent legislation, must be approved by MAFF and adhere to the conditions of the approval.

"Birds, domestic and wild animals should be kept out of feed stores. Avoid dry feed becoming wet as any contaminating salmonella may multiply rapidly in damp conditions.

"All feed bins and delivery pipes for dry feed should regularly be thoroughly dry cleaned. Cleaning and disinfection of wet feed delivery systems should form part of a regular routine."

THE CONTROLS DIRECTIVE 95/53 (EC, 1995 a)

Fixes the principles which govern the organisation of official inspection in the field of animal nutrition. It is under this Directive that EU officials can monitor what controls the member states have on the production of animal feed.

Proposed changes

For some time the Commission has been trying to get a safeguard clause adopted. This would enable the Commission to take action without reference to member states in the event that an imported feedingstuff posed a serious risk to human health, animal health or the environment. This is opposed by some member states because they do not wish to pass control to the Commission. The EP supports the Commission proposal and the matter is, at the moment, subject to the conciliation process. The most likely outcome appears to be to allow the Commission to invoke the clause, but to make it subject to confirmation by the Standing Committee within 10 days of the Commission action. The safeguard clause is supported by FEFAC.

It is further proposed to introduce a rapid alert system for feedstuffs, similar to that for food. This would require the industry to inform their national authority of any incident involving feed which potentially endangers human health, animal health or the environment. The national authority would then have to inform the Commission, who would then inform other member states. There is obviously a danger that this, in some circumstances, might lead to over-reaction. FEFAC believes that professional organisations should be involved in any rapid alert system in order to try to prevent over-reaction, but at the same time to ensure that the industry is informed of any problems as soon as possible.

The third proposal is that member states should carry out a monitoring programme for contaminants in feedingstuffs. This will form part of an EU harmonised feed control programme. The EP has called for a legal right for the Commission to carry out on the spot checks without prior notice. This amendment has been accepted by the Commission in relation to operators, providing the member states' authorities are notified in advance. This arrangement is currently being brought into operation.

New regulations proposed in the White Paper

The White Paper published by DG SANCO proposes the introduction of two new Regulations.

The first will lay down a framework for official controls on all aspects of feed and food safety, along the chain from feed to food. It aims to merge and complete existing rules for national and community controls and inspection, and will include border controls and third countries. It will integrate monitoring and surveillance schemes from farm to table, and will merge existing rules on mutual co-operation. Finally, it aims to create a Community approach towards financial support for official controls. What this means for the cost of inspection etc. of compounders in the UK is not yet known. It is the Commission's aim to finalise this Regulation during 2001.

The second Regulation on feed has the objective of establishing animal and public health as the primary objective of EU feed legislation. Its aim will be to ensure implementation of HACCP and traceability with efficient controls and enforcement. In theory it will recast all existing measures on feedingstuffs so as to create a comprehensive legislative tool to increase transparency and legal security. FEFAC supports simplification of EU feed legislation on feed/food hygiene and safety, including traceability, but the authors doubt whether the Commission will be prepared to eliminate detailed controls and stop responding to each problem that occurs by the introduction of new legal controls.

Following the White Paper, the Commission has now committed itself to the setting up of a European Food Standards Agency. Current proposals are that this agency, with its own scientific committees, would be consulted on all aspects of animal feeding. However the writing and implementation of legislation would remain as a responsibility of DG SANCO.

Changes in UK Legislation

The Feedingstuffs (Sampling and Analysis) (Amendment) Regulations 2000 – FSA consulted on these draft regulations in October. They implement Council Directive 2000/45/EC setting out statutory methods for vitamins A and E, and tryptophan.

The regulations also concern amendments to the Feedingstuffs (Enforcement) Regulations 1999 and the Feedingstuffs (Establishments and Intermediate) Regulations 1999.

Conclusion

The existing and proposed EU legislation has a common objective – that is the production of safe animal feed that will not be harmful to the health of animals or humans consuming the food from these animals, nor to the environment in which these animals are kept, or on which their manure is spread. Unfortunately in meeting this objective the legislation is becoming more extensive and complex. What the industry requires is simple rules which can be understood, i.e. are transparent, by all operators in the farming community, the feed industry and, most importantly, by the consumers of animal products. All feed, whether from farm or commercially-produced, should be subject to the same rules in order to ensure safety. Safe feed equals safe food.

References

EEC 1970 Council Directive 70/524/EEC concerning additives in feedingstuffs. OJ No.L270/1 of 14.12.1970

EEC 1979 Council Directive 79/373/EEC on the marketing of compound feedingstuffs. OJ No.L86/30 of 6.4.1979

EEC 1982 Council Directive 82/471/EEC concerning certain products used in animal nutrition. OJ No.213/8 of 21.7.1982

EEC 1986 Council Directive 86/362/EEC on the fixing of maximum levels for pesticide residues in and on cereals. OJ No.L221/37 of 7.8.1986

EEC 1990(a) Council Directive 90/167/EEC laying down the conditions governing the preparation, placing on the market and use of medicated feedingstuffs in the Community. OJ No.L92/42 of 7.4.90

EEC 1990(b) Council Directive 90/220/EEC on the deliberate release into the environment of genetically modified organisms. OJ No.L117/15 of 8.5.90

EEC 1990(c) Council Directive 90/667/EEC laying down the veterinary rules for the disposal and processing of animal waste, for its placing on the market and for the prevention of pathogens in feedstuffs of animal or fish origin. OJ No.L363/51 of 27.12.90

EEC 1991 Commission Decision 91/516/EEC establishing a list of ingredients whose use is prohibited in compound feedingstuffs. OJ No.L281/23 of 9.10.91

EEC 1992 Council Directive 92/117/EEC concerning measures for protection against specific zoonoses and specific zoonotic agents in animals and products of animal origin in order to prevent outbreaks of food-borne infections and intoxications. OJ No.L62/38 of 15.3.93

EEC 1993 Council Directive 93/74/EEC on feedingstuffs intended for particular nutritional purposes. OJ No.L237/12 of 22.9.93

EC 1994 Commission Decision 94/381/EC concerning certain protection measures with regard to bovine spongiform encephalopathy and the feeding of mammalian derived protein. OJ No.L172/23 of 7.7.94

EC 1995(a) Council Directive 95/53/EC fixing the principles governing the organisation of official inspections in the field of animal nutrition. OJ No.L265/17 of 8.11.95

EC 1995(b) Council Directive 95/69/EC, laying down conditions and arrangements for approving and registering certain establishments and intermediaries operating in the animal feed sector. OJ No.L332/15 of 30.12.95

EC 1996 Council Directive 96/26/EC on the circulation of feed materials. OJ No.L125/35 of 23.5.96

EC 1999 Council Directive 1999/29/EC on the undesirable substances and products in animal nutrition. OJ No.L115/32 of 4.5.99

EC 2000(a) White Paper on Food Safety. Commission of the European Communities. 12.1.2000 COM (1999) 719 final. Brussels, Belgium

EC 2000(b) Commission Decision 2000/418/EC regulating the use of material presenting risks as regards transmissible spongiform encephalopathies. O.J.No. L158, 76-82 of 30.6.00

EC 2000(c) Commission Regulation No. 2437/2000 concerning permanent authorisation of an additive and the provisional authorisation of new additives in feedingstuffs. O.J. No.L280, 28-36 of 4/11/2000.

EC 2000(d) Commission Regulation No.2697/2000 concerning the provisional authorisations of additives in feedingstuffs. O.J.No.L319, 1-59 of 16/12/2000.

MAFF 2000 The Feeding Stuffs Regulations 2000 (S.I. 2000 No.2481). HMSO, London, UK.

UKASTA 2000 UKASTA Feed Assurance Scheme. UKASTA Code of Practice for the Manufacture of Safe Compound Animal Feedingstuffs. Edition 2 – July 2000. UKASTA, 3 Whitehall Court, London SW1A 2EQ, UK

3

E-COMMERCE – OPPORTUNITIES FOR COMPOUNDERS AND THE FEED INDUSTRY

A. Chappell and K. Richards
Computer Applications Limited, Rivington House, Drumhead Road, Chorley, Lancs PR6 7BX

Introduction

E-commerce, and the World Wide Web in general, are set to revolutionise the way that people do business. This change has, in fact, already started and is fairly well advanced in some cases. For example, traders and material buyers can already get information on their own Personal Computer (PC), via the web, direct from Reuters (www.reutersinform.com), without the need for the previous heavy investment in satellite dishes and dedicated hardware infrastructure.

A wealth of information relating directly to Agriculture is published on the Internet via web pages. A web page can be thought of as a very sophisticated document. Users can find information in a number of ways, but the commonest method is via search engines and there are many of these available (yahoo, lycos, alta vista, excite - to name but a few). You select the one that gives you the best results. Examples of the benefits and pit falls of search engines will be described later.

What is the current position regarding the Internet, e-commerce and Agriculture in general?

It has been stated by many companies whose livelihoods depend upon e-commerce that over 65% of UK farmers have PCs that are capable of connection to the web. As an industry, we need to be confident that (a) this figure is increasing and (b) farmers are using the Internet. One thing is clear - with the advent of WAP phones and Internet access through the TV screen, the use of the Internet is only going to increase. The Internet is definitely a powerful method of getting to your customers – it can act as a technical consultant, order predictor and taker, information provider, cash collector and account manager - at a fraction of the cost of traditional methods of customer service.

Business to business e-commerce is not very well developed in Agriculture. What there is tends to be along the lines of publishing product catalogues and allowing customers to buy – a shopping cart approach. There is very little true business to business where the Internet forms the bridge and the systems have true integration to the back-end commercial systems. What usually happens is that orders may be captured and may be sent electronically to the company, but then they are often printed, verified and manually entered into the commercial software and the first time the customer is aware that his order has been accepted is when the product arrives.

"Inteletrade" from Computer Applications Limited (CAL) is an exception to this rule. It allows for complete on-line trading and order management and provides end users with direct, on line access to corporate data. It has given proven benefits to the companies who deploy it.

The Internet and e-commerce are here to stay and companies who take advantage of them will reap the benefits at the expense of those who ignore them. We will now look at what is available out there in 'cyber space'

Search Engines

A search engine is a software tool, which is accessed via your internet browser, that allows you to enter a key word or phrase and which will then search the web for sites, documents etc that contain those key words or phrases.

When the Internet was new, and there was little information available, the search engines would invoke 'web crawlers' that would scan the pages for the words, index the results an publish them to you

Search engines now work in a different way and it is important that you to know this, especially if you are planning, or have created, your own web site. A search engine looks for words that are stored in the web page in the informational content of the page and not the HTML. In general, the engine retrieves words from between the title tags, body tags, alt tags and meta tags and indexes them according to some algorithms. These algorithms are designed to ignore sites where keywords are repeatedly entered as a means to increase the frequency of finds (so called 'spamming'). This means that when people are designing web pages they ensure that they have inserted the correct keywords into the correct places. It is estimated, however, that only about 15% of web sites are currently indexed correctly

We are now starting to see the emergence of more specialised search engines (currently, there is not one for agriculture – or even a 'Usenet' or chat room!). There are many search engines available in the Internet world today.

www.yahoo.co.uk
www.altavista.co.uk
www.lycos.co.uk
www.excite.co.uk
www.askjeeves.co.uk – this actually can give some remarkable
results if you ignore the standard response format

Whatever search engine is used, it is important to remember a few things about
the service they offer and how to use them:-

a)   The speed of response and performance depends on the size of the 'pipe'
      connected to the web. You need to get the biggest pipe that you can afford.
      ISDN connection to the web is ideal and a 28kb dial up modem is barely
      adequate. This will determine how fast things are downloaded to you.
b)   Once connected to the site the speed of presentation to you depends on the
      power and speed of your computer (the bigger the better) and how the site
      was designed in the first place. As a recent example (using a P400 PC with
      128Mb memory connected via ISDN) connection to www.lastminute.com
      took less that 3 seconds to load but connection to www.comet.co.uk took
      over a minute – imagine that performance over a 28kb dial up line!
c)   You need to be specific in what you ask for. Searching for Agriculture
      brings back over 450,000 sites; searching for 'conjugated linoleic acid' brings
      back over 1800 sites; changing this to 'conjugated linoleic acid in animals'
      reduces to 580 and 'conjugated linoleic acid dale bauman' reduces to 18.
      Note the use of the quotes in this search.
d)   Be careful what you ask for. Trying to book a rail journey by searching for
      'virgin' gives some very revealing sites.
e)   Remember that spelling counts – search engines are unforgiving on spelling
      mistakes. It has been said that a lot of pornography promoters have
      registered frequently misspelled words as porn sites! Use the security settings
      of your internet browser to control access to the net.
f)   Remember that you are connected to the net via telecommunications and
      these can pose problems. If a search seems to be taking ages, simply press
      'stop' and 'refresh' on the browser to re-start the search.
g)   Everything published on the web is in the public domain, but do not be
      fooled into thinking that everything on there is authentic and true – check
      the reliability of information that you get from the net before using it
      extensively in your organisation. Most information is accurate, but there are
      people out there who enjoy providing misinformation.
h)   It is ESSENTIAL that you build in adequate 'firewalls' to protect your
      systems from the Internet. Just as you are connected to the web and can

see what is out there, other people can see you. If you do not have firewalls then your core business systems are wide open and are at risk – business information has been scrambled in the past.

i)   Ensure that ALL downloaded information is scanned by adequate virus software (e.g. Norton Anti Virus) before it is opened and used (this also applies equally, if not more, to email). If any viruses are found they should be killed immediately. It is a simple task to configure this to be automatic.

j)   Finally, surfing the net can be a time consuming activity and is definitely 'more-ish' - you may need to consider limiting access to the net in your organisation (i.e. use the security and firewall to prevent access to sport channels). You should consider putting clauses in staff's contracts regarding misuse of the Internet.

WEB Pages

Most companies in the British Agricultural Supply industry now have web pages – why?

One of the major reasons people have put up web pages is that "everyone else is doing it - so we should". As a result of this naivety, the design, look and feel, navigation and usability of the sites often suffer. If we consider the majority of web sites out there we find that:-

a)   Most sites consist of static pages and are often a simple reproduction of the company brochure. Whilst intrinsically there is little wrong with this approach it does not give any incentive for users to revisit the site.

b)   Little thought is given as to "why will users visit the site?". All web sites should be customer-oriented; they should address what the customer wants and not what you as a company want – these are usually different objectives. You have to provide what customers want to see on the site or they will not come back – think of the web site as an extension of your service to them – if you don't have the product they want, they will go somewhere else for it. It is the same on the web.

c)   Most web sites do not change. A survey of the feed industry would find that most sites have not been updated in the last month, let alone this week. Sites must change regularly to keep up the level of interest. If the site does not change there is no incentive for visitors to come back.

d)   The web page is an opportunity to provide an on line service to your customers on a 24x7 basis. You can think of the web site as a customer service desk, a technical rep, a chat room, and (ultimately) a supermarket for the farmer.

e)   The Internet is a global service and the site can be accessed from anywhere in the world so the site should indicate areas of service provision.

When you are building your web site, or updating or refurbishing it, you need to consider the following:-

a)   Why do I need a web presence?
b)   Who is it aimed at?
c)   How will I attract new visitors?
d)   What are the key words that I should use to attract visitors and where should they be placed?
e)   How will I monitor the effectiveness of the site? Like any other marketing tool, you need to monitor its success not only in the number of hits but in terms of increased revenue.
f)   When designing the site you need to get the balance of usability and technology right. It should look good (whilst avoiding technical gimmicks), be easy and simple to navigate and should be quick to download. A useful tip would be to give your web developers the slowest PC and the slowest modem on which to design and deploy the site.
g)   When you design and build the site your company's image and philosophy will come through, so be professional. It can be the flagship and banner of your company and can often act as the first point of contact with your company.
h)   You must ensure that the web and web presence form a part of your overall business strategy – it cannot be outside your overall business goals.

E-Commerce

What is the state of e-commerce in British Agriculture? If you were to have asked that question in March 2000 you would have got the answer 'not a lot'. At that time there were a few sites that offered business to business e-commerce in the form of providing a product catalogue and allowing users to purchase via a shopping basket approach (e.g. FOL, Farming Express). Such sites were aimed at farmer customers and as such are classed as Business to Customer.

These however did not have any real-time integration into a customer's commercial back-end database.

There was only one product at that time that offered a customer the ability to interrogate their position and account, on-line and via the Internet. This was developed by ourselves (CAL) for the Trident Feeds arm of ABN.

A number of new trading ventures have sprung up since that time (e.g. Globalfarmers, Grainnet, Grainman, Cigrex, etc) that act as trading platforms. These products provide a mechanism where sellers can register products for sale (mostly grain) and buyers can register an interest to purchase. The software puts the two together via matching software and emails both parties and takes a fee for doing it. There is no integration provided to either party's back-end system, apart from email messages. They are, in effect, acting like on-line brokers and have provided a second layer of administration and are, therefore, increasing the cost of the transaction.

One product that offers true business-to-business functionality, with direct integration to back-end databases, is Inteletrade from CAL. This product is available for registered users and allows them to interrogate on the current contract position with Trident. It allows them to cancel deliveries, do price enquiries and place new orders – on line and with direct access to the back-end database.

To ensure the integrity of data, the whole product is wrapped in a security system called Securit-e. This allows the owners of the Inteletrade product to completely control who has access to what data and what functionality is made available to them. Our security system will soon allow end users to tailor the screens to match their own requirements!

Providing access to your own corporate database for your end users to view and manipulate selected information is only the start of the e-commerce revolution. With the advent of XML messaging and the adoption of BASDA standards the ability to share information between corporate databases, via the Internet, is getting ever nearer. These messaging protocols and standards allow one system to send a message that can be interpreted by another machine (You could also include people as a type of machine!). The ability to enter (for example) a purchase order in one system and that to automatically become a sales order in a different system can only reduce the cost of a transaction.

E-commerce is technically very simple. The difficulty always arises in the politics of business:-

a)   Agreeing standards for the information transmitted (this has largely been resolved and has been pulled in from other business sectors where e-commerce is more prevalent).

b)   Getting agreements between trading partners that orders received via the web are real and will be acted on.

c)   Ensuring that businesses are geared to respond to the receipt of orders from a different source (e.g., for electronic orders, how are mill managers informed of new demand?).

d)   Business needs to ensure that the system is available 24x7. This is a radically different working pattern than is normal in the agricultural industry. This means that business must invest in strategies to ensure that the systems are secure, available, backed up and are always relevant.

e) Businesses often fail to decide upon an e-commerce strategy and incorporate it into their business plans and more importantly fail to follow through and put enough resources into marketing the solution and monitoring its effectiveness.

f) In the e-commerce world there have been a number of spectacular failures (e.g. boo.com) and surveys suggest that 30% of e-commerce transactions go unfilled. This is generally because companies who have embarked on e-commerce were retail businesses (e.g. Tesco Home Deliveries, Toys 'R' Us) or, even worse, start-up businesses. They were not geared for delivering products to customers at short notice. The animal feed industry knows all about receiving orders this morning for delivery this afternoon, so the delivery infrastructure is in place.

Conclusion

The Internet and e-commerce are here to stay – those investing in e-commerce now are starting to see real benefits in their business with decreased costs, new revenue streams and new means of getting to market. The Forrester group are predicting that, in the US alone, the e-commerce market will reach $1,400 billion by 2003, compared with $251 billion in 2000.

To develop a successful e-commerce presence you should be asking yourselves:-

a) Can you afford to ignore the Internet?
b) What is my objective for e-commerce?
c) How does this fit in with my business goals?
d) Do you understand your target audience?
e) Is my company geared for the internet revolution?
f) What are your competitors doing?
g) If you don't do it, some one else will!
h) Remember, plan it, do it, review it and take it to the bank!

Useful Web Addresses

The following table is a non-exhaustive list of web sites which might be of use.

www.abm.org.uk   Associated British Meat

www.abpi.org.uk	Association of the British Pharmaceutical Industry
www.adas.co.uk	ADAS
www.bsi.org.uk	British Standards Institution
www.bva.co.uk	British Veterinary Association
www.caluk.com	Computer Applications Ltd.
www.cbot.com	Chicago Board of Trade (Market prices and analysis)
www.cheap-flights.co.uk	To search for and book cheap flights
www.environment-agency.gov.uk	Environment Agency
www.gafta.co.uk	GAFTA
www.hgca.co.uk	Home Grown Cereals Authority
www.ib-uk.gov.uk	Intervention Board
www.maff.gov.uk	MAFF
www.nfu.org.uk	National Farmers Union
www.noah.co.uk	National Office of Animal Health (NOAH)
www.nottingham.ac.uk	Nottingham University
www.nottingham.ac.uk/feedconf	Nottingham University Feed Conference
www.open.gov.uk/hse/press/press.htm	HSE Inspectorate (Information & press releases)
www.open.gov.uk/vmd/vmdhome	Veterinary Medicines Directorate (difficult to get into)
www.reuters.com	General News and limited price information
www.reutersinform.com	Live market price feeds and market analysis
www.rpsgb.org.uk	Royal Pharmaceutical Society of Great Britain (RPSGB)
www.ukasta.co.uk	UKASTA

4

SPECIALITY DIETS FOR HORSES

Anne Rodiek
California State University, Fresno, California, USA

Introduction

Many products are marketed in, for example, both the United Kingdom and the United States for horses with different dietary needs or in different physiological states. This trend appears to have followed the companion-animal food industry, with foods specially designed to meet the nutrient needs of growing, mature and older animals. Diets for growing horses, mature horses, and "senior" horses have been available for several years. More recently, diets to meet more specific dietary needs have become available, including various speciality diets for performance horses and diets for horses with respiratory ailments or with hyperkalemic periodic paralysis. Some of the claims made regarding the attributes of these "designer diets" appear to be well supported by scientific research, some appear to be logical extensions of scientific research, and some appear not to be related to science at all.

The scientific knowledge of horse nutrition is not as advanced as for other farm species, or even many companion animals. This is due partly to lack of funding for equine research, but is also due to the lack of easily measurable parameters of success. Horse rations that promote maximal rate of growth are easier to formulate (and their efficacy proven) than rations that promote "optimum or paced" growth rate, that provide "calm energy", or that improve athletic performance, because all of these desirable outcomes are difficult to assess quantitatively. Because of the difficulty in scientifically evaluating the effects of rations on somewhat nebulous characteristics, it is tempting to accept testimonials as evidence of efficacy in place of scientific evidence. Scientifically, this is a dangerous practice. However, it is probably not wrong to extrapolate, in at least some cases, from research in other species, or to broaden the application of specific research in the formulation of rations that attempt to meet real needs in the horse industry. In all cases, both the feed manufacturer and the customer would be well advised to understand how nutrients/feed/rations promote their claimed effects and to separate the science-founded claims from the non-science claims. It is the objective of this chapter to summarise some of the generally accepted, science-based, nutritional concepts that can be used to formulate or assess some of the speciality diets popular today, with particular emphasis on energy requirements.

Feeding older horses

Little research has been conducted on "old" horses. Even the age at which horses should be considered old is highly variable from horse to horse. In general, old horses that have trouble maintaining their weight have problems with improper worming, poor dentition or debilitating disease. Older horses also frequently have decreased ability to digest fibre, protein and phosphorus compared with young horses. Malabsorption of nutrients may be caused by chronic parasite damage or other alterations in the digestion processes in the small or the large intestine. Older horses may benefit from diets that are highly digestible and more concentrated in nutrients than young horses. Diets formulated for weanlings or yearlings have higher levels of nutrients and may better meet the needs of older horses with decreased digestion or absorption capabilities. Recommended nutrient concentrations/ kg of diet include: 120-140 g crude protein, 6 – 8 g calcium and 3 – 4 g phosphorus (Ralston, 1997). Feeds should be highly digestible and protein should be high quality. Raw materials such as soyabean meal, sugar beet pulp, processed grains and vegetable oil are more easily digested than others. Complete feeds that are pelleted and extruded and contain forage as well as concentrate (complete feeds) may increase intake. Coarse or poor quality hays should be avoided or minimised. Forage intake should not be discouraged, as gut fill is important. From a management perspective, the body condition of old horses should be carefully managed so that horses do not become too fat or too thin. Obesity will probably exaggerate arthritic conditions. Thin horses have little body reserves in case of illness and will probably suffer more in the cold. Adequate water intake is important.

Specific disease conditions may require specific dietary changes. Horses with pituitary dysfunction are frequently glucose intolerant and should not be given diets high in soluble carbohydrate (grains). They also often have low blood levels of vitamin C and therefore may benefit from vitamin C supplementation.

Horses with renal failure should receive diets low in protein (less than 120 g/ kg), calcium (less than 6 g/kg) and phosphorus (less than 3 g/kg).

Microbial supplements, excess vitamins and herbs are frequently added to rations for older horses, but their scientific merit is largely unsubstantiated.

Feeding athletic horses

Unlike the older horse, a large amount of research has been devoted to optimising nutrition for the athletic horse. However, because horses compete in such a wide variety of athletic activities, the research has been broad and sometimes difficult to compare, and the goals of various diets are different and sometimes even conflicting.

It is well accepted that equine athletes require rations of high energy concentration. The sources of the energy are generally either carbohydrates (cereal grains) or fats (vegetable oils or animal fat). The digestion, absorption and metabolism of these components are quite different, and the timing and form of the absorbed energy-yielding components may be factors that affect energy availability for different athletic endeavours.

High-fat diets (up to 150 g/kg or more) provide concentrated energy, not only to fuel athletic performance, but also to maintain body weight during training. Fats, both from dietary or adipose tissue sources, can only be metabolised aerobically and cannot supply energy directly during anaerobic work. However, some research indicates that dietary fat not only enhances the efficiency and amount of fat utilised during work, but also promotes preferential selection of fat as a metabolic fuel over glycogen during aerobic work. This fat preference may reduce glycogen utilisation until very high levels of work are required, and thus enhance glycogen availability during high intensity, anaerobic work. Several studies on fat supplementation to horses in training and performing high intensity work have been inconclusive. One study showed that fat supplementation (70 g added fat/kg diet) to a cereal-grain/forage diet increased both pre-exercise muscle glycogen levels (20.31 vs. 24.06 mg muscle glycogen/g wet tissue, for unsupplemented and fat-supplemented horses, respectively, 21 days into a feeding regime) and total muscle glycogen utilisation (8.05 vs. 11.30 mg muscle glycogen/g wet tissue in unsupplemented and fat-supplemented horses, respectively) during anaerobic work (Hughes, Potter, Greene, Odom and Murray-Gerzik, 1995). Another study reported no difference in either parameter in fat-supplemented horses (Eaton, Hodgson, Evans, Bryden and Rose, 1995). Similarly, inconclusive results have been shown in studies designed to examine the effects of fat supplementation on lactic acid production during anaerobic work, although there is some tendency to believe generally that, when fat utilisation is increased, lactic acid accumulation is diminished.

Horses performing low intensity, prolonged work appear to be most likely to benefit from high fat supplementation. Endurance horses are frequently fed diets high in fat, low in soluble carbohydrate, low in protein and high in fibre. Vegetable oil, sugar beet pulp and high-quality grass hays are popular, supplemented with electrolytes to balance the losses in sweat, particularly in hot climates or during competition.

Horses involved with high intensity work (racehorses in particular) may be fed fat, but generally are also fed liberal amounts of cereal grains. There are two schools of thought about how race horses should best be "fuelled". Some believe that high-fat diets are best because of their overall higher energy concentration and because of their "glycogen-sparing" effect. Others believe that fat is not a suitable energy substrate for horses as fat cannot be catabolised anaerobically. These latter individuals believe that carbohydrate is the better source, particularly during sprint work, because glucose can be metabolised anaerobically, making it available for the very intense, short-duration work of racing. Much of the difference in opinion about optimum metabolic fuel is caused by different opinions or findings on the limiting factors in various types of work. In very short duration, high intensity work, depletion of muscle glycogen stores, or even depletion of glycogen in fast twitch muscle fibres, may be the limiting factor in racing performance. In somewhat longer duration, but still quite intense work, accumulation of lactic acid and the feeling of pain and fatigue associated with its accumulation may be the first factor that causes work intensity to be diminished. Of course, all of these factors depend not only on diet, but also on fitness, muscle fibre type distribution and stride attributes of the horse, as well as numerous other factors, some which remain to be identified.

In the future, there may be enough nutritional information to warrant the production of diets specially formulated for horses competing in high intensity, short duration (primarily anaerobic) work and for horses competing in low intensity, long duration (primarily aerobic) work. There are already some supplements marketed that allegedly contain complex, concentrated carbohydrates for quick boosts in energy or for improved recovery after strenuous exercise. These supplements may provide readily available energy for some types of exercise. They should be used conservatively, however, to determine their most useful application, and also to guard against potentially harmful carbohydrate-overload.

In all diets for athletic horses, the supply of adequate levels of forage is necessary to maintain gut health and to maintain adequate water and electrolytes in the large intestine.

It is generally agreed that high protein levels do not enhance athletic performance. However, legume hays should not be discounted entirely as they are higher in energy and some minerals, and often more palatable than grass hays, so may make a greater contribution to the energy and nutrient requirements of horses.

Not only ration content, but also timing of a meal prior to exercise may affect a particular athletic performance, or at least affect the availability of energy substrates during work performance. Different feeds produce different blood concentrations of glucose and free fatty acids over different time courses after consumption. Several studies have shown that blood glucose (and insulin) levels peak approximately 2 to 3 hours after eating (Stull and Rodiek, 1988; Hintz, Argenzio and Schryver, 1971; and Glade, Gupta and Reimers, 1984) and that different feeds produce different glycaemic (blood glucose) responses. Maize, and a combination of maize and alfalfa, produced greater peak concentrations of glucose than iso-energetic amounts of alfalfa hay or a combination of maize and maize oil in resting horses (Stull and Rodiek, 1988). If exercise starts when glucose levels are high, glucose is used to a greater extent to fuel the exercise. If exercise starts when glucose levels are low, fatty acids are mobilised to a greater extent. Stull and Rodiek (1995) reported that exercising horses showed greater declines in blood glucose concentrations when exercised 1 or 4 hours after a meal (supplying approximately 17.2 MJ of digestible energy) of maize than after a meal of alfalfa hay or when fasted. During exercise, glucose concentrations in maize-fed horses fell below pre-feeding concentrations, while glucose in alfalfa-fed and fasted horses gradually rose through exercise. Blood insulin concentrations mirrored the glucose responses, being higher in the maize-fed horses. Lactic acid concentrations tended to be higher in horses given maize than fasted horses or horses fed alfalfa hay. Free fatty acid levels responded approximately inversely to the glucose levels, showing highest levels during exercise in the fasted horses and lowest levels in the maize-fed horses (Figures 4.1 and 4.2).

When exercise is of short duration and intense, the goal may be to provide quickly-available, high blood glucose levels. When exercise is of long duration, however, having large amounts of glucose being catabolised and causing a drop in blood glucose levels early in work may not be ideal. This is a relatively new area of nutritional research but, in the future, feed labels may contain information not only about amounts of feed to be offered, but also about the timing of meals before exercise (soon before or long before) to enhance further the ability of the diet to optimise athletic performance.

Feeding laminitis-prone horses (feeding "natives")

In contrast to the equine athletes, where maximising energy intake is often the goal, there are other types and breeds of horses and ponies that are the "air ferns" of the equine world. They are the ultimate "good doers," seemingly able to maintain themselves on nothing and get fat on anything, no matter how meagrely they are fed. The metabolic differences in these animals have not been established conclusively, but many animals that maintain weight very easily also seem to be more sensitive to laminitis when overfed even relatively minimally. Some evidence points to pituitary or thyroid dysfunction, associated with detrimental alterations in glucose metabolism, as predisposing factors for laminitis.

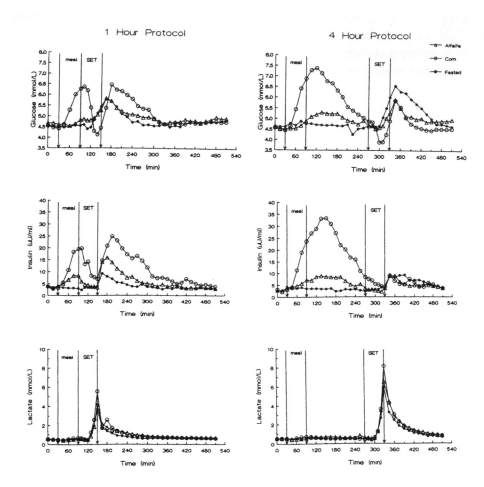

Figure 4.1. Mean plasma glucose, serum insulin and plasma lactate concentrations of fasted, maize-fed, or alfalfa-fed horses exercised (SET) 1 or 4 hours after eating.

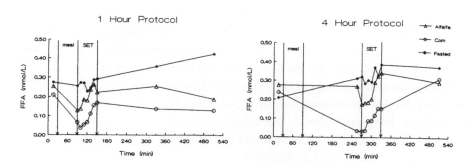

Figure 4.2. Mean free fatty acid concentrations of fasted, maize-fed, or alfalfa -fed horses exercised (SET) 1 or 4 hours after eating.

Soluble carbohydrates are strongly associated with laminitis, particularly if they pass to the large intestine and are fermented by the microbes therein. Microbial fermentation of soluble carbohydrates results in acid production (primarily lactic acid), damage to the mucosa of the large intestine, and escape of toxins from the digestive system into the circulation. More than one theory has been proposed regarding how blood-borne toxins from the digestive system cause laminitis in the hooves, but the concept of an ultimate cause being carbohydrate-overload in the large intestine is generally accepted. Grain is not the only feed material that supplies soluble carbohydrates; pasture grasses can also cause problems because of their high concentration of fructans (polysaccharides of fructose) that largely by-pass the small intestine and are fermented by microbes in the large intestine. Longland, Cairns and Humphreys (1999) showed that fructan concentration varied greatly in a common species of United Kingdom pasture grass (*Lolium perenne*), due to variety as well as seasonal and diurnal variation. Fructan content of *Lolium perenne* varied throughout the year, with lowest levels (100 to 140 g/kg dry matter) in April and highest levels (300 to 420 g/kg dry matter) in July. Time of day and environmental conditions were also shown to influence fructan concentration. Fructan levels were lowest during the evening and morning hours (0 to 140 g/kg dry matter) and highest about midday (140 to 300 g/kg dry matter). Warm, sunny weather produced higher fructan concentrations than cool, rainy weather.

Formulating low energy, high fibre and palatable feeds is both the key and the challenge. Horses should not be given rations where the feed allocation is less than 15 g/kg body weight, since a minimum of dietary bulk is required for optimum health of the digestive tract. Rations must also meet needs for vitamins and minerals, which may mean supplementation is needed if the ration is based on grass hay (particularly for trace minerals and vitamin E and possibly also for protein). Long hay, coarse mixes or cubes are preferred over pelleted diets as the latter are usually consumed more readily, and pass through the digestive tract faster, than diets comprised of forages with larger particle size. Newer methods of manufacturing cubes compress the hay pieces along the long axis of the cube instead of the short axis of the cube, further increasing the size of the hay pieces incorporated into each cube. "Softer" cubes are preferred to "hard, glassy" cubes as they are easier to chew. Slow rates of eating pacify the horse for longer periods of time and are also more natural, considering the "nibbler" vs. "meal eater" evolution of the horse. A slower rate of passage increases the amount of time that feed is retained in the digestive tract. This enhances microbial digestion and also increases the feeling of satiety.

Palatability of grass hays can sometimes be increased by the addition of molasses (sparingly) or other feed flavourings; horses generally like apple or aniseed flavouring. On the one hand, it is not economical to pay for processing and bagging of hay when long-stem, baled hay can be fed (and is a superior source of long fibre) but, on the other hand, a specially blended feed for at-risk animals is more likely to contain the proper amounts and distribution of supplemental vitamins and minerals throughout the mix. If long-stem hay is preferred, a supplement of vitamins and minerals of low dietary energy concentration can be mixed with alfalfa hay or wheat bran as a relatively low energy, low soluble-carbohydrate, palatable carrier.

From a management perspective, overweight horses, ponies of almost all descriptions, and previously-foundered horses/ponies should be fed as continuously as possible. This means that several small meals should be offered daily instead of only two "bolus" feedings. Clarke, Roberts and Argenzio (1990) summarised the physiological changes that occur in the digestive tract and the circulatory system when ponies were fed a single large meal. It was reported that, when ponies ingested a large single meal (as opposed to several smaller meals), blood volume decreased by approximately 15% (approximately 1 litre in a 160 kg pony) within one hour of feeding. The post-feeding reduction in blood volume was regained within two hours of eating; however, a second, smaller decrease was detected six hours after eating, attributable to colonic secretion during microbial fermentation in the caecum and colon. Meal feeding also promoted a faster rate of passage, allowing a greater proportion of the ration, particularly a greater proportion of soluble carbohydrates, to reach the large intestine. This resulted in intense fermentation, as demonstrated by decreased luminal pH and increased organic acid production, similar to what is observed during carbohydrate overload that results in laminitis. These cyclic changes in blood volume and rate of microbial fermentation may not be harmful *per se* but, when these stresses are compounded by other stressors, they may push a horse or pony over the "brink" into colic or laminitis.

Another management consideration is housing of overweight or laminitis-prone animals. Like should be housed with like, and group-feeding of horses with widely different nutritional needs should be minimised. An older or larger horse or pony can easily bully younger or smaller pasture-mates and over-consume dangerous amounts of feed. Free exercise is generally considered to be beneficial to burn excess energy consumed and to enhance circulation in the lower legs.

Palatability of feeds

Rations formulated for horses require not only that feed ingredients meet the nutrient and energy requirements of the horses for which they are formulated, but also that diets are millable and palatable. Horses generally prefer maize oil to other plant and animal fat sources. One study ranked palatability of common horse feeds in order, based on time of consumption (D'Ambrosi, 2000). Table 4.1 shows the relative palatability ranking of 16 different common equine feeds, categorised by feed type, with oats as a standard in each category.

Table 4.1. Relative palatability ranking of common horse feeds*.

Ranking of feeds	Cereal grains	Hays	Other feeds
1	Maize, steam-rolled	Oats, whole	Oats, whole
2	Oats with molasses (100 g/kg)	Alfalfa, baled	Wheat bran, loose
3	Barley, steam-rolled	Alfalfa, cubed (large)	Alfalfa with molasses (100 g/kg)
4	Wheat, whole	Timothy, baled	Sugar beet pulp, loose
5	Oats, whole	Winter forage (various cereals, primarily oats)	Carrots, fresh, sliced
6	Oats with vegetable oil (100 g/kg)	Bermuda grass, baled	Rice bran, pelleted

(D'Ambrosi, 2000)

*   Maize: *Zea mays*
   Oats: *Avena sativa*
   Barley: *Hordeum aestivum*
   Wheat: *Tricium aestivum*
   Alfalfa: *Medicago sativa*
   Timothy: *Phleum pratense*
   Winter forage, primarily oats: *Avena sativa*
   Bermuda grass: *Cynodon dactylon*
   Wheat bran: *Triticum aestivum*
   Sugar beet pulp: *Beta saccharifera*
   Carrots: *Daucus ssp*
   Rice bran: *Oryza sativa*

Feeding the 'hot' horse

This is an area in which science-based nutrition appears to be left behind. No studies, to date, have been published that show the effects of specific feeds or diets on measurable behaviours. The concept of "grain high", while popular, is without scientific substantiation. It makes sense that a horse with excess energy intake may display more exuberant behaviour than a horse that is deficient in energy, but the idea that grain makes horses behave badly, or that fat supplementation promotes better behaviour, is based on testimonial evidence only. The same must be said about herbal supplements, although the knowledge of the author on this subject is limited.

Summary

The demand for speciality diets for horses exceeds the scientific knowledge of the special nutritional, physiological and management needs of these horses. While it is tempting to formulate rations that "might" improve various conditions, it is probably better to educate horse owners about which nutritional formulations are science-based and which are not. It makes owners better able to care for their horses and helps them to understand the potential and the limitation of nutrition to influence the health and well being of their horses.

References

Clarke, L.L., Roberts, M.C. and Argenzio, R.A. (1990) Feeding and digestive problems in horses: Physiologic responses to a concentrated meal. In *The Veterinary Clinics of North America. Equine Practice – 1990*, pp 433-450. Guest edited by H.F. Hintz. W.B. Saunders Co. Philadelphia, PA, USA

D'Ambrosi, R.D. (2000) Personal communication.

Eaton, M.D., Hodgson, D.R., Evans, E.L., Bryden, W.L. and Rose, R. J. (1995) Effect of a diet containing supplementary fat on the capacity for high intensity exercise. In *Equine Exercise Physiology 4 – 1995 (Proceedings of the Fourth International Conference on Equine Exercise Physiology)*, pp 353 –356. Edited by N.E. Robinson. Equine Veterinary Journal Limited, Newmarket, Suffolk, UK.

Glade, M.J., Gupta, S. and Reimers, T.J. (1984) Hormonal responses to high and low planes of nutrition in weanling thoroughbreds. *J. Animal Sci.*, 59: 658-665.

Hintz, H.F., Argenzio, R.A. and Schryver, H.F. (1971) Digestion coefficients, blood glucose levels and molar percentage of volatile acids in intestinal fluid of ponies fed varying forage-grain ratios. *J. Animal Sci.*, 33: 992-995.

Hughes, S. J., Potter, G.D., Greene, L.S., Odom, T.W. and Murray-Gerzik, M. (1995) Adaptation of thoroughbred horses in training to a fat supplemented diet. In *Equine Exercise Physiology 4 – 1995 (Proceedings of the Fourth International Conference on Equine Exercise Physiology)*, pp 349 - 352. Edited by N.E. Robinson. Equine Veterinary Journal Limited, Newmarket, Suffolk, UK.

Longland, A.C., Cairns, A.J. and Humphreys, M.O. (1999) Seasonal and diurnal changes in fructan concentration in *Lolium perenne*: implications for the grazing management of equines pre-disposed to laminitis. *Proceedings of the 16th Equine Nutrition and Physiology Symposium*, Raleigh, North Carolina. pp 258-259.

Ralston, S.L. (1997) Geriatric Horse Nutrition. In *The Veterinarian's Practical Reference to Equine Nutrition – 1997*, pp 27 – 31. Edited by Kent N. Thompson. Published jointly by The American Association of Equine Practitioners and Purina Mills, Inc., St. Louis, MO, USA.

Stull, C. and Rodiek, A. (1988) Responses of blood glucose, insulin and cortisol concentrations to common equine diets. *J. Nutrition,* 118: 206-213

Stull, C. and Rodiek, A. (1995) Effects of post prandial interval and feed type on substrate availability during exercise. In *Equine Exercise Physiology 4 – 1995 (Proceedings of the Fourth International Conference on Equine Exercise Physiology)*, pp 362 - 366. Edited by N.E. Robinson. Equine Veterinary Journal Limited, Newmarket, Suffolk, UK.

5

FORMULATING FEEDS FOR SOWS – FEEDING GESTATING SOWS WITH HIGH FIBRE DIETS

R.DURÁN GIMÉNEZ – RICO

Technical Department, Trouw Nutrition España, Nutreco, Ronda de Poniente, 9, 28760, Tres Cantos, Madrid, Spain.

Introduction

"Prediction is very difficult, especially about the future" (Niels Bohr)

Hunger in pregnant sows is considered a major welfare problem, especially in the Northern European countries. Pregnant sows are fed restrictively; the biggest effect of feeding during pregnancy is on the body weight of the sow and excessive dietary energy intake leads to heavier sows with increased maintenance requirements. Furthermore, those animals might show a decrease feed intake in the following lactation and, most probably, an excessive loss of body condition at weaning. On the other hand, under-feeding should be avoided, as this is associated with a risk of impairing reproductive performance and poor body condition. Pregnant sows are fed once or twice per day and, with these allowances, their maintenance requirements and those necessary to support a small amount of maternal growth and the growth of conceptus are met. However, these conditions might not fulfil the feeding motivation needs of these animals; this means that satiety level is not reached and sows will develop stereotyped activities (Meunier-Salaün, 1999). The latter are characterised by behavioural patterns performed repetitively in a fixed order and without any evident function. Abnormal behaviours (including stereotypes) are indicators of poor welfare (Vestergaard, 1997).

Fibre is suggested as a means of increasing bulkiness of the diet and acts as an energy diluent. The influence of fibre in the diet will depend on the inclusion rate, the type of fibre used and the non-starch polysaccharide (NSP) content. Increased bulkiness may be beneficial in pregnant sows that are currently fed well below their appetite potential. In this way, the utilisation of fibre in diets for pregnant sows and the improvement of their welfare by means of a reduced stereotypic behaviour observed in some recent studies (Brouns, 1994-95; Vestergaard, 1997; Ramonet, 1999) has gained popularity in countries such as The Netherlands and

Denmark. The former decided to make it compulsory to formulate diets for pregnant sows with a minimum of 140 g CF/kg diet or 340 g OOS("Overige Organische Stoff", or Other Organic Material) / kg diet. Within the Dutch Agricultural Legislation of September 1998, a list of fibre-containing ingredients was included as possible materials for feeding to pregnant sows ("Varkensbesluit", Sept., 1998). Moreover, the above mentioned 140 g CF/kg level was also adopted as a way of enabling the formulation of complete diets without the need for extra dry or green roughage.

At the same time, the new evaluation system of net energy for sows (Noblet, 1993-1994) was the subject of study by the main feed compound industries in The Netherlands: a useful link between fibre-containing ingredients – new NE system for sows – and welfare of sows was established.

This chapter describes a detailed programme of work conducted by Nutreco in the Netherlands. The overall objective was to obtain a practical solution to feeding pregnant sows high levels of fibre without impairing reproductive performance.

Utilisation of energy from ingredients

The Gross Energy (GE) of ingredients is not fully available to animals. Part of that energy is lost in faeces, in urine and in gaseous products resulting from fermentation in the hindgut. The final metabolic utilisation of the energy from feed implies also a loss of energy called heat increment. Different ratios are calculated in order to appreciate the efficiencies at each level. Mean energy losses, expressed as proportion of GE, together with mean efficiencies, were proposed by Noblet (1994) (Figure 5.1).

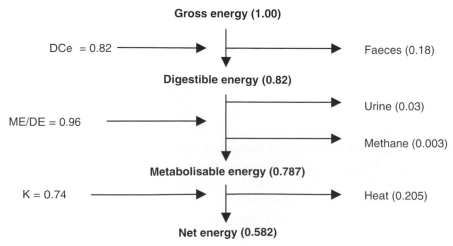

Figure 5.1. Energy utilisation in pigs (Noblet, 1994)

For most pig diets, the coefficient of energy-digestibility (DCe) varies between 0.70 and 0.90 (Noblet, 1996); for individual ingredients, variations are greater (0 to 1.00). These variations are due to differences in faecal digestibility of the constituents of organic matter. Starch and sugars, both soluble carbohydrates, are highly digestible (>0.95); protein and crude fat between 0.60 and 0.95, while most of the variation in DCe is associated with the presence of fibre, referred to subsequently as OOM (from the Dutch OOS).

OOM DEFINITION

OOM is a dietary category that includes all carbohydrates, including lignin, but with the exception of starch and sugar. Essentially, OOM represents that proportion of a feed and/or ingredient that might be potentially fermented in the hindgut, or even partially fermented in the lower parts of the small intestine (e.g. OOM from sugar beet pulp).

Though seen as an artefact, OOM is defined as (OM – CP – EE – Starch – Sugar), where OM=organic matter; CP = crude protein; EE = ether extract. It has become a practical tool for formulating high-fibre diets for gestating sows (SRC, Nutreco 1997-2000) while designing and performing the trials on sow feeding. Differences between the NSP and OOM contents of several feedstuffs are shown in Table 5.1.

Table 5.1. Comparison of crude fibre (CF) and non-starch polysaccharide (NSP) contents of some commonly used ingredients (g/kg DM; Longland and Low, 1995). OOM values in g/kg DM (CVB, 1999 and CVB Verkorte tabel, 1994)

	CF	NSP	OOM
Barley	53	147	221
Wheat bran	85	360	474
Soya hulls	354	868	758
Soyabean meal	85	190	284
Rapeseed meal	120	286	392
Sugar beet pulp	174	635	736
Citrus pulp	118	415	600
Wheat straw	400	632	877

Differences in NSP and OOM are as predicted if fibre is defined in accordance with Low (1985), who described fibre as "the non-starch polysaccharides (NSP) plus lignin". In other words, OOM is the NSP plus the lignin content of a given ingredient or complete feed.

DIGESTIVE UTILISATION OF FIBRE

In agreement with Noblet and Shi (1993), the digestibility coefficient of energy in diets decreased linearly when the NDF content increased. According to the previous definition of OOM, the neutral detergent fibre (NDF) fraction is included in OOM; therefore, it is assumed that the more OOM in the diet, the less digestible it becomes for pigs. Additionally, sows showed a higher capacity to digest fibrous diets than growing pigs (Shi and Noblet, 1993). The cell wall components present in the dietary fibre fraction (NDF or NSP or OOM) show much lower digestibility coefficients than other components of ingredients or diets. Ramonet (1999, Table 5.2) showed differences in DC of fibre between growing pigs and sows for various ingredients. While the NDF fraction of sugar beet pulp is well digested by both pigs and sows, the difference for maize fibre was 0.42. Pectins in sugar beet pulp are water soluble and more digestible than lignin, in which maize fibre is very rich. Consequently, variability in DCe for different ingredients is associated with fibre origin as well as cell wall content (lignin content).

Table 5.2. Coefficient of digestibility of fibre fractions in growing pigs and sows (adapted from Ramonet, 1999).

	Wheat bran		*Maize fibre*		*Sugar beet pulp*	
	Pig	*Sow*	*Pig*	*Sow*	*Pig*	*Sow*
NDF	0.48	0.51	0.39	0.81	0.84	0.89
ADF	0.26	0.33	0.38	0.82	0.89	0.92
NDF – ADF	0.58	0.62	0.38	0.82	0.88	0.93
Cellulose	0.25	0.32	0.38	0.82	0.87	0.91
NSP	0.46	0.54	0.36	0.82	0.89	0.92

NDF = neutral-detergent fibre; ADF = acid-detergent fibre; NSP = non-starch polysaccharides.

The major site of fibre digestion is the large intestine, in which microbial fermentation takes place. The volatile fatty acids produced here, as a result of fermentation processes, can meet between 0.05 and 0.25 of the energy needs of pigs. The proportions of volatile fatty acids produced in the large intestine can be modified by fibre level. In accordance with results of Noblet and Shi (1993), and also with those presented in Table 5.2, a higher contribution of fibre to the final energy utilisation of pigs can be assumed when considering sows.

ENERGY VALUE FOR SOWS

In 1997, a programme was started to investigate digestibility of feed ingredients for sows. Twenty-five ingredients were evaluated. Barley was considered as the reference ingredient; therefore its faecal-energy digestibility was determined three times by including it in the basal diet at 980 g/kg. The remaining 24 ingredients were mixed at different inclusion rates in combination with the basal diet (barley+ marker + premix) and their digestibility coefficients were calculated by difference from the one measured for barley (average of three determinations; 0.835 ± 0.0082). The inclusion levels (g/kg) of the ingredients in the basal diet ranged from 250 (citrus pulp, soya hulls, lucerne) to 400 (wheat bran, sugar beet pulp, soyabean meal). Diets were evaluated with non-pregnant sows fed 2 kg daily, in two feeds, for 15 days (10 days acclimatisation; 5 days total faecal collection).

RESULTS AND FURTHER APPROACH (Net Energy for sows: NEsows).

Data for DCe are presented in Table 5.3, in comparison with data published by Noblet (1997).

Table 5.3. Coefficient of digestibility of gross energy in 11 feed ingredients given by INRA (Noblet, 1997) and SRC (1998, internal communication).

	Growing pigs	*Sows*	*Sows (SRC)*
Wheat	0.875	0.892	0.908
Barley	0.826	0.835	0.836
Maize	0.889	0.916	0.917
Peas	0.808	0.871	0.908
Soyabean meal	0.845	0.894	0.914
Wheat bran	0.585	0.646	0.712
Maize gluten feed	0.686	0.755	0.716
Sugar beet pulp	0.698	0.764	0.795

For a given ingredient, energy digestibility is higher as live weight increases and as feeding level falls (Noblet, 1997). Growing pigs might represent finishing pigs, but they do not represent sows. The DE data obtained were converted to NE using the conversion factors identified in Figure 5.1, with an assumption that NE was 0.77 of ME (Table 5.4).

The NE values for sows proposed by Noblet (1997) differ slightly from those proposed by SRC (1998) when considering cereals. However, differences already existed between values given by CVB (1998) and Noblet (1997).

Table 5.4. NE values (MJ/kg) proposed for sows (SRC, 1998).

	NE pigs (CVB, 1998)	NE sows (SRC, 1998)	NE sows (Noblet, 1997)
Wheat	9.76	9.91	11.02
Barley	9.13	9.13	10.28
Maize	10.70	10.01	11.33
Peas	9.38	9.92	10.25
Soyabean meal	8.06	9.98	9.43
Wheat bran*	5.97	7.77	7.89
Maize gluten feed	6.93	7.82	7.44
Sugar beet pulp	9.02	8.68	8.28

(*):NE sows = (DCe Wheat bran / DCe Barley) * NE pigs Barley

The conclusion from this phase of the programme was that DE values obtained for growing- finishing pigs will underestimate the energy contribution of a number of raw materials currently used in practical formulations of diets for sows. Therefore, for NE (and also for DE), it is advisable to assign two energy values to each ingredient: NEpigs and NEsows.

EFFECT OF FIBRE ON WELFARE

A comprehensive review on the effect of fibre on welfare of gestating sows was recently published by Meunier-Salaün (1999). The current chapter will summarise the major findings and present additional data from Roelofs (1998).

Stereotypies can be described as regularly repeated movements that are identical and without any obvious function (Ödberg, 1978, cited by Roelofs, 1998). A stereotypic behaviour develops as follows:

1.   Environmental factors cause frustration;
2.   It results in a burning "desire" to do something;
3.   This leads to obvious behaviour, but this behaviour is not fulfilling;
4.   The behaviour is repeated;
5.   The behaviour becomes automated, simplified and inherent.

Stereotypies probably arose from the behaviour that was typical at the outset (in this case, lack of feed). It is not impossible that, in performing these ritual behaviours, the animal helps to lessen the seriousness of the frustration. However, animals that show stereotypic behaviour have serious welfare disorders. It is beneficial for the individual animal to express stereotypies because it is a compromise for what they are missing (more feed in order to chew…). Nevertheless, for the whole system, it is a sign that animal welfare is seriously compromised (Vermeer, 1997).

The pig possesses considerable drives both to forage for feed over long periods of time (and distances) and to manipulate thoroughly what it finds in its mouth. The natural situation can be contrasted with modern systems of production where the full daily allowance is provided in a trough and can be consumed in a feeding time of 20 minutes or less (Roelofs, 1998). This shows that it is impossible for sows to perform their natural behaviour in modern systems of production. In other words, the low current feeding level of pregnant sows is probably related to the stereotypic behaviour of sows housed in tethers and/or cages.

Vestergaard (1997), Ramonet (1999) and Meunier-Salaün (1999) have recently published very convincing data showing the benefit of dietary fibre on the welfare of pregnant sows. The figures presented here are from the results of Roelofs (SRC, 1998). The objective was to study the effect of feeding high-fibre diets, combined with different feed presentations, on the stereotypic behaviour of mature non-pregnant Hypor sows. Twenty sows, varying from the third to the eleventh parity, were fed during two periods of three weeks a diet containing either a standard (65 g CF/kg) or high level of fibre (155 g CF/kg), in pellet or meal form. The total daily intake of energy was calculated to be the same for both diets (maintenance level). Daily allowances of the standard and the high-fibre diets were, respectively, 2.0 kg and 2.4 kg. Feed was delivered once per day during a three-week experimental period (two weeks of adaptation). The animals were closely observed for five days, during the last week, from 7.30 a.m. until 10.00 a.m. Eating time (total time eating the feed allowance) increased with the high-fibre diet (P=0.001). It took approximately 2.4 minutes/kg longer (5 minutes extra/daily allowance) to consume the high-fibre feed. When the standard diet was compared with the high-fibre diet, active behaviour decreased by 13% (P=0.003) with the high-fibre diet, lying behaviour increased by 15% (P=0.009; Figure 2) and oral stereotypic behaviour decreased by 10% (P=0.007). When the pellet form was compared with the meal form, active behaviour decreased by 5% (P=0.027) with the meal form and lying behaviour decreased by 16% (P=0.002; Figure 5.3). The meal form showed the largest differences between the two fibre levels.

These results indicate that high-fibre diets may have beneficial effects on the welfare of sows.

The sows fed on the standard diet were more active and spent less time lying down than the sows fed on the high-fibre diet (Table 5.5), confirming the observations of other authors (Broom and Potter, 1984; Robert, et al., 1993; Haaksma, 1994; Brouns, et al., 1994; Ramonet, et al, 1997). This increase in resting time and decrease in active behaviour for the sows fed on the high-fibre diet has been associated

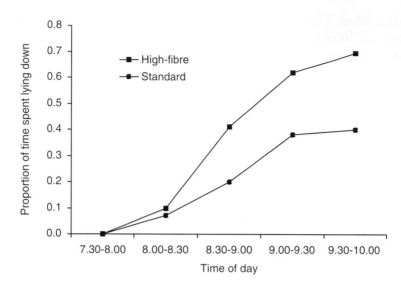

Figure 5.2. Effect of dietary fibre on sow behaviour

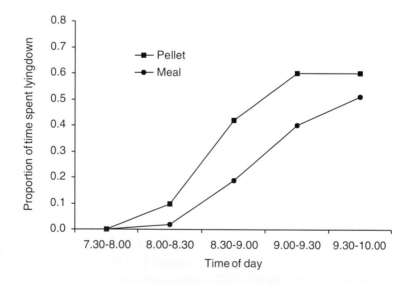

Figure 5.3. Effect of diet form on sow behaviour.

with a decrease in feeding motivation, as a consequence of gut fill (probably due to the fibre content of feed) controlling appetite. However, feeding motivation will only stop if the diet fulfils the minimal metabolic needs of the animal (Robert, *et al*, 1993).

Table 5.5. Effect of diet on sow behaviour: proportion of time spent on various activities (8.00 – 10.00 a.m.) (Roelofs, 1998)

| Behavioural category | Standard | | High fibre | | |
	Mean	S.E.	Mean	S.E.	P
Oral stereotypic	0.65	0.259	0.55	0.278	0.007
Stereotypic lying	0.18	0.240	0.24	0.268	0.077
Stereotypic standing/sitting	0.47	0.324	0.31	0.267	<0.001
Stereotypic 'rooting'	0.20	0.179	0.18	0.109	NS
Active	0.86	0.182	0.73	0.270	0.003
Lying	0.30	0.296	0.45	0.282	0.009

It was interesting to note the effect when pellet and meal feeding were compared (Table 5.6), possibly associated with physical characteristics of the food and their influence on eating behaviour. The meal form may not have provided enough oral stimulation (a minimum chewing activity) and this might have led to increased feeding motivation and increased restlessness. It was concluded, however, that neither feed form was adequate in terms of satisfying foraging behaviour, and therefore improving welfare. Although sows fed the pellet form spent a greater proportion of their time lying down (0.45 vs 0.29), they also performed more stereotypic activities while seated/lying (0.28 vs 0.13). However, it is presumed that when the sows lie down, they feel calmer and welfare is improved.

Table 5.6. Effect of diet form on sow behaviour: proportion of time spent on various activities (8.00 – 10.00 a.m.) (Roelofs, 1998)

| Behavioural category | Pellet | | Meal | | |
	Mean	S.E.	Mean	S.E.	P
Oral stereotypic	0.60	0.277	0.59	0.270	NS
Stereotypic lying	0.28	0.268	0.13	0.220	<0.001
Stereotypic standing/sitting	0.32	0.279	0.46	0.315	<0.001
Stereotypic 'rooting'	0.21	0.171	0.17	0.118	NS
Active	0.82	0.227	0.77	0.247	0.027
Lying	0.45	0.289	0.29	0.285	0.002

It could be concluded that high-fibre diets have a positive, though variable, effect on welfare in three ways: decreasing oral/rooting activity, reducing total activity and increasing lying time. The latter effect might be variable, but it was assumed to reduce stereotypic behaviour. Feed form had no clear effect, although differences in oral stereotypic behaviour were greater between the two CF levels when the sows were fed in meal form (differences of 0.17 vs 0.03); which, although not significantly different, indicated a clear trend (Table 5.7).

Table 5.7. Interaction between feed form and CF level on behaviour of mature sows: proportion of time spent on various activities (8.00 – 10.00 a.m.) (Roelofs, 1998)

	Pellet form				Meal form			
	Standard CF		High CF		Standard CF		High CF	
	Mean	S.E.	Mean	S.E.	Mean	S.E.	Mean	S.E.
Oral stereotypic	0.61[ac]	0.291	0.58[ab]	0.265	0.68[c]	0.222	0.51[b]	0.288
Lying	0.39[ac]	0.293	0.51[a]	0.276	0.20[b]	0.268	0.38[c]	0.275
- Stereotypic	0.26[a]	0.266	0.31[a]	0.270	0.10[b]	0.184	0.17[b]	0.249
- Passive	0.13[a]	0.180	0.21[ab]	0.254	0.10[a]	0.160	0.22[b]	0.225
- sham chewing	0.25[a]	0.266	0.29[a]	0.265	0.10[b]	0.179	0.16[c]	0.249

[a,b,c]: Different superscripts in the same row indicate differences among treatments ($p<0.10$).

AD LIBITUM FEED INTAKE

Gestating sows are fed below their maximal feed-intake capacity, thus it is well accepted that their natural rooting and chewing behaviour is not fully expressed, which compromises welfare. On the other hand, sows cannot be fed *ad libitum,* as this would not allow maintenance of optimum body condition and live weight to be achieved during gestation. This would result in reproductive failures due to lower feed intake during lactation, excessive weight loss at weaning and an increased interval from weaning to oestrus. Sows are currently fed in tethered in cages; in this way, feeding to achieve defined body condition is not a real challenge for the farmer, other than time taken to distribute the daily allowances. However, in many European countries concerns about animal welfare have already been reflected by legislation. Countries like the UK, The Netherlands and others ban the use of tether and stall housing for dry sows. One existing clause in European legislation has major implications for sow feeding practice: EU directive 91/630 on The Welfare of Pigs states that *"all pigs must be provided with a diet appropriate to their age, weight and behavioural and physiological needs, to promote a positive state of health and well-being".* The *"Codes of recommendations for the welfare of livestock, pigs"*(UK, 1994), *"Act on indoor keeping of gestating sows and gilts"* (Denmark, 1994) or the *"Varkensbesluit"* (The Netherlands, 1994), are clear and well defined examples of a commitment towards an improved welfare of pigs and sows. In September 1998, in The Netherlands, a minimum level of either 140 g CF/kg or 340 g OOM/kg of diet became compulsory when feeding gestating sows with no access to any other fibre source (also clearly defined by law in their quality and quantities/sow/day) additional to the daily standard diet allowance.

The same EU directive, 91 630/EU, states that from the 1st of January 2006 keeping dry sows in tethers will not be allowed. They can be kept in pens (limited space and movement) or in groups. Expensive housing designs, which offer individual feeding facilities in group-housing systems, have been developed to prevent aggression due to feed competition. However, the problem still exists and sows still suffer from hunger express stereotypic behaviours.

Both the need to feed *ad libitum* in order to reduce the development of stereotyped oral behaviour and the need to house sows in groups make the utilisation of high fibre diets an interesting approach to feeding dry sows. Moreover, precise nutritional evaluation of the fibre-like ingredients will allow nutritionists to formulate the most appropriate diets for these animals.

When diets were offered *ad libitum*, Brouns et al. (1995) showed how, depending on the NSP/OOM origin, the effect varies considerably, especially in the case of sugar beet pulp; levels of unmolassed sugar-beet pulp as high as 650 g/ kg were used. This could be a simple effect of reduced palatability, as the organoleptic properties of such a diet may be very different from those of wheat, maize or barley-based diets. However, the other diets used could be also considered to have low palatability (Table 5.8): diets formulated with barley straw, oat husks, malt culms, rice bran or wheat bran were also extremely high in CF or NSP/ OOM.

Table 5.8. Mean voluntary feed intake (last two weeks of experimental period) of diets by gestating sows (Brouns *et al.*, 1995).

	Inclusion level (g/kg)	*DE (MJ/d)**	*VFI (kg/d)**	*CF (g/d)*	*OOM (g/d)*	*DOOM (g/d)***
Sugar beet pulp	650	18	2.3	294	1060	867
Straw	360	42	6.4	1029	2450	802
Oat husks	370	58	7.7	1088	3047	809
Malt culms	460	52	6.8	624	2171	797
Rice bran	610	52	7.6	1485	2810	806
Wheat bran	670	54	7.1	590	2204	738

*: *p<0.001.*
**: DOOM refers to the digestible (fermentable) OOM.

Materials particularly high in soluble fibres (sugar beet pulp) had the greatest effect on restricting feed intake. According to a number of Danish studies (National Committee for Pig Breeding, Health and Production, Annual Report, 1998) the long-term feeling of satiety is achieved by means of dry complete feed, with large amounts of sugar beet pellets, or wet complete feed with sugar beet waste or maize silage. The awareness of the adverse effects of hunger on abnormal behaviour, competition for food and overall aggression has led to attempts to devise practical and cheap systems whereby group-housed pregnant sows could be fed *ad libitum*. The system has been proven to be effective when based on diets with a high inclusion of sugar beet fibre (Noordman, 1999): results from the survey on 37 Dutch farms with this *ad libitum* system concluded, among other things, that the mean gestational feed intake was 3.5 kg/sow/day (sugar beet pulp-based). Because the energy content was calculated to be approximately 10.7 MJ DE/kg, sows would have eaten on average 37.5 MJ DE/d, which matches the 35 MJ DE /d of standard feed (2.7 kg/d) given to sows on some reference farms. The report commented on the possibility that water holding capacity and the viscosity properties of sugar beet pulp could explain these effects.

There seems to be a relationship between daily intake of DOOM and total feed consumption (Table 5.8): in other words, once a sow has consumed a certain quantity of DOOM (750-900 g/d), maximum intake capacity is also attained. In this way, by including high amounts of soluble fibre (thus high amounts of DOOM) in the diet, voluntary feed intake might be "naturally" restricted (Figure 5.4).

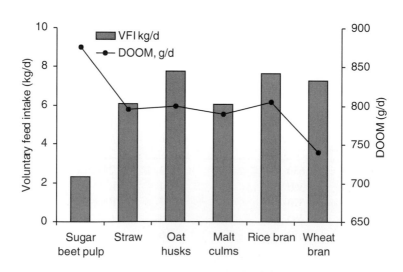

Figure 5.4. VFI (kg/d) and DOOM (g/d) from Brouns *et al.* (1995).

A number of experiments with mature barrows were conducted at the Nutreco experimental research unit in The Netherlands. Recent results showed a similar trend to those observed by Brouns *et al.* (1995). The combination of sugar beet pulp (400 g/kg diet) and citrus pulp (200 g/kg diet) gave intakes of 2.28 kg/d and 765 g DOOM/d, whereas intakes of 2.82 kg/d and 760 g DOOM/d were observed when citrus pulp (400 g/kg), combined with soya-bean hulls (200 g/kg), was offered *ad libitum* to mature entire males (194 kg live weight) kept in groups of five animals per pen. The difference was not significant (p=0.13), but suggest a requirement for a certain amount of fermentable fibre (DOOM) before the animal stops eating when offered free access to feed.

Reproductive performance with high-fibre diets

The high intake capacity of dry sows limits any attempts at offering a conventional diet *ad libitum* because of concomitant obesity. However, diets with a high inclusion level of fibrous ingredients (soluble fibre type) might be able to restrict nutrient intake to acceptable levels. Moreover, the interest in fibrous ingredients – especially in The Netherlands - for gestating-sow diets has increased for three reasons already mentioned: firstly, Dutch regulations (September 1998) established a minimum level of 140 g CF/kg (or 340 g OOM/kg) in all gestation diets (a trend that might be followed in other EU countries); secondly, the low nutrient requirements of gestating sows make the use of bulky diets diluted with fibrous ingredients possible and economically attractive, while minimizing feed restriction, thus improving welfare; thirdly, during the past decade, research performed by INRA in France indicated that high-fibre raw materials are more efficiently utilised by sows when compared with growing-finishing pigs. Therefore, because of the higher digestibility and utilization of dietary fibre in sows, there seems to be a great potential for using various fibrous feeds.

Introduction of high-fibre feeds, especially those high in soluble fibre (those that are readily fermented: sugar beet pulp, citrus pulp), during gestation seems practical because no adverse effects on reproductive performance (i.e., pigs farrowed, birth weight, mortality) were observed (Varel and Yen, 1997). Even slight improvements were observed in some studies (Matte et al., 1994): indeed, the use of bulky diets diluted with fibrous ingredients has beneficial effects on body condition and on some measures of reproductive performance in sows. The growth of the litter was increased by approximately 20% during the second parity when a bulky diet, based on wheat bran and maize cobs, was fed to sows during the first and second parities. This improvement may be linked to the beneficial effect of this diet on sow well-being, which was also reported. In previous experiments, Robert and Matte (1993) found an increase of 12.8% (more than 3 hours/d) in resting time, and a reduction of at least 50% in incidence of stereotypies during the second gestation was observed when the same diet was fed. This altered behaviour could have decreased the energy requirements for maintenance and consequently left more energy available for sow and conceptus growth.

Furthermore, positive effects on development of the digestive tract could be expected which, in turn, could increase voluntary feed intake capacity of sows during subsequent lactation. Voluntary feed intake of sows during lactation is frequently insufficient to meet nutrient demands for maintenance and maximal milk production. Insufficient feed intake during lactation results in sows losing large amounts of weight. Excessive weight loss has been associated with several common reproductive problems, including an increased interval from weaning to oestrus (King and Williams, 1984 cited by Noblet et al., 1998) and reduced litter size in subsequent litters (Kirkwood et al., 1987, cited by Williams, 1998). The effect of *ad libitum* feeding on the gastrointestinal tract of young sows has been examined (National Committee for Pig Breeding, Health and Production, Annual Report 1998, Denmark). Although no data are given in the report, they reported a difference in the weight of stomach, small and large intestines plus rectum from five young sows fed with an ordinary feed mix, compared with the weight of the gastrointestinal tract from five young sows fed given a diet containing large amounts of sugar feed waste.

Vestergaard (1997) showed that when nutritional requirements were met, sows had a large capacity to increase their total feed intake during pregnancy, without influencing on body weight and litter size (Tables 5.9 and 5.10). Dietary fibre increased the gut fill of sows, leading to a greater live-weight gain during pregnancy, but also to a larger weight loss at farrowing, when a standard lactation diet was offered. Thus, the true body-weight gain was the same for all three groups of sows.

Table 5.9. Chemical analysis and calculated net energy content of diets (Vestergaard, 1997).

	Standard	*SBP[1]*	*Mixed[2]*	*Lactation*
Dry Matter (DM; g/kg)	889	888	891	884
Crude Protein (g/kg DM)	154	158	144	203
Ether Extract (g/kg DM)	55	55	62	75
Crude Fibre (g/kg DM)	46	127	135	36
Net Energy (MJ/kg DM)	9.03	8.49	6.72	9.88

[1] SBP: 500 g sugar beet pulp + 315 g barley
[2] Mixed : 415 g barley + 150 g grass meal + 150 g wheat bran + 200 g oat hulls.

Feed intake during pregnancy (kg/d) was highest for the mixed diet and lowest for the standard diet. However, net energy intake (MJ/d) was the lowest for the sugar beet pulp diet. Mean piglet weight at birth was lower for the sugar beet pulp diet but, at weaning, the mean weights of piglets were the same for all three treatments (Table 5.10). According to these results, it is possible to feed pregnant sows with high levels of fibre; the sows showed capacity to reach the same energy intake without detrimental effects on performance.

Table 5.10. Effect of three diets (Table 9) on intake and reproductive performance (Vestergaard, 1997)

	Standard	*SBP*	*Mixed*	*P-value*
Feed Intake during gestation				
- mean for three cycles (kg)	281.1[a]	289.7[b]	368.2[c]	<0.001
- mean for three cycles (MJ NE)	2296[a]	2216[b]	2282[a]	<0.001
Feed intake during lactation				
- mean for three cycles	1450[a]	1528[b]	1438[a]	0.002
Performance (mean of three cycles)				
- pregnancy live-weight gain (kg)	58.7[a]	61.6[a]	67.9[b]	<0.001
- live-weight loss at farrowing (kg)	21.4[a]	26.7[b]	25.9[b]	<0.001
- live-weight loss during lactation (kg)	8.0	6.3	8.0	0.270
Litter performance (mean of three cycles)				
- number of piglets born alive	10.8	10.9	10.7	0.860
- number of piglets weaned	9.5	9.3	9.3	0.900
- live weight at birth (kg)	1.62[a]	1.47[b]	1.65[a]	<0.001
- live weight at weaning (kg)	8.4	8.1	8.4	0.550
- litter weight at weaning (kg)	77.2	73.7	75.8	0.350

[a,b,c] means in same row with different superscripts bare significantly different.

A similar experiment (Durán, 1998) was conducted to measure the effect of bulky diets, based on soluble fibre (163 g CF/kg with sugar beet pulp, 450 g/kg and soyabean hulls, 200 g/kg) or insoluble fibre (170 g CF/kg with grass meal, 230 g/kg, wheat bran, 100 g/kg, lucerne, 60 g/kg and straw, 80 g/kg), compared with a standard diet (70 g CF/kg), on performance of sows during gestation and their following lactation. Though no significant differences were found, the trends were similar to those observed by previous researchers (Matte, 1994, Vestergaard, 1997, Ramonet, 1999). Results shown in Table 5.11 were obtained from 93 Hypor sows fed with the diets described; a standard vs. two bulky ones – insoluble and soluble fibre types. Mean feed allowances were 2.4, 3.0 and 2.9 kg respectively, leading to similar intakes of NE (21MJ/d), more than 325 g CP/d and around 10 g ileal-digestible lysine/d; sufficient to meet maintenance requirements and allow for maternal gain and conceptus growth.

Table 5.11. Performance of sows fed on diets differing in fibre content, with similar intakes of net energy (Durán, 1998)

	Standard diet	Insoluble-fibre diet	Soluble-fibre diet
Parity number	3.9	4.6	4.5
Initial weight (kg)	193.4	190.8	186.5
Live-weight gain during gestation (kg)	51.1	49.6	54.1
Net maternal gain during gestation (kg)	34.0	32.1	37.7
[1]NEpigs intake (MJ/d)	21.0	21.0	21.0
[1]NEsows intake (MJ/d)	22.3	23.7	23.2
Lactation			
- live-weight loss (kg)	-15.2	-15.9	-17.5
- litter birth weight (kg)	17.5	17.2	16.5
- litter live-weight gain after 24 days (g/d)	2332	2269	2269
- Feed intake during lactation (kg/d)	7.6	7.6	7.8

[1] NEpigs and NEsows: net energy values for growing pigs or sows, as given in Table 4.

Fibre-based diets were efficiently used by multiparous sows, but there was no positive effect of fibre on litter birth weight due to, as suggested by Matte (1994), an energy-sparing effect of fibre in the feed. However, there was a higher net weight gain in those sows fed the soluble fibre. It is debatable whether differences in weight arose from an increased hind-gut mass (weight), a gut-fill effect of diet (larger amounts of digesta in the hind-gut), or to an "extra energy intake" by the fibre-fed sows because of the higher intake of energy when the latter was expressed as NE sows (previously defined). Litter performance, birth weight, growth rate and weaning weight did not differ among treatments. Although sows fed the soluble fibre showed a higher feed intake during lactation, this was not reflected in any litter improvement.

Conclusions

It is possible to improve the welfare of sows by correct feeding of fibrous ingredients, without impairing reproductive performance.

It may be predicted that although this practice is already being used in some countries, the utilization of fibrous diets for gestating sows will keep increasing across the rest of the EU members.

Acknowledgements

I am grateful to all the staff at SRC, especially Jos Weerts, who taught me all that is not in the books about management and feeding of sows: *Hartelijk bedankt Jos !!*. Also to Yannick Ramonet and Rosil Lizardo, from INRA: *merci beaucoup !!*. Thanks to C. van der Peet-Schwering, from Rosmalen.

References

Broom, D.M. and Potter, M.J., (1984) In: Factors affecting the occurrence of stereotypes in stall-housed dry sows. J. Unshelm, G. van Putten and K. Zeeb (Editors), Proceedings of the International Congress on Applied Ethology in Farm Animals. pp. 229-231. Kiel, Germany. KTBL, Germany.

Brouns, F., Edwards, S.A. and English, P.R. (1994) Effect of dietary fibre and feeding system on activity and oral behaviour of group housed gilts. *Applied Animal Behaviour Science*, **39,** 215-223.

Brouns, F., Edwards, S.A. and English, P.R. (1995) Influence of fibrous feed ingredients on voluntary intake of dry sows. *Animal Feed Science and Technology*, **54,** 301-313.

CVB, Veevoedertabel 1998 (1998).

CVB, Veevoedertabel 1999 (1999).

CVB, Verkorte tabel 1994 (1994). Centraal Veevoederbureau, cvb-reeks no. 15, August 1994.

Durán, R. (1998) Effect of bulky diets based on beet pulp/soyabean hulls or grass/ straw on gestation performance of sows and their subsequent lactation . *Report S0297 – 03. SRC, Nutreco.*

Haaksma, J. (1994) Bietenpulp: Voederwaarde en het effect op het gedrag bij varkens. *Bergen op Zoom (Instituut voor rationele suikerproduktie).*

Longland, A.C. and Low, A.G. (1995) Prediction of the energy value of alternative feeds for pigs. In *Recent Advances in Animal Nutrition - 1995*, pp 187-209. Edited by P.C. Garnsworthy and D.J.A. Cole. Nottingham University Press, Nottingham.

Manteca, J. (1998) El bienestar animal en la granja del 2000. *II Jornadas Técnicas de porcino de Nanta, S.A.*

Matte, J.J., Robert, S., Girard, C.L., Farmer, C. and Martineau, G.-P. (1994) Effect of bulky diets based on wheat bran or oat hulls on reproductive performance of sows during their first two parities. *Journal of Animal Science,* **72,** 1754-1760.

Meunier – Salaün, M-C. (1999) Fibre in diets of sows. In *Recent Advances in Animal Nutrition - 1999,* pp 257-273. Edited by P.C. Garnsworthy and J. Wiseman. Nottingham University Press, Nottingham.

National Committee for Pig Breeding, Health and Production, Annual Report, 1998.

Noblet , J. Etienne, M. and Dourmad, J.-Y. (1998) *Energetic efficiency of milk production.* The Lactating sow, Wageningen Pers, 1998, pp: 113-130

Noblet, J. (1996) Digestive and metabolic utilization of dietary energy in pig feeds: comparison of energy systems. In *Recent Advances in Animal Nutrition - 1996,* pp 207-232. Edited by P.C. Garnsworthy, J. Wiseman and W. Haresign. Nottingham University Press, Nottingham.

Noblet, J. (1997) Two energy values for each ingredient. *Pig International,* **27,** *no.10, 25-26.*

Noblet, J. and Shi, X.S. (1993) Comparative digestible of energy and nutrients in growing pigs fed ad libitum and adult sows fed at maintenance. *Livestock Production Science,* **34,** 137-152.

Noblet, J. and Shi, X.S. (1994) Effect of body weight on digestive utilization of energy and nutrients and diets in pigs. *Livestock Production Science, 37, 323-338.*

Noordman, M. (1999) Onbeperkt voeren van dragende zeugen. *Hendrix UTD VSN*

Ramonet, Y. (1999) *Réponses comportementales, digestives, hormonales et métaboliquesá la restriction énergétique chez la truie gestante: effects d´une incorporation de fibres alimentaires dans la ration.* Thesis, University of Rennes.

Ramonet, Y.; Meunier-Salaün, M-C.; Dourmad, J.Y. (1997) Effets d'une incorporation de parios végétales dans la ration alimentaire sur l'activité comportementale des truies gestantes. *Journées Rech. Porcine en France, 29, 167-174.*

Robert, S., Matte, J.J., Farmer, C., Girard, C.L. and Martineau, G.-P. (1993) High fibre diets for sows: effects on steretypies and adjunctive drinking. *Applied Animal Behaviour Science, 37,* 297.

Roelofs, J. (1998) Effect of feed form and crude fibre level on the oral stereotypic behaviour of tethered sows. *Report S1197 – 02. SRC, Nutreco.*

Shi, X.S. and Noblet, J. (1993) Contribution of the hindgut to digestion of diets in growing pigs and adult sows: effect of diet composition. *Livestock Production Science,* **34,** 237-252.

Van der Peet-Schwering, C. (2000) *The use of welfare diets during gestation.* Nutrition of sows during gestation and lactation. Wageningen, Tuesday March 28, 2000.

Varkensbesluit (1994). Helpdesk Dierenwelzijn sector Varkenshouderij Nr 5.

Vermeer, H.M.; Ekkel, E.D.; Groot, J.S.M. de; Klooster, C.E. van 't; Peet, G.F.V. Van der Swinkels, J.W.G.M (1997), Welzijn van varkens: van verzorgingsvoorschriften naar verzorgingsmaatregelen. *Proefverslag nummer P1.173, Rosmalen, Nederland.*

Vestergaard, E-M. (1997) Phd Thesis: *The effect of dietary fibre on welfare and productivity of sows.* Institut for Husdyrbrug og Husdyrsundhed, Veterinary School, Copenhagen, Denmark.

Vestergaard, E-M. and Danielsen, V. (1998) Dietary fibre for sows: effects of large amounts of soluble and insoluble fibres in the pregnancy period on the performance of sows during three reproductive cycles. *Animal Science,* **68**, 355-362.

Williams, I.H. (1998) Nutritional effects during lactation and during the interval from weaning to oestrus. In *The Lactating Sow,* pp: 159-181. Wageningen Pers, Wageningen.

6

HIGH-ENERGY DIETS FOR POULTRY - EFFECTS OF DIET COMPOSITION ON PERFORMANCE AND CARCASS QUALITY

JULIAN WISEMAN

University of Nottingham, Sutton Bonington Campus, Loughborough, Leics LE12 5RD

Introduction

It is widely recognised that performance and carcass quality of broilers is influenced considerably by variations in energy and nutrient inputs during growth. However, addition of fats to diets to achieve required levels of apparent metabolisable energy (AME), together with high dietary levels of AME *per se*, may promote excess lipid accretion in fast-growing broiler chickens fed under *ad libitum* conditions. The consequences of such high levels of body fat relate both to evisceration losses and to the fat content of saleable meat. These have been the subject of many previous investigations examining the effects of nutrition, in particular manipulation of the energy:protein ratio (E:P), on carcass quality (e.g. Bartov *et al.* 1974a). It is generally considered that AME cannot be regarded in isolation from E:P. This chapter will discuss, initially, the consequences of changing E:P (under a variety of circumstances) for performance and carcass characteristics, although a consideration of protein quality is outside the scope of this discussion. Similarly, discussion of actual AME values of high-energy ingredients may be found elsewhere (e.g. Wiseman, 2001 for fats; Short *et al* 1999; Wiseman 2000; Wiseman *et al* 2000, for wheat, which is the major energy-yielding commodity for poultry diets in much of Europe and other regions of the World).

Maintaining fat deposition at acceptable levels, whilst producing chicken meat economically from current commercial strains of broiler chickens, will be a compromise between achieving a satisfactory carcass and realising the potential of the bird for maximum live-weight gain. However, fat deposition is an integral part of broiler growth and its extent and distribution throughout the carcass is an important aspect of meat quality. Hakansson *et al.* (1978a,b) examined the distribution of lipid in tissues within broiler carcasses, as influenced by dietary nutrient concentration and age, but no data for individual fat depots were reported.

A subsequent area that will be considered is 'catch-up' growth. It has been suggested that feed (and dietary energy) restriction early in the life of the broiler, followed by a refeeding period, may improve efficiency of feed utilization and reduce fat content of the carcass (Kubena *et al.* 1974; Plavnik & Hurwitz 1985; Jones & Farrell 1992*a*), although it is likely that this response might be accompanied by a smaller finisher body weight (Jones and Wiseman 1985). Such early feed restriction has been considered in the context of prevention of metabolic diseases (e.g. ascites) associated with growth rates that are too rapid. Other studies have failed to confirm this catch-up following early feed restriction (Deaton *et al.* 1973; Griffiths *et al.* 1977). Susbilla *et al.* (1994) demonstrated that early feed restriction did improve subsequent growth rate, although no differences were observed for feed conversion ratio and carcass fat content; Jones and Farrell (1992*b*) examined body fat growth. However, the majority of these studies have tended to consider the consequences of diet changes at one point in time (slaughter) rather than over time up to slaughter, although Susbilla *et al.*, (1994) examined the growth pattern of internal organs.

There has been considerable interest recently in the manipulation of fatty acid profiles of broiler meat (e.g. see review by Leskanich and Noble, 1997). Notwithstanding the current trend (at least in the UK) for dietary fat levels to be reduced (in favour of increasing cereal starch so as to maintain overall AME), change in carcass fatty acids, in response to dietary changes, is an important area for study. It allows accurate dietary regimes to be developed to produce carcasses with a specific fatty acid profile.

Developments in carcass quality are generally assuming greater importance as broiler meat markets in many regions move from a volume-dominated to a quality-controlled position.

Effect of E:P on performance and carcass quality

Maximisation of lean-tissue growth rate is one of the fundamental objectives of any meat-producing animal enterprise. Lean-tissue (muscle) accretion is an energy-demanding process, which suggests that dietary protein (P - assumed to be of optimum quality) inputs can only result in lean tissue gain if there is adequate dietary energy (E) intake. E:P may be altered by varying E at constant P, maintaining E and varying P or by altering both simultaneously.

Narrowing E:P by reducing AME whilst maintaining P (quantitatively and qualitatively) was studied (Trial 1 - Jones and Wiseman, 1985; Trial 2 - University of Nottingham unpublished data) as the literature appeared to offer conflicting evidence on the effect of this variable during starter (0 to 23 days of age) and finisher (24 to 49 days of age) time periods. For example, Griffiths *et al.* (1977) concluded that the starter regime imposed had no effect on abdominal fat pad (AFP) weight at eight weeks of age, in contrast to the observations of Kubena *et al.* (1974) and March and Hansen (1977). The AFP is commonly employed as an indicator of total carcass fat content; correlations between the two are usually high.

Trial 1 was based on constant P at varying E and the second (Trial 2) employed varying P at constant E. Dietary regimes are presented in Tables 6.1A and 6.1B respectively. The transfer between starter and finisher diets was at 23 days of age. Performance, together with basic carcass responses from Trial 1 are presented in Table 6.2A. The data indicate that a low-energy starter diet was associated with less carcass fat at 24 days of age and that this effect persisted until slaughter at 49 days of age. However, this was associated with a smaller bird and a poorer feed conversion ratio. Consequently, an economic appraisal of such a regime would not favour any approach that compromised the two major performance indicators, which are rate and efficiency of live-weight gain. Nevertheless, if financial incentives for carcass quality were introduced, as in other livestock production systems, then modifying carcass fat content may be an attractive proposition.

Table 6.1A. Dietary Regimes Employed in Trial 1 (Jones and Wiseman, 1985)

Diet	*AME levels (CP constant)*					
	L		*M*		*H*	
	AME (MJ/kg)	*CP (g/kg)*	*AME (MJ/kg)*	*CP (g/kg)*	*AME (MJ/kg)*	*CP (g/kg)*
0 - 23 d						
Starter	10.78	230	12.78	230	14.78	230
E:P (kJ AME/g CP)	46.9		55.6		64.3	
24 - 49d						
Finisher	10.78	200	12.78	200	14.78	200
E:P (kJ AME/g CP)	53.9		63.9		73.9	

Table 6.1B. Dietary Regimes Employed in Trial 2 (University of Nottingham, unpublished data)

Diet	*CP levels (AME constant)*					
	L		*M*		*H*	
	AME (MJ/kg)	*CP (g/kg)*	*AME (MJ/kg)*	*CP (g/kg)*	*AME (MJ/kg)*	*CP (g/kg)*
0 - 23 d Starter	12.75	190	12.75	230	12.75	270
E:P (kJ AME/g CP)	67.1		55.4		47.2	
24 - 49d Finisher	12.75	170	12.75	203	12.75	235
E:P (kJ AME/g CP)	75.0		62.8		54.3	

Table 6.2A. Selected Performance and Carcass Data from Trial 1 - L, M and H refer respectively to dietary levels of AME (from Jones and Wiseman, 1985)

Starter	*FI (kg/b)*	*LWG (kg/b)*		*FCR Period*	*Overall*	
L	1.09	0.50		2.180		
M	1.11	0.63		1.762		
H	1.09	0.73		1.493		

	Finisher	*FI (kg/b) Period*	*LWG (kg/b) Period*	*LWG (kg/b) 0 - 49d*	*FI (kg/b) 0 - 49d*			*Carcass Fat (g/kg DM)*
	L	4.17	1.48	1.98	5.26	2.818	2.657	326
L	M	3.52	1.58	2.08	4.61	2.228	2.216	362
	H	3.29	1.68	2.18	4.38	1.958	2.009	399
	L	4.11	1.62	2.25	5.22	2.537	2.320	364
M	M	4.16	1.68	2.31	5.27	2.476	2.281	387
	H	3.49	1.72	2.35	4.6	2.029	1.957	412
	L	4.08	1.51	2.24	5.17	2.702	2.308	379
H	M	4.27	1.52	2.25	5.36	2.809	2.382	403
	H	3.73	1.65	2.38	4.82	2.261	2.025	438

Table 6.2B. Selected Performance and Carcass Data from Trial 2 - L, M and H refer respectively to dietary levels of CP (University of Nottingham, unpublished data)

Starter	FI (kg/b)	LWG (kg/b)		FCR Period	Overall
L	1.30	0.64			2.031
M	1.28	0.74			1.730
H	1.29	0.82			1.573

Finisher		FI (kg/b) Period	LWG (kg/b) Period	LWG (kg/b) 0 - 49d	FI (kg/b) 0 - 49d			Carcass Fat (g/kg DM)
	L	3.24	1.32	1.96	4.54	2.455	2.316	362
L	M	3.03	1.26	1.90	4.33	2.405	2.279	346
	H	3.53	1.31	1.95	4.83	2.695	2.477	346
	L	3.47	1.27	2.01	4.75	2.732	2.363	384
M	M	3.45	1.34	2.08	4.73	2.575	2.274	303
	H	3.63	1.46	2.20	4.91	2.486	2.232	320
	L	3.7	1.25	2.07	4.99	2.960	2.411	368
H	M	3.78	1.36	2.18	5.07	2.779	2.326	335
	H	3.29	1.25	2.07	4.58	2.632	2.213	281

Differences between treatments in Trial 2 were also important. Although dietary protein affected performance, in terms of live-weight gain and food conversion ratio, there was a suggestion that the very high protein diet was used less efficiently. This implies that the excess protein was catabolized rather than contributing to increased lean-tissue deposition. Interestingly, this response was associated with a reduction in carcass fat, suggesting that dietary energy was used in catabolism of excess protein rather than being deposited as fat. This therefore represents a further means of altering carcass quality, although feeding high (supra-optimal) protein diets would be associated with increased nitrogen excretion (which would have serious environmental consequences) and may be metabolically challenging to the bird as it attempts to accommodate excessive protein degradation.

The consequences of changing nutrient regimes can also be viewed over time following modelling of data. Table 6.3 presents data from a programme in which three starter diets were fed from 0 to21days of age, all with E:P of 57.2 (achieved by varying both E, MJ/kg, and P, g/kg; 11.39/199, 12.77/224, 14.15/248 for H, C and L respectively) and three finisher diets were fed from 22 to 70 days of age, with E:P of 64.8 (11.85/183, 13.07/202, 14.29/220 for H, C and L respectively). This gave five treatment combinations - HH, HL, CC, LH and LL. Representative

Table 6.3. Estimates of time (d) to reach specific live weight (g) and weight of carcass components (g) at this live weight

Diet	Live weight (g)								
	1000	*1250*	*1500*	*1750*	*2000*	*2250*	*2500*	*2750*	*3000*
Time to reach Live weight (d)									
HH[*]	24	29	32	35	39	42	46	50	55
HL	27	32	36	40	44	47	52	56	61
CC	27	31	35	38	42	46	49	53	58
LH	30	33	36	39	43	46	49	53	57
LL	32	36	40	44	48	53	57	61	66
Carcass Component									
Carcass Protein (g)									
HH	166	214	264	314	365	417	469	521	573
HL	170	226	284	342	400	456	510	563	612
CC	178	230	282	334	385	436	484	532	577
LH	171	225	279	333	386	438	488	536	581
LL	181	239	296	350	402	451	497	538	577
Carcass Fat (g)									
HH	93	124	159	198	240	286	337	394	456
HL	82	113	147	181	217	252	288	322	355
CC	88	121	158	196	238	277	317	358	399
LH	97	134	174	217	262	309	359	410	463
LL	82	106	131	159	189	222	258	299	345

birds were slaughtered at regular intervals, allowing assessment of responses through use of the Gompertz equation (Wiseman and Lewis 1999). Performance (in terms of time taken to reach a specific live weight) was influenced considerably by diet, as was carcass composition (in terms of fat and protein content)

CONTROL OF FEED INTAKE

In the studies of Leeson *et al* (1996a and b) conflicting evidence was presented on the ability of the broiler to compensate feed intake for differing dietary AME concentrations. Whilst feed intake increased with decreasing AME concentration, intake of AME (MJ over the experimental period studied; 35-49d) was reduced (from 35.3 to 28.0 MJ and from 34.7 to 28.8MJ) irrespective of E:P (Leeson *et al* 1996a). Intake of AME actually increased when the experimental period was 0-49d (Leeson *et al* 1996b). Bartov (1998) subsequently observed that increasing dietary AME had no effect on feed intake but, consequently, was associated with an increase in AME intake (54.4 to 58MJ) over the experimental period (from 7-49d). The hypothesis that the bird is able to control its energy intake is therefore not tenable.

General observations (see Table 6.4) tend to support the principle of a modest control of feed intake (to an extent determined by many factors, including E:P), but this 'compensatory mechanism' is insufficient, so increasing dietary AME concentration is associated with increasing AME intake.

Table 6.4. Influence of dietary AME (MJ/kg) on feed intake (FI, kg/bird) and AME intake (MJ/bird)

Data are from:

T1 Jones and Wiseman, 1985; T2 University of Nottingham, unpublished data; T3 Wiseman and Lewis, 1999

	FI *(kg.bird)*	*AME* *(MJ/kg)*	*AME* *consumed* *(MJ/bird)*
T1	4.17	10.78	45.0
	3.52	12.78	45.0
	3.29	14.78	48.6
	4.11	10.78	44.3
	4.16	12.78	53.2
	3.49	14.78	51.6
	4.08	10.78	44.0
	4.27	12.78	54.6
	3.73	14.78	55.1
T2	3.24	12.75	41.3
	3.03	12.75	38.6
	3.53	12.75	45.0
	3.47	12.75	44.2
	3.45	12.75	44.0
	3.63	12.75	46.3
	3.70	12.75	47.8
	3.78	12.75	48.2
	3.29	12.75	41.9
T3	3.74	14.29	53.4
	3.76	11.85	44.5
	3.49	13.07	45.7
	3.59	14.29	51.2
	3.69	11.85	43.7

COMPENSATORY GROWTH RATE

Data from HH and LH (Wiseman and Lewis, 1999) were used to examine the effects of early feeding of diets with low nutrient and energy concentration, followed by a return to a more adequate diet, on the ability of birds thus treated to achieve rates of performance comparable with those fed adequately throughout. Gompertz values for 'M' (the age in days at which growth is maximum) were 33.3 and 38.0 for HH and LH respectively, indicating that maximum growth rate was achieved earlier on the HH treatment. Similarly, A + C (asymptote) values (which represent mature weight) were higher for HH (4.40 kg) than for LH (4.15 kg), suggesting a more persistent growth rate. Whilst these data and analyses suggest that HH birds were always heavier than LH, these differences became progressively smaller with increasing age. Rogers *et al.* (1987), with a more acute restriction of early feed, also observed that Gompertz asymptote values were always below those of the control group. These observations on the influence of dietary treatment support conclusions on time taken to reach specific live weights (as presented in Table 6.3).

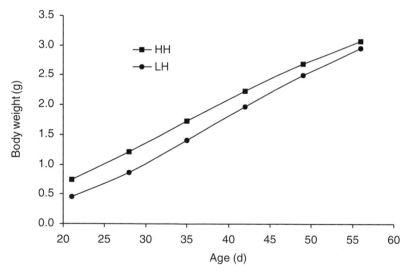

Figure 1. Comparison between fully fed (HH) and early-feed restriction followed by re-alimentation (LH) on broiler growth (from Wiseman and Lewis, 1999)

Early feed restriction in broilers has been investigated because of the possible reductions in body fat, but not overall body weight, that might result. There was the suggestion (Wiseman and Lewis, 1999) that LH birds gradually approached the live weight of HH (see Figure 6.1), a response that was associated with increasing feed intake by LH (but which did not reach that of HH). Comparisons between treatments for carcass protein and fat content require separation on the basis of whether time or live weight is the independent variable. LH birds had slightly higher carcass protein and fat contents when comparisons were at the same live weight (Figure 6.2); on the basis of time (Figure 6.3) carcass fat and protein contents for LH approached, but did not reach, those for HH.

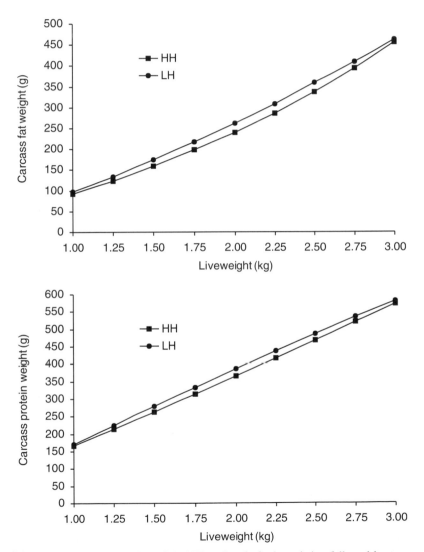

Figure 6.2. Comparison between fully fed (HH) and early-feed restriction followed by re-alimentation (LH) on composition of live weight in broilers based on live weight (from Wiseman and Lewis, 1999)

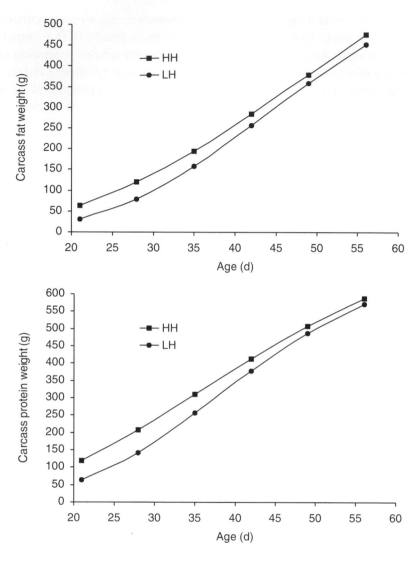

Figure 6.3. Comparison between fully fed (HH) and early-feed restriction followed by re-alimentation (LH) on composition of live weight in broilers based on time (from Wiseman and Lewis, 1999)

Plavnik & Hurwitz (1985) observed that there was a lower proportion of abdominal fat (the depot most often employed to characterize overall fatness of broilers) following early feed restriction, which may also have contributed to the better feed efficiency of these birds. It was further argued that reduced fatness was attributable to impaired hyperplasia of adipocytes (as observed by March & Hensen, 1976). Jones & Farrell (1992b) concluded that the success of early feed restriction in reducing carcass fat (but no other component) was attributable to a temporary delay in fat deposition. However, the older the birds, the greater the likelihood that this advantage would disappear. The results of the study of Wiseman and Lewis (1999) support this latter observation, although LH birds were always fatter at a given weight than HH when total fat content was considered. However there was variation with individual fat depots where LH birds were less fat. This suggests differential responsiveness of individual fat depots to nutrient and energy intake. Hargis and Creger (1980) reduced nutrient and energy intake for the first 7 days of life whereas Arafa *et al.* (1983) reduced energy intake during the last 10 days of the growth period following a commercial starter. Both studies reported a reduction in carcass fat at 49 days of age.

The ability of birds to compensate for early nutrient and energy restriction in terms of lean tissue weights is, according to Jones & Farrell (1992b) dependent upon the severity of this restriction. Susbilla *et al.* (1994) observed that the weight of breast and thighs at 40d of age was not influenced by feed restriction. It is likely that the study of Wiseman and Lewis (1999) was more severe in this context as the content of white (i.e. breast) muscles from birds fed HH was always, albeit marginally, greater at the same live weight than birds fed on LH. The effect of feeding diets of low energy and nutrient content early in life was different when considering dark meat, where the small differences recorded were in favour of LH, particularly at higher weights.

LIPID:PROTEIN RATIOS

The composition of lipids is considerably more variable than that of other components in broilers. Whilst there may be a genetic maximum body lipid composition, the level of this component is subject to a number of variables; those associated with diet are essentially the provision of dietary energy relative to the bird's energy and nutrient (specifically protein / amino acid) requirements. Thus, if protein accretion is at a genetic maximum, additional dietary energy supplied beyond the requirement to support such accretion will result in increased lipid deposition. Conversely, energy intake below requirement will risk lipid catabolism and a consequent reduction in lipid content (a further complication is the dietary E:P as considered earlier). Lipid:protein ratios have been employed to separate genotypes (e.g. Hancock *et al* 1995) which would, accordingly, require different energy and nutrient regimes in order for them to grow optimally. In the study of Wiseman and Lewis (1999), lipid:protein ratios ranged from 1.27, 0.62, 0.78, 1.02 to 1.65 respectively for HH, HL, C, LH and LL diet combinations which is further demonstration of the significant influence of diet on carcass composition. Changes in lipid:protein ratios, as influenced by diet within the same genotype, are presented in Table 6.5 (on the basis of either live weight or time).

Table 6.5. Carcass lipid:protein ratios (Wiseman and Lewis, 1999)

a. At specific live weight

Live weight (g)	Diet Combinations				
	HH	HL	C	LH	LL
1000	0.56	0.48	0.49	0.57	0.45
1250	0.58	0.50	0.53	0.60	0.44
1500	0.60	0.52	0.56	0.62	0.44
1750	0.63	0.53	0.59	0.65	0.45
2000	0.66	0.54	0.61	0.68	0.47
2250	0.69	0.55	0.64	0.71	0.49
2500	0.72	0.56	0.66	0.73	0.52
2750	0.76	0.57	0.67	0.77	0.55
3000	0.80	0.58	0.69	0.80	0.60

b. At specific time

Time (d)	Diet Combinations				
	HH	HL	C	LH	LL
21	0.54	0.46	0.44	0.48	0.54
28	0.58	0.49	0.50	0.55	0.47
35	0.63	0.51	0.56	0.61	0.45
42	0.68	0.54	0.61	0.67	0.45
49	0.75	0.56	0.65	0.73	0.47
56	0.81	0.57	0.68	0.79	0.51

EFFICIENCY OF USE OF AME FOR GAIN

It is evident from the three Nottingham trials reported that there are major differences between diets in terms of both growth performance and carcass quality. However, although food conversion ratios were also influenced by diet, it remains to be established whether the efficiency of utilisation of AME varied between treatments.

Accordingly, data were analysed further by regressing AME intake on weight gain to derive the efficiency of utilisation of dietary energy (e.g. g weight gain / MJ AME consumed) and results are presented in Table 6.6A, 6.6B and 6.6C

Table 6.6A. Feed intake and efficiency of utilisation of AME for growth (Trial 1, from Jones and Wiseman, 1985)

Starter	Diet	Food Intake kg/bird	Calculated AME MJ/kg as fed	AME intake	Efficiency of AME utilisation - lwt gain (g) / AME intake (MJ)*
	L	1.09	10.78	11.75	42.6
	M	1.11	12.78	14.19	44.4
	H	1.09	14.78	16.11	45.3
Finisher					
L	L	4.17	10.78	44.95	32.9
	M	3.52	12.78	44.99	35.1
	H	3.29	14.78	48.63	34.5
M	L	4.11	10.78	44.31	36.6
	M	4.16	12.78	53.16	31.6
	H	3.49	14.78	51.58	33.3
H	L	4.08	10.78	43.98	34.3
	M	4.27	12.78	54.57	27.9
	H	3.73	14.78	55.13	30.0

* Live-weight gain presented in Table 6.2A

Table 6.6B. Feed intake and efficiency of utilisation of AME for growth (Trial 2; University of Nottingham, unpublished data)

Starter	Diet	Food Intake kg/bird	Calculated AME MJ/kg as fed	AME intake	AME utilisation - lwt gain (g) / AME intake (MJ)*
	L	1.30	12.75	16.58	38.6
	M	1.28	12.75	16.32	45.3
	H	1.29	12.75	16.48	50.0
Finisher					
	L	3.24	12.75	41.31	32.0
L	M	3.03	12.75	38.63	32.6
	H	3.53	12.75	45.08	29.1
	L	3.47	12.75	44.24	28.7
M	M	3.45	12.75	43.98	30.5
	H	3.63	12.75	46.28	31.5
	L	3.7	12.75	47.18	26.5
H	M	3.78	12.75	48.20	28.2
	H	3.29	12.75	41.95	29.8

* Live-weight gain presented in Table 6.2B

Table 6.6C. Feed intake and efficiency of utilisation of AME for growth (Trial 3; from Wiseman and Lewis, 1999)

	Cumulative:		Weekly:		Efficiency of AME utilisation for weight gain	
	AME intake MJ	Weight gain g	AME intake MJ	Weight gain g	g gain/MJ intake	Week
HH	10.12	470	10.12	470	46.5	4
	22.78	983	12.67	513	40.5	5
	37.45	1486	14.66	503	34.3	6
	53.37	1940	15.93	454	28.5	7
	69.79	2327	16.42	387	23.6	8
	86.00	2643	16.21	316	19.5	9
	101.47	2893	15.46	250	16.2	10
HL	8.01	371	8.01	371	46.3	4
	18.31	789	10.30	418	40.6	5
	30.63	1227	12.32	438	35.5	6
	44.49	1659	13.86	432	31.2	7
	59.32	2066	14.83	407	27.4	8
	74.53	2436	15.21	369	24.3	9
	89.59	2762	15.06	326	21.6	10
CC	9.09	398	9.09	398	43.8	4
	20.19	851	11.09	453	40.8	5
	33.21	1327	13.02	475	36.5	6
	47.98	1793	14.78	467	31.6	7
	64.25	2230	16.27	436	26.8	8
	81.70	2622	17.44	392	22.5	9
	99.97	2964	18.27	342	18.7	10
LH	9.18	413	9.18	412	44.9	4
	21.06	944	11.88	531	44.7	5
	35.28	1511	14.23	567	39.9	6
	51.26	2042	15.98	531	33.2	7
	68.29	2496	17.03	454	26.7	8
	85.66	2860	17.37	364	21.0	9
	102.73	3139	17.08	279	16.3	10
LL	7.64	316	7.64	316	41.4	4
	17.61	704	9.97	387	38.9	5
	29.79	1131	12.19	427	35.1	6
	43.89	1566	14.10	435	30.9	7
	59.46	1982	15.57	416	26.7	8
	75.97	2361	16.51	379	23.0	9
	92.91	2694.0	16.94	333	19.6	10

respectively for Trials 1, 2 and 3. Although there is the suggestion that starter diets with higher nutrient and energy concentrations might be associated with greater efficiencies, no such trend was apparent for the finisher diets introduced from 24 days of age.

A similar conclusion could be drawn from Trial 3. Mean efficiencies of utilisation of AME for weight gain were 36.9, 37.9, 37.1, 39.0 and 35.5 g/MJ respectively for HH, HL, CC, LH and LL over all time periods; these efficiencies are somewhat lower than in Trials 1 and 2 because Trial 3 extended to 10 weeks - a point when efficiencies are declining rapidly. The reduction in efficiency is probably associated with changing composition of live-weight gain with time where the more efficient muscle growth is progressively replaced by the less efficient fat growth. Differences between these means were not significant (P=0.827) confirming that the influence of diet on growth performance is based substantially on energy intake and not on efficiency of utilisation of energy and nutrients *per se*; with the qualification that nutrients are provided in adequate amounts.

Altering fatty acid profiles of broiler meat

This is an increasingly topical area as advice to the human consumer continues to advocate an increase in the intake of polyunsaturated fatty acids (specifically the n-3 or omega 3 families, e.g. COMA, 1984; DHSS 1994). Numerous studies have demonstrated how closely carcass fatty acids of poultry reflect those of the diet. Time over which changes take place has received some attention. Bartov *et al.* (1974b) recommended a withdrawal of polyunsaturated fat source between 3 and 4 weeks prior to slaughter and Lopez-Ferrer *et al.* (1999), together with Gonzalez-Esquerra and Leeson (2000), confirmed that the response of carcass to dietary fatty acids was greater with longer feeding periods. However, the rate at which carcasses respond to dietary changes has not been studied in detail.

Information on the rate of change would be valuable to those who wish to increase the degree of unsaturation of fatty acids in broiler meat. Such information might also be used to reduce the risks of lowering the melting point of adipose tissue, which has been associated with excessive degrees of fatty acid unsaturation. Finally, because of the well-established effect of age on the AME values of fats (e.g. Wiseman *et al.*, 1999) it is common to feed relatively unsaturated fats early in life (when the differential between AME of unsaturated and saturated fats is greatest) to be replaced subsequently by relatively saturated sources. It is important to point out that degree of saturation of carcass lipid is generally not thought to be associated with the 'oily bird' syndrome. Although Hammershoj (1997) observed that birds fed on diets containing soya bean oil had a greater skin tearing frequency and reduced skin elasticity, most other studies have implicated impaired collagen structure and cross-linking (e.g. Ramshaw *et al* 1986; Granot *et al*, 1991).

In a programme of work conducted at the University of Nottingham (unpublished data), broilers were fed iso-energetic and iso-nitrogenous diets based on fats and oils of varying unsaturation before being transferred (at 21 days of age) to a diet based on the considerably more saturated tallow. A serial-slaughter approach allowed modelling of the rate of change of carcass fatty acids based on the function (Salmon and O'Neil, 1973):

$y =$ $A + BR^X$, where y is an increase or decrease in fatty acid concentration at a given value of X,

$R =$ $\exp(-k)$, an estimate of the rate at which fatty acid levels change in the asymptotic region of the response,

$A =$ an estimate of the fatty acid concentration towards which each response tends to approach (the asymptote),

$B =$ an estimate of the rate of change of fatty acid in the initial portion of the curve,

$X =$ number of days between the baseline and slaughter weights.

The data (Figure 6.4) indicate that the pattern of change in carcass fatty acids is substantially complete within approximately 20 days of dietary change (although it is not possible to determine categorically whether these changes are simple dilution of the existing fatty profile or actual replacement of one fatty acid by another). Intriguingly, the data also suggest that different carcass components respond differently. Thus, the abdominal fat pad (one of the major storage sites of adipose tissue in the broiler) is much more responsive than the thigh meat. Breast meat is very unresponsive to dietary changes; as fat extracted from this region will be predominantly functional lipid, as opposed to storage lipid, the lack of response is presumably due to homeostatic mechanisms.

Conclusions

It is evident that alteration of carcass composition of broilers is achievable through a number of dietary routes, including changes to dietary energy: protein ratios (by changing one or both). There can be no optimum diets recommended as these will depend on the circumstances under which they are fed and the desired product. What is intriguing, however, is that analysis of responses can be reduced substantially to a consideration of feed intake. Finally, for those markets wishing to identify a specific fatty acid profile, diet is a means whereby this may be achieved although, similarly, definitions of what is optimum are not possible.

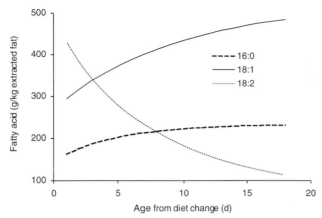

(a) Fatty acid profile of abdominal fat pad (AFP); 16:0, 18:1 and 18:2 are, respectively, palmitic, oleic and linoleic acids.

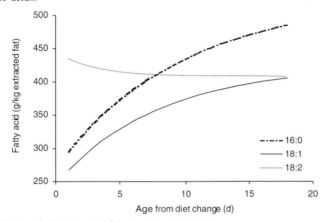

(b) Pattern of change of 18:1 in specific carcass components.

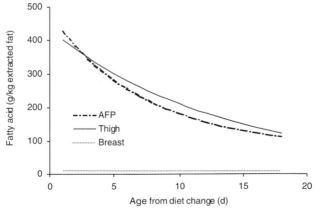

(c) Pattern of change of 18:2 in specific carcass components.

Figure 6.4. Rate of change of carcass fatty acids in response to a dietary change (from safflower oil to tallow) at 21 days of age (day zero in responses; University of Nottingham unpublished data).

References

Arafa, A.S., Boone, M.A., Janky, D.M., Wilson, M.R., Miles, R.D. and Harms, R.M. (1983) Energy restriction as a means of reducing fat pads in broilers. *Poultry Science,* **62,** 314-320.

Bartov, I. (1998) Lack of interrelationship between the effects of dietary factors on food withdrawal on carcase quality of broiler chicks. *British Poultry Science,* **39**, 426-433.

Bartov, I, Bornstein S. and Lipstein, B. (1974a) Effect of calorie to protein ratio on the degree of fatness in broilers fed on practical diets. *British Poultry Scienc*e, **15,** 107-117.

Bartov, I., Lipstein, B. and Bornstein, S., (1974b) Differential effects of dietary acidulated soybean oil soapstock, cottonseed oil soapstock and tallow on broiler carcass fat characteristics. *Poultry Science,* **53,** 115-124.

COMA (1984) Department of Health and Social Security, Diet and cardiovascular disease. Report on Health and Social Aspects No. 28. London HMSO.

Department of Heath and Social Security (1994) Report of Health and Social Subjects No. 46. Nutritional aspects of cardiovascular disease. London HMSO

Deaton, J.W., Reece, F.N., Kubena, L.F., Lott, B.D. and May, J.D. (1973) The ability of the broiler chicken to compensate for early growth depression. *Poultry Science,* **52,** 262-265

Gonzalez-Esquerra, R. and Leeson, S. (2000) Effects of menhaden oil and flaxseed in broiler diets on sensory quality and lipid composition of poultry meat. *British Poultry Science,* **41,** 481-488.

Granot I., Pines, M., Plavnik, I., Wax, E., Hurwitz, S. and Bartov, I. (1991) Skin tearing in broilers in relation to skin collagen - effect of sex, strain, and diet. *Poultry Science,* **70,** 1928-1935

Griffiths, L. Leeson, S. and Summers, J.F. (1977) Fat deposition in broilers: effect of dietary energy to protein balance and early life restriction on productive performance and abdominal fat pad size. *Poultry Science,* **56,** 638-646.

Hakansson, J., Eriksson, S. and Svensson, S.A. (1978*a*) The influence of feed energy level on chemical composition, growth and development of different organs of chicks. Report No.57. Swedish University of Agricultural Science, Department of Animal Husbandry.

Hakansson, J., Eriksson, S. and Svensson, S.A. (1978*b*) The influence of feed energy level on chemical composition of tissues and on the energy and protein utilisation of broiler chicks. Report No. 59. Swedish University of Agricultural Science, Department of Animal Husbandry.

Hammershoj, M. (1997) *Acta Agriculturae Sscandinavica,* **47,** 197-205

Hancock, C.E., Bradford, G.D., Emmans, G.C. and Gous, R.M. (1995) The evaluation of the growth parameters of six strains of commercial broiler chickens. *British Poultry Science,* **36,** 247-264.

Hargis, P.H. and Creger, C.R. (1980) Effects of varying dietary protein and energy levels on growth rate and body fat of broilers. *Poultry Science,* **59,** 1499-1504.

Jones, G.P.D. and Farrell, D.J. (1992*a*) Early-life restriction of broiler chickens. II. Methods of application, amino acid supplementation and the age at which restriction should commence. *British Poultry Science,* **33,** 579-587.

Jones, G.P.D. and Farrell, D.J. (1992*b*) Early-life restriction of broiler chickens. II. Effects of food restrictions on the development of fat tissue. *British Poultry Science,* **33,** 589-601.

Jones, R.L. and Wiseman, J. (1985) Effect of nutrition on broiler carcase composition: influence of dietary energy content in the starter and finisher phases. *British Poultry Science,* **26,** 381-388.

Kubena, L.F., Chen, T.C., Deaton, J.W. and Reece, F.N. (1974) Factors influencing the quantity of abdominal fat in broilers. 3. Dietary energy levels. *Poultry Science,* **53,** 974-978.

Leeson, S., Caston, L. and Summers, J.D. (1996a) Broiler response to energy or protein dilution in the finisher diet. *Poultry Science,***75,** 522-528.

Leeson, S., Caston, L. and Summers, J.D. (1996b) Broiler response to diet energy. *Poultry Science,* **75,** 529-535.

Leskanich, C.O and Noble, R.C (1997) Manipulation of the n3 polyunsaturated fatty acid composition of avian eggs and meat. W*orlds Poultry Science Journal,* **53,** 155-183.

Lopez-Ferrer, S., Baucells, M.D., Barroeta, A.C. and Grashorn, M.A. (1990) N-3 enrichment of chicken meat using fish oil: alternative substitution with rapeseed and linseed oils. *Poultry Science,* **78,** 356-365.

March, B.E. and Hansen, G. (1976) Lipid accumulation and cell multiplication in adipose bodies in White Leghorn and broiler-type chicks. *Poultry Science,* **56,** 886-894.

Plavnik, I. and Hurwitz, S. (1985) The performance of broiler chicks during and following a severe feed restriction at an early age. *Poultry Science,* **64,** 348-355.

Ramshaw, J.A.M., Rigby, B.J., Mitchell, T.W.and Nieass, A. (1986)Changes in the physical and chemical properties of skin collagen from broiler-chickens exhibiting oily bird syndrome. *Poultry Science,* **65,** 43-50.

Rogers, S.R., Pesti, G.M. and Marks, H.L. (1987) Comparison of three nonlinear regression models for describing broiler growth curves. *Growth,* **51,** 229-239.

Salmon, R.E. and O'Neil, J.B., 1973. The effect of dietary fat and storage temperature on the storage stability of turkey meat. *Poultry Science,* **52,** 314-317.

Short, F.J., Wiseman, J. and Boorman, K.N. (2000) The effect of the 1B/1R translocation and endosperm texture on amino acid digestibility in near-isogenic lines of wheat for broilers. *Journal of Agricultural Science,* **134,** 69-76.

Susbilla, J.P., Fraankel T.L., Parkinson, G. and Gow, C.B. (1994) Weight of internal organs and carcase yield of early food restricted broilers. *British Poultry Science,* **35,** 677-685.

Wiseman, J. (2000) Correlation between physical measurements and dietary energy values of wheat for poultry and pigs. *Animal Feed Science and Technology,* **84,** 1-11.

Wiseman, J. (2001) The quantitative contribution of fat to metabolisable energy. In "26th Poultry Science Symposium", edited by J.M. McNab, CAB International, Wallingford (in press)

Wiseman, J. and Lewis, C.E. (1999) Influence of dietary energy and nutrient concentration on the growth of body weight and of carcass components of broiler chickens. *Journal of Agricultural Science,* **131,** 361-371.

Wiseman, J., Nicol, N.T. and Norton, G. (2000) Relationship between apparent metabolisable (AME) values and *in vivo / in vitro* starch digestibility of wheat for broilers. *Worlds Poultry Science Journal,* **56,** 1-14.

Wiseman, J. Powles, J. and Salvador, F. (1998) Comparison between pigs and poultry in the prediction of the dietary energy value of fats. *Animal Feed Science and Technology,* **71,** 1-9.

7

BACTERIOLOGICAL ASPECTS OF THE USE OF ANTIBIOTICS AND THEIR ALTERNATIVES IN THE FEED OF NON-RUMINANT ANIMALS

K. HILLMAN

Microbiology unit, SAC Aberdeen, Ferguson building, Craibstone, Aberdeen AB21 9YA, UK

Introduction

The prophylactic use of antibiotics in animal feeds has made intensive farming possible, and has shown improved feed conversion in these animals. However, the continuous feeding of low levels of antibiotics over many years has resulted in the formation of a gut microflora which is, overall, resistant to these antibiotics. Incoming pathogens may not be, so the inclusion of these antibiotics in feed should still act prophylactically against the spread of many pathogens. When antibiotics are withdrawn, the intensive nature of modern farming could result in the rapid spread of infective agents.

It may be worth mentioning the distinction between the effects of antibiotics on bacteria when these are used as disease therapy or as growth promoters. Many bacterial populations will contain a few individual cells which are resistant to a particular antibiotic, usually by the production of an enzyme which inactivates the antibiotic. Therapeutic dosing is designed to overwhelm this small resistance, so that the resistant cells do not have the opportunity to re-establish as a resistant population. In the presence of low levels of antibiotic, the resistant cells survive and grow, producing an antibiotic-resistant population (Eichner and Gravitz, 2001). As all the cells in this population can now produce the antibiotic-inactivating enzyme, much higher doses of antibiotic are required to overcome the resistance - often, the dose required would be toxic to the animal.

Antibiotics in therapy can also cause serious intestinal problems, particularly when pathogens such as *Clostridium difficile*, which show a natural resistance to many antibiotics, are present. The disruption of the normal microflora by therapeutic doses of antibiotics provides such pathogens with an opportunity to proliferate and infect (Nord and Edlund, 1990).

This paper intends to examine the influence of antibiotic use on the gut microflora, and the possible effects of some of the alternatives to antibiotics currently available.

Antibiotic origins

The discovery of antibiotics is credited to Alexander Fleming, although the use of antibiotic-containing preparations (eg as "mouldy-bread poultices") on infected wounds dates back much further. The important term here is "discovery", not "invention". Antimicrobials, including antibiotics, are produced by microorganisms as part of the natural competition between species, and it is likely that such compounds were being produced by microorganisms well before the emergence of the human species. When these compounds were discovered, they were already in an advanced stage of development by microbes. Consequently, the mechanisms of resistance must have been around for almost as long. We cannot now say whether the presence of antibiotic resistant bacteria has been brought about by human activity or whether these microbes have been there all along, although human activity has certainly increased the prevalence of resistance.

Development of resistance

Continued exposure of a population of bacteria to low levels of antibiotics will result eventually in the development of resistance. As an observation, this is entirely correct, although the mechanism of development of resistance is not dependent on the presence of the antibiotic. The bacterial genome is easily mutated, and can be modified by the incorporation of DNA from the environment, processes which occur continuously. Occasionally, mutation, plasmid transfer or incorporation of DNA will result in the acquisition of resistance to an antibiotic. Even if we consider this to be a rare event, perhaps one in a thousand million, the density of the microbial population of the gut exceeds this number of microorganisms per gram of contents. One antibiotic-resistant bacterium in this population is irrelevant in the absence of the antibiotic, but in the presence of a low concentration of the appropriate antibiotic, this one cell can rapidly proliferate. The antibiotic is therefore not involved in the mechanism of development of the resistance, but represents the selective pressure that allows the proliferation of the resistant strains.

As an example, penicillin interferes with cell wall formation in Gram-positive bacteria. If a population contains a few penicillin-resistant individuals, this is normally due to the formation of a penicillin-degrading enzyme (ß-lactamase, penicillinase). Therapeutic doses of penicillin will overwhelm the penicillinase produced by these individuals so that the population is unable to replicate, making it a simple matter for the immune system to remove the infection.

However, if the population is treated with a low continuous dose of penicillin, the enzyme-producers are unaffected and are able to proliferate while the rest of the population is inhibited. Eventually, the entire population is composed of penicillin-resistant individuals. Now, the total production of penicillinase is such that therapeutic doses of penicillin are simply brushed aside. Further, a large population of a penicillin-resistant species in a complex ecosystem such as the gut could effectively "protect" other, normally penicillin-sensitive, pathogens. Antibiotic resistance is not purely a pathogen-linked problem: even where the antibiotic-degrader is a harmless species, the effectiveness of that antibiotic against an infection may be reduced by virtue of its inactivation by the resistant population.

Antibiotic resistant bacteria thrive in environments with constant exposure to antibiotics because the environmental antibiotics effectively remove those bacteria's competitors. Therefore, if we remove the antibiotic from the microbial environment, then the antibiotic resistant species should decline.

Removal of resistant strains

The occurrence of antibiotic resistant bacteria in pigs in Denmark is now lower than in those countries where antibiotics are still routinely used in feeds. However, antibiotic resistant bacteria are still present in Danish pigs. In the UK, pigs which have been raised organically still retain some antibiotic resistant bacteria within their guts. It is possible to accelerate the decline of antibiotic resistant strains in the gut by the introduction of an antibiotic-sensitive microflora, but it is not currently possible to eradicate resistance altogether. Are the remaining resistant bacteria a residue of the previous, highly resistant population or do they represent a natural gut population? It may not be possible to answer this question, as this would require access to farmed pigs which have not been fed with antibiotics for many generations, and it is unlikely that such pigs currently exist among domesticated herds. It is therefore difficult to establish a baseline level, where we could say that the resistance caused by human intervention has been removed.

Development and composition of the intestinal microflora

Until the moment of birth, all animals are microbially sterile. However, the animal is quickly colonised by bacteria from its immediate environment, particularly the mother's urinogenital tract and faeces. All skin surfaces, including the gut, will bear a measurable microbial population within a few hours of birth. Close skin and mouth contact between mother and offspring aid the transfer of parental microflora, while the lactose content of mammalian milk encourages the growth of lactic acid bacteria in the neonate intestine. Selectivity of the neonatal microflora is also enhanced by immunoglobulins in the milk, so that the gut of the newborn effectively remains under the protection of its mother until weaning.

In the case of poultry, the sterile hatchling derives its microflora by pecking at ground contaminated by adult faeces. As poultry do not subsequently undergo the intestinal disruption caused by weaning, the initially derived microflora effectively determines the microbial status of the gut throughout the bird's life (Nurmi and Rantala, 1973). As the gut microflora is derived from the parent animal, it will contain considerable antibiotic resistance from the moment it begins to develop.

In the nonruminant mammal, abrupt weaning results in a sudden change in the composition of the nutrients provided to both the animal and the intestinal microflora. This dramatic change results in equally dramatic effects within the microflora, as the complex population struggles to adapt to the new environment. There are also considerable changes in the structure of villi and crypts over the weaning period (Ewing and Cole, 1994). These physical changes, along with changes in substrate, result in an unstable intestinal microflora over the weaning period as the microflora, like the host animal, change from infant towards adult structure. The reduction in villus height at weaning can also result in reduced absorption from the small intestine, allowing an increase in fermentable carbohydrate reaching the large intestine, with corresponding increase in fermentative microbial activity (Pluske, Williams and Aherne, 1996a).

The lactic acid bacteria that dominate the infant gut are gradually overgrown by strictly anaerobic bacteria such as *Bacteroides* (Ewing and Cole, 1994), as the gut lumen enlarges and the feed becomes more solid, allowing the formation of oxygen-free microbial environments. It is notable that the transition from milk-fed to solid-fed microflora is not a smooth one, with wild daily fluctuations in the composition of the microflora for some 7-14 days after weaning (Reid, 1999). Abrupt weaning in piglets may also be associated with a 100-fold drop in the numbers of lactobacilli in the intestine, and a 50-fold increase in the numbers of *Escherichia coli* (Huis in't Veld and Havenaar, 1993). The process of change in the intestinal microflora continues throughout life, albeit more gradually, with the gut tending to favour a higher proportion of strictly anaerobic bacteria as the animal ages. (Maczulak, Wolin and Miller, 1989; Piattoni, Demeyer and Maertens, 1996). The disruption of the gut at weaning often results in intestinal infection, requiring treatment with antibiotics which would favour the retention of antibiotic-resistant strains. So, even where the young animal has never been directly fed with antibiotics, it may still harbour a significant antibiotic-resistant microbial population within its gut.

The composition of the intestinal microflora thus depends on the initial colonisation at birth, on the current state of the gut physiology, on the type of feed consumed by the animal, and also on its genetic predisposition to the maintenance of some of the component species. The genetic composition of the host animal has been identified as an important component in certain cases, notably in pigs in susceptibility to the various substrains of *Escherichia coli* K88 (Bijlsma, de Nijs, van der Meer and Frik, 1982; Vogeli, Kuhn, Kuhne, Obrist, Stranzinger, Huang, Hu, Hasler-Rapacz and Rapacz, 1992). A less well known example exists in the genetic predisposition to harbouring methane-producing bacteria shown by certain animal species (Hackstein, van Alen, op ten Camp, Smits and Mariman, 1995: Hackstein, Langer and Rosenberg, 1996).

More recently, data has been collected which suggests that intestinal lactobacilli with antipathogenic activity are not evenly distributed between pigs: some farms show a high prevalence of these active lactobacilli in pigs, while others show little or no antipathogenic activity within their lactobacilli. A preliminary study (Table 7.1; Hillman and Robertson, 2001) has shown a wide variation in the inherent antipathogenic activity of intestinal lactobacilli in pigs from 19 farms. In this study, the ability of isolated *Lactobacillus* spp to inhibit the growth of *Escherichia coli* K88 was assessed (Hillman and Fox, 1994) and the isolates were grouped by the significance of the difference between the growth rates of *E. coli* in the presence or absence of the lactobacillus. Although only a small part of the total lactobacilli within the intestine could be sampled in this study, those tested were isolated from high dilutions of colonic contents and should therefore represent the predominant groups within the colonic population of these pigs. The data obtained contained examples of both zero and 100% probiotic activity within the colonic *Lactobacillus* population of pigs, with a range of activities detected within the samples from most farms. The implication is that this distribution could affect perceived efficacy of many added antibacterials, particularly probiotics, since adding an antimicrobial to a population with a high indigenous antimicrobial activity is likely to be less effective than adding it to a population which shows little or no such activity. Whether this variation is connected with the genetics of the animals is not known, but the pattern of distribution within herds rather than as a random distribution among individuals suggests that this possibility may be worthy of consideration.

Table 7.1. Numbers of isolates of *Lactobacillus* spp from 19 farms, grouped by significance of inhibition of *E. coli* K88.

	Farm no.																		
	1	2	3	4	5	6	7	8	9	10	11	12	13	14	15	16	17	18	19
ns	6	4	6	4	7	6	3	8	6	3		4	8	10	7	10	3	4	39
$P<0.05$			3	2	1		2	1		1							1	4	4
$P<0.01$	4	2		2		2	4	1		3	2	1					4		9
$P<0.001$		4								3	5	2					1	2	2
Total	10	10	9	8	8	8	9	10	6	10	7	7	8	10	7	10	9	10	54

ns, no. of isolates showing no significant inhibitory activity. *P* values, no. of isolates showing significant inhibitory activity at the levels indicated. Total, total no. of isolates per farm. Farm 19 was the SAC Tillycorthie experimental farm, while farms 1-18 were commercial piggeries within Scotland.

Seasonal variation in Campylobacter carriage by poultry has been demonstrated (Wallace, Stanley, Currie, Diggle and Jones, 1997), indicating that seasonal effects on the microbial population of the gut may also occur in other cases. Our own studies have produced indications of seasonal changes in the populations of *Lactobacillus* spp, *Bifidobacterium* spp, coliform bacteria and *Clostridium* spp, though not in total anaerobic bacteria or *Bacteroides* spp, in the colonic digesta of pigs (Stefopoulou and Hillman, unpublished). Although the experimental work is currently incomplete, the data obtained so far suggest that the effects of any attempt at gut manipulation may depend on the time of year at which it is applied. This would be particularly true in the case of feed additives which include, or which are intended to affect, the coliform or lactic acid bacteria.

These variations in the intestinal microflora, together with other known sources of variation such as animal age, herd and individual differences and dietary effects, would probably be entirely predictable if their causes were better understood. At the moment, it is not possible to take account of these variabilities within animal trials (more importantly, in comparing data from different trials) because the influence of many of these causes of variation have not been fully studied. It is, however, important to be aware that replication of experimental work will be influenced by all of these factors, and possibly by many other factors not yet identified. It is therefore possible that two separate studies into the effects of a feed additive on the growth of an animal species may give different results, although both datasets would be correct within the current state of all these variables.

Stability of the intestinal microflora and its importance to health and growth

Where a mixed microbial population is in a state of constant change, opportunities arise for the introduction of new species into the population as niches become temporarily accessible. However, when the same mixed population is stable, *ie* its nutrient supply and environmental conditions do not change rapidly, then all available niches within the system are occupied by one or more species of microorganism. Put simply, this means that any incoming microorganism would have to compete for a niche by displacing its current occupant. For this reason, a stable intestinal microflora is inherently more resistant to pathogenic infection than an unstable one (Van der Waaij, Berghuis de Vries and Lekkerkerk van der Wees, 1971; Nurmi and Rantala, 1973; Hill, Fadden, Fernandez and Roberts, 1986). Much of the resistance of the intestine to invading bacterial species can be attributed to direct competition and inhibition of the invader by the existing microflora (*eg* Wilhelm, Lee and Rosenblatt, 1987, Hillman, Murdoch, Spencer and Stewart, 1994). For the same reasons, an incoming antibiotic-sensitive bacterial species would have to compete with the established antibiotic-resistant strain of the same species in order to remove it. It should be noted that viruses, and many protozoan pathogens, do not take part in the competition for niche and nutrients between the bacteria and so are able to bypass this level of pathogen-resistance.

The health of the gut has a direct bearing on the growth and productivity of livestock animals, since the gut comprises the body's largest organ and represents a considerable part of the animal's protein and energy requirements (Edmunds, Buttery and Fisher, 1980; Murumatsu, Takasu, Furuse, Takaski and Ikumura, 1987). Therefore, an inefficient gut will result in an increased drain on these resources, resulting in less of the feed being converted to muscle tissue. At the same time, a diseased intestine will result in general ill health of the animal, leading to reduced feed intake which would exacerbate the inefficiency of feed conversion. So, providing a stable, disease-resistant microflora within the gut would lead to improved productivity by the animal.

What is actually required in a production system is not a "normal" intestinal microflora, but the development of an "optimal" microflora, which would provide the maximum protection from infection while minimising the microbial use of the animal's feed. There is no reason to suppose that a normal, or wild-type, microflora would be optimised for growth of the animal. In the wild state, animals whose intestines are optimised for disease-resistance would fare better than those optimised for growth, even if enhanced disease-resistance resulted in a higher microbial load for the gut. It is possible that improved growth and improved disease resistance may not be mutually exclusive, although a reduction in numbers or activity of the indigenous microflora would be expected to result in a reduction in its anti-pathogenic activity.

Alternatives to antibiotics

Developing an alternative to antibiotics is hampered by one important question - what is it that antibiotics do to the gut to improve animal performance? This has never been clear, although some plausible theories have been put forward, such as an improvement in vitamin B_{12} assimilation by reducing the activity of those bacteria which utilise B_{12} while leaving the B_{12}-producers unaffected (Postgate, 1992). The idea that antibiotics reduce total bacterial load in the gut is, however, not realistic. While a reduction in total bacteria may be observed within the first few days of treatment, the intestinal bacteria would take very little time to replace the inhibited groups with either new, resistant strains or with bacterial groups which display a natural resistance to the antibiotic. The total bacterial load of any environment is determined by the availability of nutrients and suitable niches for bacterial growth. If we kill one strain of bacteria in an environment like the gut, another will quickly replace it.

So, in the absence of a defined mechanism of action for an antibiotic growth promoter, it is futile to look for something which "works in the same way". Rather, we should look for something that produces the same end result, ie. faster-growing and more cost-efficient animals. Unfortunately, this can only be reliably done by the use of lengthy and expensive animal feeding trials.

It is beyond the scope of this work to provide a comprehensive discussion of all available potential alternatives. The following paragraphs consider a few of the alternatives available, as well as some potentially useful future products which are still at the research stage.

DIETARY MANIPULATION

Possibly the least controversial method of adjusting the intestinal microflora is by adjusting the composition of the diet: this involves no additives of any kind and can be argued to be a completely "natural" form of controlling gut bacteria. Significant differences in the composition of the gut microflora can be achieved by dietary alterations, which may be as simple as changing the source or treatment of the starch component of the diet (Reid and Hillman, 1999). The amylose/amylopectin content of the starch, the structure of the granule and the degree of pre-feed treatment, influences both the extent of its digestion within the small intestine and the nature of the fermentation of the undigested residue by intestinal bacteria (Reid, 1999; Reid, Hillman and Henderson, 1998). The effects of three starches, singly and in combination, on the microflora of the piglet colon are shown in Table 7.2 (McFarland, 1998). In this study, potato starch was found to significantly suppress coliform and *Escherichia coli* numbers in relation to the other two starches. Mixtures of these starches produced intermediate effects in comparison with individual starches, suggesting that this form of manipulation should have the potential to accurately control microbial populations within the gut.

Table7.2. Selected bacterial counts (per gram wet weight) in the proximal colon of weaned piglets fed meal diets containing different starches.

	Potato	*Waxy maize*	*Maize*	*Potato/ waxy maize*	*Potato/ maize*	*Maize/ waxy maize*	*SED*
Total anaerobes	9.82[a]	8.66[b]	8.88[b]	10.05[a]	9.86[a]	9.56[ab]	0.43
Coliform bacteria	1.90[a]	6.93[b]	7.88[b]	6.12[b]	4.24[ab]	7.12[b]	1.81
Escherichia coli	1.73[a]	6.62[bc]	7.39[c]	5.99[bc]	3.64[ab]	6.52[bc]	1.62
Lactobacillus spp.	9.62[a]	8.62[b]	8.76[ab]	9.67[a]	9.67[a]	9.17[ab]	0.42
Enterococcus spp.	5.20[a]	6.44[ab]	7.43[b]	6.11[ab]	4.69[a]	6.19[ab]	0.90
Bacteroides spp.	7.16[a]	7.43[ab]	7.16[a]	8.27[bc]	8.54[c]	7.17[a]	0.44

Values within a row bearing different superscript letters differ significantly (P<0.05). Data represent the means of determinations from three piglets on each diet.

However, there is an important disadvantage in reliance on dietary manipulation. Starch, and other feed components, are natural products, and are therefore inherently variable. In the case of starch, the precise amylose/amylopectin content, granule size and structure, gelatinisation temperature and other factors are dependent on the growing conditions of the plant from which the starch was extracted. Therefore, the effects obtained in this one trial cannot be reliably extrapolated to other situations, because the starches and their effects will vary from one year to the next. Additionally, starches from the same plant species grown in different climates will show differences in response when fermented by intestinal bacteria. Although the general principles should remain consistent (ie, potato starch should result in lower coliform numbers than maize starch), the actual values obtained may show considerable variation. When this is superimposed on other sources of variation inherent in bacteriological studies of the gut (discussed earlier), it is clear that dietary manipulation is currently an imprecise tool in the control of the gut microflora. It is possible that, in the future, accurate and comprehensive chemical analysis of feed components would allow for a more accurate control by dietary manipulation, but at the present time this method is not sufficiently reproducible for commercial application.

ENZYMES

Although the addition of supplementary enzymes to diets does not directly affect the intestinal bacteria, there is an indirect effect analogous to that observed as a result of dietary manipulation. Feed components such as phytates and non-starch polysaccharides cannot normally be digested by the host animal and therefore become available for fermentation in the colon (pigs) or caeca (birds). Where enzymes are used to render normally indigestible components of the diet susceptible to small intestine digestion, there will be a resultant reduction in the proportions of these components reaching the fermentative bacteria of the lower gut (Ferraz de Oliveira, Hillman and Acamovic, 1994). There may therefore be considerable indirect effects on the intestinal microflora as a result of enzyme treatment of animal diets. Whether or not these effects will all be bacteriologically advantageous is not clear at this time, although a reduction in the concentration of non-starch polysaccharide reaching the lower gut is desirable, as this has been implicated as an initiator of porcine dysentery (Pluske, Pethick and Mullan, 1998).

ORGANIC ACIDS

Eidelsburger, Roth and Kirchgessner (1992) found that live-weight gain in piglets was increased by the addition of formic acid to the feed, although buffering the feed with sodium bicarbonate reduced this effect. The addition of sodium bicarbonate to a negative control reduced live-weight gain, possibly due to buffering of stomach acids. The improvement in feed conversion ratio obtained with formic acid was not affected by buffering the feed, although the protection against diarrhoea afforded by the acid was abolished.

Feeds with high buffering capacity have adverse effects on animal performance, probably by buffering of stomach acids with a consequent reduction in pepsin activity (Eidelsburger, Kirchgessner and Roth, 1992). Reduced stomach efficiency may result in increased feed retention with consequent reduced intake. The reciprocal of this would be that feed acidification would improve the efficiency of digestion in the stomach, resulting in a shorter retention time and improved feed intake.

The most likely mechanisms by which organic acid growth promoting feed additives improve feed conversion efficiency include acidification, either of the feed or of the gut, or selective inhibition of particular intestinal microorganisms. They may also be absorbed directly by the animal, providing an additional nutrient source. The reason for the reported protection from diarrhoea by certain acids is not immediately apparent. Although reduced gut pH would inhibit most pathogens, this effect would be required in the lower gut. Acidification has not been reported to extend this far.

The stomach of the young pig does not secrete sufficient hydrochloric acid to deal with the solid feed it receives at weaning (Ravindran and Kornegay, 1993). Acid secretion in the piglet stomach takes some 3-4 weeks after weaning to reach appreciable levels, so that acidification of feed for weaned piglets may be effective in reducing gastric pH. Inefficient digestion of the feed allows undigested material to reach the colon, and the presence of large quantities of fermentable material in the colon may allow the proliferation of pathogenic microorganisms, resulting in colitis and diarrhoea (Pluske *et al*, 1998). It is therefore possible that the reported reduction in diarrhoea in animals fed acidified feeds is due to a reduction in colonic fermentation, as a consequence of improved digestion in the stomach and small intestine. If this is the case, reducing the proportion of poorly digested carbohydrate in the diet of the piglet would result in a similar reduction in diarrhoea to that observed with acidification.

Although the principal areas of microbial fermentation in the pig are the caecum and colon, there is considerable microbial activity in the small intestine, and bacteria are present throughout the digestive tract of all animals (Ewing and Cole, 1994). In poultry, most microbial activity is found in the caeca. It is difficult to define the role of poultry caecal microflora since surgical removal of these organs seems to have little effect on the health or growth of the bird (Lewis and Swan, 1971). However, Gasaway (1976) suggested that up to 10% of the energy requirements of birds may be provided by volatile fatty acids produced by the microflora of the caeca, while the microbial fermentation in the large intestine of pigs may produce enough volatile fatty acids to account for up to 20% of the maintenance energy requirement for the piglet (Freind, Nicholson, and Cunningham, 1964). Since a considerable proportion of the energy requirements of birds and pigs can be obtained from the organic acids produced by intestinal microbial fermentation, the possibility that the provision of additional organic acids in the feed may act as a rapidly-absorbed energy source cannot yet be ruled out. Although the increase in growth observed appears out of proportion to the energy content of the added acids, this may be a contributory factor in the observed response.

PREBIOTICS

Most dietary sugars are digested and absorbed in the small intestine. However, complex oligosaccharides may survive small intestinal digestion and reach the lower gut intact, where they are fermented by intestinal bacteria. With careful selection, oligosaccharides have the potential to selectively enhance specific groups of bacteria.

The term "prebiotic" is used to define those compounds which selectively enhance particular microorganisms in the intestinal tract. The prebiotic may be fed alone, or as a selective substrate in combination with a suitable probiotic bacterium. Normally, compounds are used which are not digested by the host animal, although the use of the term "prebiotic" is occasionally vague, sometimes referring to the prefermentation of food by lactic acid bacteria (Bengmark, 1998) and may also be interchanged with other terms such as "biobiotic" or "synbiotic". The choice of term is perhaps unfortunate, as "prebiotic" has long been used to refer to the chemical soup from which life first developed. This poses considerable difficulties in researching the literature.

Studies *in vitro* have demonstrated the selective enhancement of the growth of certain groups of intestinal bacteria by oligosaccharides, obtained by the hydrolysis of oat beta-glucan and xylan hydrolysates (Jaskari, Kontula, Siitonen, Jousimies-Somer, Mattila-Sandholm and Poutanen, 1998). Fructooligomers increased the growth of *Bacteroides*, *Clostridium* and *Escherichia* as well as *Lactobacillus* and *Bifidobacterium*; while beta-glucooligomers and xylooligomers enhanced the *Lactobacillus* and *Bifidobacterium* strains preferentially but not significantly.

Sucrose caramel, containing a high proportion of oligofructose, can improve growth rate and feed conversion in poultry, and can increase the proportions of *Lactobacillus* and *Bifidobacterium* in the caeca (Manley-Harris and Richards 1994; Orban, Patterson, Sutton and Richards, 1997). However, this caramel had no effect when fed to pigs (Orban, Patterson, Adeola, Sutton and Richards, 1997).

Galactosyl-lactose was found to produce improved live-weight gain, comparable to the use of oxytetracycline/neomycin, in week-old calves (Quigley, Drewry, Murray and Ivey, 1997). These sugars appear to work well in both poultry and cattle, although no reports of their successful use in pig diets has been found. It is possible that they act as fermentation-modifiers and are suited more to the low-oxygen environments of the rumen and of the caeca of hens.

In pigs, mannose (mannan)-oligosaccharides have been shown to produce improvements in live-weight gain and feed conversion efficiency comparable to that obtained with olaquindox, or with a combination of Zn-bacitracin and toyocerin (Bolduan, Schuldt and Hackl, 1997). Piglets fed mannan-oligosaccharides in combination with a probiotic (*Enterococcus faecium* M74) showed significant improvements in live-weight gain, feed conversion efficiency and fibre digestibility (Kumprecht and Zobac, 1998). As the surface structures of the intestinal mucosa carry mannose residues which are used as receptors for attachment by many bacteria, and by dietary lectins, mannose and manno-oligosaccharides are important in intestinal toxicity and pathogenicity. Manno-oligosaccharides may therefore directly inhibit the activity of certain lectins by competitive binding (Oda and Tatsumi, 1993) and may also interfere with pathogen attachment.

Since much of the microflora of the intestine cannot currently be isolated and cultured, it is difficult to determine with certainty which species are making use of prebiotic substances. Additionally, making significant alterations to the numbers or activity of one genus of intestinal bacteria may have consequences for many other genera in terms of the substrates used and produced by the enhanced population. For example, the numbers of *Bifidobacterium* spp. can be increased in the gut by the use of inulin - but *Bifidobacterium* has complex nutrient requirements, including a requirement for vitamin K. Therefore, increasing that population 10-fold will result in a 10-fold increase in the requirement for vitamin K within the gut lumen, possibly depriving other genera of this and other nutrients. The gut microflora is a complex system, and the whole system must be considered in attempts at manipulation.

PROBIOTICS

The intended uses of probiotics are as a prophylactic for the prevention of intestinal disease or to accelerate re-stabilisation of the microflora after illness or antibiotic therapy (Fuller, 1992). They are generally of limited use in the treatment of active infection, since it is difficult for any bacterium to remove an established pathogen. The growth-promoting and other effects reported as a result of probiotic application are outwith the original intention of the products, but may arise as a natural consequence of improvements to the health and efficiency of the gut. Some probiotic preparations have been shown to produce antibiotic-like compounds, which have direct effects on particular groups of bacteria within the gut. Whether these compounds constitute the primary mode of action of the probiotic is unknown; the effects observed under *in vitro* conditions may not correspond exactly with the action of these cultures *in vivo*. It is therefore difficult to assess whether the mode of action of the probiotic growth promoter is similar to that of the antibiotic growth promoter, since little is known of the modes of action of either.

Furthermore, since poultry do not undergo the disruption of weaning, this microflora can persist throughout the life of the bird. The disruption of weaning results in the removal of any probiotic protection which may have been applied to monogastric or ruminant animals at birth, so that continued dosing with large numbers of probiotic bacteria is required throughout the weaning period.

A probiotic preparation may consist of a pure culture of one species, a mixture of known species in known proportions, or an undefined general gut microflora. An example of the latter would be the familiar Nurmi principle, in which the dosing of newly-hatched chicks with a healthy, *Salmonella*-free adult microflora can lead to the establishment of a *Salmonella*-resistant microflora in the adult chickens (Nurmi and Rantala, 1973).

The bacterial species in current use as probiotics for farmed animals comprise principally *Lactobacillus* and *Enterococcus* species, sometimes *Bacillus*, occaisionally *Escherichia* (Goerg and Schlorer, 1998) species. Unusual species have been tried, such as the diatom algae *Scenedesmus* and the cyanobacteria *Spirulina* and *Aphanizomenon* (Kay, 1991) although these do not appear to have found commercial application. Organisms which are to be active in the intestine must be able to survive in that environment, so they require, for example, resistance to bile salts (Chateau, Deschamps and Sassi, 1994) and acid tolerance (*eg* Nemcova, Laukova, Gancarcikova and Kastel, 1997) and an ability to inhibit pathogenic bacteria is desirable. They should also prove capable of competing effectively within the gut microflora (*eg* Kontula, Jaskari, Nollet, DeSmet, vonWright, Poutanen and Mattila-Sandholm, 1998). Diatoms and cyanobacteria are unlikely to fulfil these requirements.

Yeast and mould probiotics, often using strains of *Saccharomyces cerevisiae* and *Aspergillus oryzae*, can produce improvements in live-weight gain, feed conversion efficiency and milk production in cattle (reviewed by Stewart, Hillman and Maxwell, 1995). Both appear to work by enhancing rumen fermentation. These probiotic fungi do not colonise the rumen, neither are they typical inhabitants of the non-ruminant intestine. It is therefore unsurprising that yeast populations were not maintained in the intestine when a yeast-based probiotic was used in pig feeds (Mathew, Chattin, Robbins and Golden, 1998), although these authors reported an improvement in daily weight gain in pigs with this probiotic. There were no detectable effects on the intestinal microflora.

Yeast-based probiotics do not appear to have significant beneficial effects on the intestinal microflora of the pig, although an effect which may be related to the mechanism of action in the rumen has been observed in the caeca of hens (Andruetto and Doglione, 1993).

Table 7.3. Comparisons of bacterial probiotics for pigs with antibiotic feed additives.

Additive	Growth	Feed conversion efficiency	Authors
Virginiamycin	+	-	Harper *et al*, 1983
Freeze-dried *Lactobacillus* *acidophilus*	-	-	
Tylosin	+	+	Pollman *et al*, 1980
ASP-250	+	+	
Lincomycin	+	+	
L. acidophilus	+	+	
Streptococcus faecium C68	-	—*	
Furazolidone	+	+	Zani *et al*, 1998
Bacillus cereus ("CenBiot")	+	+	
Zn-bacitracin	+	++	Adami *et al*, 1997
B. coagulans	++	+	
Carbadox	+		Kreuzer, 1994
B. licheniformis + *B. subtilis* ("BioPlus 2B")	+		
Carbadox + BioPlus 2B	++		
Zn-bacitracin	+		Vassalo *et al*, 1997
B. toyoi	+		
L. acidophilus + *S. faecium* + *Saccharomyces cerevisiae*	+		

* In this case, the organism resulted in reduced feed conversion efficiency.

A selection of studies comparing the efficacy of probiotic growth promoters with feed antibiotics in pigs is shown in Table 7.3. These studies are, on the whole, supportive of the use of certain probiotic bacteria as replacements for growth-promoting antibiotics in pig feeds. However, while the results of probiotic administration can equal that of antibiotics, the probiotics appear to be less consistent in activity than antibiotics in both pigs and poultry (reviewed by Thomke and Elwinger, 1998). It is therefore the consistency of activity, rather than the degree of activity, which needs to be improved in the probiotic growth promoters.

A probiotic *Sporolactobacillus* sp. increased apparent absorption of glucose, galactose, and amino-nitrogen in growing pigs. The effects were lost immediately on withdrawal of the supplement, so that the presence of live bacteria was important (Rychen and Nunes, 1993). These same authors later reported that three probiotics (*Sporolactobacillus*, *Bacillus cereus* and a mixture of *Lactobacillus acidophilus*, *Lactobacillus fermentum* and *Lactobacillus brevis*) showed no effect on glucose, galactose or lactic acid absorption, although the absorption of amino-nitrogen was increased (Rychen and Nunes, 1995). So, variable results have been found by the same authors within a single probiotic preparation.

There have been a number of other reports concerning the efficacy of probiotics as growth promoters in pigs, although as these have not included an antibiotic as a positive control so that direct comparison of their effectiveness in relation to a feed antibiotic is not possible (Table 7.4). Most reports show improvements in comparison with a control, but most gains due to probiotic application occur in the first few weeks after weaning.

Studies on probiotics for poultry show results similar to those observed in pigs. Often, the probiotic produces enhancements in growth and/or feed conversion comparable to those observed with an antibiotic (Cavazzoni, Adami and Castrovilli, 1998; Zuanon, Fonseca, Rostagno and Silva, 1998). These effects are most marked in the first few weeks of growth (Yeo and Kim, 1997; Jin, Ho, Abdullah, Ali and Jalaludin, 1998; Istvan, Zoltan, Dalma and Laszlo, 1995). It appears that the benefits of probiotics are restricted to the young animal. It is possible that these improvements, where they occur, are a consequence of improved function in the immature gut. The gut of the fully mature animal may not require the additional protection of extraneous lactobacilli. Additionally, a much larger gut would require a correspondingly larger dose of probiotic in order for any effects to become apparent.

Despite the variability in activity of many probiotic preparations, and the apparent restriction of their utility to the young animal, it is possible that the use of probiotics during early growth may lead to savings on feed due to improved feed conversion efficiency. In the neonate, it is also possible that faster early growth may result in more robust young animals with improved disease resistance.

The overall conclusion must be that while probiotics can be as efficient at growth promotion in pigs as antibiotics, the consistency of their activity seems poor. This may, in part, be due to the wide range of strains and species employed as probiotic strains, and to the variability in the culture conditions used to produce the organisms. It is notable that none of the growth studies with probiotics provide the criteria used for the selection of these strains; in many cases the authors appear to have used either a randomly selected lactic acid bacterium, or one which has previously been demonstrated to have antipathogenic activity. Relatively few *Lactobacillus* species, or strains of species, exhibit probiotic activity (Hillman

Table 7.4.   Reported effects of probiotic preparations on piglets.

Organism(s)	Reported effects	Authors
"LBC" (unidentified lactic acid bacteria)	No beneficial effects on piglet performance, whether fed directly to piglets or to the sow.	Kowarz *et al*, 1994
Bifidobacterium longum	No effect on either growth or feed intake of piglets.	Brown *et al*, 1997
Bif. pseudolongum	Significant increase in piglet growth, improved FCR and reduced mortality.	Abe *et al*, 1995
Lactobacillus casei	Weight gain improved by 10% over 14 days	Danek *et al*, 1991
L. acidophilus	Significant increase in piglet growth, improved feed conversion and reduced mortality.	Abe *et al*, 1995
Enterococcus faecium M74*	Increased piglet live-weight gain by 5.75%, improved feed conversion.	Kumprecht and Zobac, 1998
Streptococcus faecium M74* + *L. casei*	Improved weight gain, reduced coliform numbers in the gut.	Tortuero *et al*, 1995
S. thermophilus + *L. bulgaricus*	Improved weight gain over the 12-21 days period only. Reduced coliform numbers in the intestine, increased phagocyte activity in the ileal epithelium.	Tortuero *et al*, 1995
Bacillus C.I.P. 32	In weaned piglets, this improved weight gain by 7 to 25%, while reducing feed intake by 8%. Improved crude protein degradation in the intestine.	Kumprecht *et al*, 1994

*The genus *Enterococcus* is taxonomically a recent subset of the genus *Streptococcus*, and the two names may be considered to be synonymous for the purposes of this discussion.

and Fox, 1994), so that random selection is unlikely to result in a useful isolate. Further variation may be attributed to the means of dosing; the efficacy of a *Lactobacillus plantarum* at inhibiting the establishment of *Listeria monocytogenes* in an *in vitro* porcine colon simulation was abolished by including the probiotic with the feed (Hillman, Khaddour and Fenlon, 1998). A more consistent means of probiotic production, together with an improved understanding of the mechanisms underlying their activity, would lead to more reliable criteria for the selection of potential strains so that consistency in results can be improved. The criteria for selection should also consider the antibiotic resistance profile of the probiotic organism.

ANTIBIOTIC RESISTANCE IN BACTERIAL PROBIOTICS

In laboratory culture, probiotic *Lactobacillus* spp. were inhibited by as little as 0.4 μg ml^{-1} of monensin, narasin, lasalocid, salinomycin or maduramycin, but in a mixture of the commercial preparation with water, the lactobacilli were resistant to up to 100 μg g^{-1} (Marounek and Rada, 1995). As the probiotic proved much more resistant to these antibiotics in a commercial mixture than in laboratory culture, the use of combined probiotic/antibiotic preparations would seem a viable option. Indeed, Mohan, Kadirvel, Natarajan and Bhaskaran (1996) showed that using a probiotic in conjunction with an antibiotic provided improved weight gain in poultry in excess of that provided by either the probiotic or antibiotic alone.

However, the principal reason for the removal of antibiotics from animal feeds is the potential for development of resistance to clinical antibiotics in the intestinal microflora. Certainly, antibiotic resistance can develop in a microbial population continually exposed to low doses of the drugs, and the genes coding for resistance can, in many cases, be passed from one bacterium to a related species by conjugation. This feature is not restricted to pathogenic bacteria. Many bacterial species can transfer genes by conjugation, including lactobacilli (Wagner and Balish, 1998; Gabin-Gauthier, Gratadoux and Richard, 1991)

Vancomycin resistance is widespread in some species of *Lactobacillus*, which may indicate a potential for direct transfer of antibiotic resistance to intestinal bacteria. Since many of these vancomycin-resistant *Lactobacillus* species are already present in the intestine, the addition of more of the same may be seen as an insignificant problem. However, the nature of this resistance is currently unclear, as is its potential for transfer. Vancomycin resistance has not been found in *L. acidophilus* or *L. delbrueckii*, but has been found in *L. rhamnosus, L. paracasei, L. plantarum, L. fermentum, L. confusus, L. salivarius* and *L. buchneri*, (Hamilton-Miller and Shah, 1998). Some of these strains are in current use as probiotics and in the preparation of human and animal fermented foods and feeds.

In the absence of antibiotic, resistant strains have no competitive advantage and their numbers reduce. However, continually adding a strain carrying an antibiotic-resistance would increase the potential for transfer of this resistance to other species. Bacteria intended for probiotic use should therefore be screened for antibiotic resistance before the product is released, to avoid the potential carriage of undesirable antibiotic resistance into the intestinal environment (Charteris, Kelly, Morelli and Collins, 1998). Furthermore, the combination of an antibiotic with a probiotic which carries resistance to that antibiotic is, in microbial terms, equivalent to applying a selective pressure while simultaneously providing the genes which confer resistance to that selection.

FERMENTED LIQUID FEED

Prefermentation of liquid feed for pigs, using *Lactobacillus* spp, has a number of theoretical advantages. The feed is acidified with lactic acid, live and possibly probiotic lactobacilli are continually fed to the animals, and the potential for contamination by pathogenic microorganisms is reduced by virtue of the low pH and high organic acid content of the feed. Unlike grass silage for ruminants, fermented liquid feed for pigs is fed immediately after fermentation. Therefore, much of the microbial population is still active, providing a continuous dosage of active lactic acid bacteria to the animals. This should, in itself, provide a beneficial influence on intestinal fermentation by continuously alleviating the physiological effects of stress on the gut.

Future possibilities

The long term future of antibiotics as treatments in both veterinary and medical disciplines is in serious doubt. Past overuse of these medications has led to the selection of highly resistant bacterial populations, including bacteria resistant to an array of antibiotics. There is currently little interest in the development of new antibiotics, and no new antibiotics active against Gram-negative pathogens are being developed (Mor, 2000). In response, there has been considerable research into alternative chemotherapeutic agents which could ultimately replace the antibiotic as the mainstay of medicine.

As is often the case, the natural world can provide the basis for new therapeutic agents. Antibacterial agents called "defensins" operate in a similar manner to gramicidin, by causing membrane perforation and ion loss from the cytoplasm. These peptide antimicrobials kill cells within seconds, unlike conventional antibiotics which are normally bacteriostatic (Mor, 2000). Within this group are the dermaseptins (secreted in the skin of frogs), and cecropin and mellitin (manufactured by moths and bees respectively). These have existed in these natural systems for some considerable time, and are still effective at preventing infections in these animals (Otvos, 2000). It is possible that the development of microbial resistance is limited by the extremely rapid activity of these compounds, and by the nature of their action. There may be insufficient time for a microorganism to produce enzymes capable of degrading these antibacterials, and the site of action is the cell membrane, a site which is difficult to alter without adversely affecting its functions. Mellitin has been found to be highly effective against spirochaetes (Lubke and Garon, 1997), while a synthetic peptide based on magainin has been demonstrated to rapidly kill *Escherichia coli* O157 (Appendini and Hotchkiss, 1999). Unfortunately, like gramicidin, the defensins would currently be suitable only for surface applications in human or veterinary medicine as they are also damaging to eukaryotic cells, particularly mammalian red blood cells.

A similar mechanism has been developed artificially using self-assembling organic nanotubes. These can be formulated in a range of diameters, and can perforate cell membranes in a similar manner to the defensins (Buriak and Ghadiri, 1997). These peptide nanotubes are formed not from helices, as are the defensins, but from stacked rings. The rings can be made stable in solution, forming tubes only when they enter cell membranes (Hartgerink, Granja, Milligan and Ghadiri, 1996). These structures can now be used to form highly effective ion channels in cell membranes (Clark, Buehler and Ghadiri, 1998) and are now being developed into specifically antibacterial compounds, so that their medicinal use should soon be realised.

If bacteria prove unable to develop resistance to these compounds, they may eventually replace antibiotics in medicine. It will be some time before they could become sufficiently cost-effective for use in animals, but the relatively simple mechanism of action and the ease with which the structures of these compounds can be manipulated would open enormous possibilities for manipulation of the intestinal microflora with highly specific antibacterials.

A further possibility for manipulation of the microflora may lie in the domain of bacterial quorum sensing, or cell-density-dependent gene expression. Originally discovered in marine bioluminescent bacteria in the 1960's (Strauss, 1999), the term relates to the activation or deactivation of certain metabolic processes within bacteria, as a direct response to the size of the population. The mechanism relies, in Gram-negative bacteria, on the production of acylated homoserine lactones (AHL; Eberhard, Burlingame, Eberhard, Kenyon, Nealson and Oppenheimer, 1981), and the concentration of these chemicals within individual cells is dependent on the number of AHL-producing cells present. A similar mechanism occurs in Gram-positive bacteria but this makes use of peptide pheromones (Kleerebezem, Quadri, Kuipers and deVos, 1997) rather than AHL's. This communication mechanism has been shown to be important in the virulence of infectious pathogens such as *Staphylococcus aureus* (Kleerebezem *et al*, 1997) and in enteropathogenic *E. coli* (Sperandio, Mellies, Nguyen, Shin and Kaper, 1999). Therefore, interfering with this extracellular communication system has the potential for control of the infective capability of these and other pathogens.

Conclusion

Reports of effective growth-promoting substances, which have included microbiological investigation of the gut microflora, show that improved gut health is linked to an increase in the numbers of lactic acid bacteria and/or a decrease in the numbers of coliform bacteria. As the coliform bacteria typically comprise about 1% of the total microbial population of the intestine, while *Lactobacillus* species account for more than 90% of the population (Khaddour, Reid and Hillman 1998, Reid and Hillman, 1999), it is not immediately apparent why a reduction in numbers of coliforms, or an increase in lactobacilli, should make much difference to the absorption of digested material by the gut?

The reason may lie in both the formation of lactic acid by lactobacilli and the sugar-utilising qualities of coliforms. The principal product of metabolism of sugars by *Lactobacillus* spp. is lactic acid, which may be absorbed and used by the animal. The degradation of feed in the gut by lactobacilli, therefore, need not result in a great deal of energy loss to the animal. On the other hand, the coliform bacteria use oxygen, which is present at appreciable quantities in the piglet intestine (Hillman, Whyte and Stewart, 1993). The coliform bacteria have a complete set of tricarboxylic acid cycle enzymes, so that the products of aerobic sugar breakdown by coliforms will be carbon dioxide and water, resulting in complete loss of the feed energy.

In this scenario, any reduction in the numbers and/or activity of coliform bacteria would be beneficial to the animal. Additionally, any increase in lactic acid bacteria in relation to the total bacterial load could also prove beneficial, since these bacteria are unable to extract more than a small fraction of the available energy from the feed. Acidification as a result of the activity of *Lactobacillus* spp. may also assist digestion by chemical hydrolysis of certain feed components, and would inhibit the activity of coliform pathogens.

Many alternatives to antibiotics are currently being made available but there seems to be little confidence in these among producers, as the efficacy of these products is considered, at best, highly variable. In fact, the efficacy of many feed antibiotics as growth promoters can be just as variable as, for example, that seen with probiotic preparations. The prophylactic effect of feed antibiotics (an effect which none of the alternatives can claim) is what really makes the antibiotics important in current production systems. The removal of antibiotics may force animal production into less intensive, more "organic" systems with a consequent reduction in productivity, at least in the short term. Unfortunately, the abrupt manner in which antibiotics are being removed from feeds means that the development of a reliable replacement is likely to take longer than the time that is left for the antibiotic feed additives.

Acknowledgements

Much of the work described was supported by SERAD or MLC

References

Abe, F., Ishibashi, N. and Shimamura, S. (1995) Effect of administration of bifidobacteria and lactic acid bacteria to newborn calves and piglets. *Journal of Dairy Science*, **78**, 2838-2846.

Adami, A., Sandrucci, A. and Cavazzoni, V. (1997) Piglets fed from birth with the probiotic Bacillus coagulans as additive: Zootechnical and microbiological aspects. *Annali di Microbiologica ed Enzimologia*, **47**, 139-149.

Andruetto, S. and Doglione, L. (1993) Modifications of poultry enteric microflora after feeding with feed containing live yeast. In *Prevention and Control of Potentially Pathogenic Microorganisms in Poultry and Poultry Meat Processing; FLAIR no. 6, Probiotics and Pathogenicity*, pp 73-76. Edited by J.F. Jensen, M.H. Hinton and R.W.A.W. Mulder, Het Spelderholt, The Netherlands.

Appendini, P. and Hotchkiss, J.H. (1999) Antimicrobial activity of a 14-residue peptide against *Escherichia coli* O157:H7. *Journal of Applied Microbiology*, **87**, 750-756.

Bengmark, S. (1998) Immunonutrition: Role of biosurfactants, fiber, and probiotic bacteria. *Nutrition,* **14**, 585-594.

Bijlsma, I.G.W., de Nijs, A., van der Meer, C. and Frik, J.F. (1982) Different pig phenotypes affect adherence of *Escherichia coli* to jejunal brush borders by K88ab, K88ac and K88ad antigens. *Infection and Immunity*, **73**, 891-894.

Bolduan, G., Schuldt, A. and Hackl, W. (1997) Diet feeding in weaner piglets. *Archiv für Tierzucht - Archives of Animal Breeding,* **40**, 95-100.

Brown, I., Warhurst, M., Arcot, J., Playne, M., Illman, R.J. and Topping, D.L. (1997) Fecal numbers of bifidobacteria are higher in pigs fed *Bifidobacterium longum* with a high amylose cornstarch than wih a low amylose cornstarch. *Journal of Nutrition*, **127**, 1822-1827.

Buriak, J.M. and Ghadiri, M.R. (1997) Self-assembly of peptide-based nanotubes. *Materials Science and Engineering C- Biomimetic Materials Sensors and Systems*, **4**, 207-212.

Cavazzoni, V., Adami, A. and Castrovilli, C. (1998) Performance of broiler chickens supplemented with *Bacillus coagulans* as probiotic. *British Poultry Science,* **39**, 526-529.

Charteris, W.P., Kelly, P.M., Morelli, L. and Collins, J.K. (1998) Antibiotic susceptibility of potentially probiotic *Lactobacillus* species. *Journal of Food Protection*, **61**, 1636-1643.

Chateau, N., Deschamps, A.M. and Sassi, A.H. (1994) Heterogeneity of bile-salts resistance in the lactobacillus isolates of a probiotic consortium. *Letters in Applied Microbiology*, **18**, 42-44.

Clark, T.D., Buehler, L.K. and Ghadiri, M.R. (1998) Self-assembling beta(3)-peptide nanotubes as artificial transmembrane ion channels. *Journal of the American Chemical Society*, **120**, 651-656.

Danek, P., Novak, J., Semradova, H. and Diblikova, E. (1991) Administration of the probiotic *Lactobacillus casei* CCM-4160 to sows - its effects on piglet efficiency. *Zivocisna Vryoba*, **36**, 411-415.

Edmunds, B.K., Buttery, P.J. and Fisher, C. (1980) Protein and energy metabolism in the growing pig. In *Energy Metabolism*, pp 129-133. Edited by L.E. Mount, Butterworths, London,

Eidelsburger, U., Roth, F.X. and Kirchgessner, M. (1992) Influence of formic acid, calcium formate and sodium hydrogen carbonate on daily weight gain, feed intake, feed conversion rate and digestibility. *Journal of Animal Physiology and Animal Nutrition - Zeitschrift fur Tierphysiologie Tierernahrung und Futtermittelkunde*, **67**, 258-267.

Eidelsburger, U., Kirchgessner, M. and Roth, F.X. (1992) Influence of formic acid, calcium formate and sodium hydrogen carbonate on dry matter content, pH value, concentration of carbonic acids and ammonia in different segments of the gastrointestinal tract. *Journal of Animal Physiology and Animal Nutrition - Zeitschrift fur Tierphysiologie Tierernahrung und Futtermittelkunde*, **68**, 20-32.

Eberhard, A., Burlingame, A.L., Eberhard, C., Kenyon, G.L., Nealson, K.H. and Oppenheimer, N.J. (1981) Structural identification of autoinducer of *Photobacterium fischeri* luciferase. *Biochemistry*, **20**, 2444-2449.

Eichner, D. and Gravitz, B. (2001) The effects of under-usage of antibiotics on bacteria. http://www.sidwell.edu/~bgravitz/bio/bio.html

Ewing, W.N. and Cole, D.J.A. (1994) *The Living Gut*, Context, Northern Ireland.

Ferraz de Oliveira, M.I., Hillman K. and Acamovic, T. (1994). The effects of enzyme treated and untreated lupins and their alkaloids on poultry gut microflora. In *Plant-associated toxins; Agricultural, Phytochemical and Ecological aspects*, pp 195-199. Edited by S.H. Colegate and P.R. Dorling, CAB International, Oxon, UK.

Freind, D.W., Nicholson, J.W.G. and Cunningham, H.M. (1964) Volatile fatty acid and lactic acid content of pig blood. *Canadian Journal of Animal Science*, **44**, 303-308.

Fuller, R. (1992) History and development of probiotics. In *Probiotics- the scientific basis*, pp 1-8. Edited by R. Fuller, Chapman and Hall, London,

Gabin-Gauthier, K., Gratadoux, J.J. and Richard, J. (1991) Conjugal plasmid transfer between lactococci on solid surface matings and during cheese making. *FEMS Microbiology Ecology,* **85**, 133-140.

Gasaway, W.C. (1976) Volatile fatty acids and metabolizable energy derived from cecal fermentation in the willow ptarmigan. *Comparative Biochemistry and Physiology*, **53A**, 115.

Goerg, K.J. and Schlorer, E. (1998) Probiotic treatment of pseudomembranous colitis: combination of intestinal lavage and oral administration of *Escherichia coli. Deutsche Medizinische Wochenschrift,* **123**, 1274-1278.

Hackstein, J.H.P., van Alen, T.A., op den Camp, H., Smits, A. and Mariman, E. (1995) Intestinal methanogenesis in primates - a genetic and evolutionary approach. *Deutsche Tierarztliche Wochenschrift*, **102**, 152-154.

Hackstein, J.H.P., Langer, P. and Rosenberg, J. (1996) Genetic and evolutionary constraints for the symbiosis between animals and methanogenic bacteria. *Environmental Monitoring and Assessment*, **42**, 39-56.

Hamilton-Miller, J.M.T. and Shah, S. (1998) Vancomycin susceptibility as an aid to the identification of lactobacilli. *Letters in Applied Microbiology,* **26**, 153-154.

Harper, A.F., Kornegay, K.L., Bryant and Thomas, H.R. (1983) Efficacy of virginiamycin and a commercially available Lactobacillus probiotic in swine diets. *Animal Feed Science and Technology*, **8**, 69-76.

Hartgerinck, J.D., Granja, J.R., Milligan, R.A. and Ghadiri, M.R. (1996) Self-assembling peptide nanotubes. *Journal of the American Chemical Society*, **118**, 43-50.

Hill, M.J., Fadden, K., Fernandez, F. and Roberts, A.K. (1986) Biochemical basis for microbial antagonism in the intestine. In *Natural Antimicrobial Systems, FEMS symposium no. 35*, pp 29-39. Edited by G.W. Gould, M.E. Rhodes-Roberts, A.K. Charnley, R.M. Cooper and R.G. Board, Bath University Press, Bath,

Hillman, K., Whyte, A.L. and Stewart, C.S. (1993). Dissolved oxygen in the porcine gastrointestinal tract. *Letters in Applied Microbiology*, **16**, 299 - 302.

Hillman, K., Murdoch, T.A., Spencer, R.J. and Stewart, C.S. (1994) Inhibition of enterotoxigenic *Escherichia coli* by the microflora of the porcine ileum, in an *in vitro* semicontinuous culture system. *Journal of Applied Bacteriology*, **76**, 294-300.

Hillman, K. and Fox, A. (1994). The effects of porcine faecal lactobacilli on the rate of growth of enterotoxigenic *Escherichia coli* O149:K88:K91. *Letters in Applied Microbiology*, **19**, 497-500.

Hillman, K., Khaddour, R. and Fenlon, D.R. (1998) Inhibition of *Listeria monocytogenes* by *Lactobacillus plantarum* is abolished by food intake, in an *in vitro* simulation of the porcine colon. *Proceedings of the British Society for Animal Science 1998*, p. 164.

Hillman, K. and Robertson, S.M. (2001) Distribution of potentially probiotic *Lactobacillus* spp. in pig farms. *Proceedings of the British Society of Animal Science* 2001, abstract no. 100.

Huis in't Veld, J.H.J. and Havenaar, R. (1993) Selection criteria for microorganisms for probiotic use. In *Prevention and Control of Potentially Pathogenic Microorganisms in Poultry and Poultry Meat Processing; FLAIR no. 6, Probiotics and Pathogenicity*, pp 11-19. Edited by J.F. Jensen, M.H. Hinton and R.W.A.W. Mulder, Het Spelderholt, The Netherlands.

Istvan, T., Zoltan, K., Dalma, M. and Laszlo, C. (1995) Effect of probiotic treatment on the body-mass gain and Salmonella excretion of growing geese. *Magyar Allatorvosok Lapja,* **4**, 197-200.

Jaskari, J., Kontula, P., Siitonen, A., Jousimies-Somer, H., Mattila-Sandholm, T. and Poutanen, K. (1998) Oat beta-glucan and xylan hydrolysates as selective substrates for *Bifidobacterium* and *Lactobacillus* strains. *Applied Microbiology and Biotechnology*, **49**, 175-181.

Jin, L.Z., Ho, Y.W., Abdullah, N., Ali, M.A. and Jalaludin, S. (1998) Effects of adherent Lactobacillus cultures on growth, weight of organs and intestinal microflora and volatile fatty acids in broilers. *Animal Feed Science and Technology,* **70**, 197-209.

Kay, R.A. (1991) Microalgae as food and supplement. *Critical Reviews in Food Science and Nutrition,* **30**, 555-573.

Khaddour, R., Reid, C-A. and Hillman, K. (1998) Maintenance *in vitro* of the microflora and fermentation patterns of the porcine intestine. *Pig News and Information,* **19**, 111N-114N.

Kleerebezem, M., Quadri, L.E.N., Kuipers, O.P. and deVos, W.M. (1997) Quorum sensing by peptide pheromones and two-component signal-transduction systems in Gram-positive bacteria. *Molecular Microbiology,* **24**, 895-904.

Kontula, P., Jaskari, J., Nollet, L., DeSmet, I., vonWright, A., Poutanen, K. and Mattila-Sandholm, T. (1998) The colonization of a simulator of the human intestinal microbial ecosystem by a probiotic strain fed on a fermented oat bran product: effects on the gastrointestinal microbiota. *Applied Microbiology and Biotechnology,* **50**, 246-252.

Kowarz, M., Lettner, F. and Zollitsch, W. (1994) Use of a microbial growth promotor in feeding sows and piglets. *Bodenkultur,* **45**, 85-97.

Kreuzer, M. (1994) Probiotic-antibiotic interactions in performance, intestinal fermentation and manure properties of piglets using a Bacillus (Bacillus licheniformis Bacillus subtilis) preparation and Carbadox. *Agribiological*

Research - Zeitschrift fur Agrarbiologie Agrikulturchemie Okologie, **47**, 13-23.

Kumprecht, I., Zobac, P. and Robosova, R. (1994) The effect of administration of Bacillus CIP-5832 on some physiological parameters and performance of piglets after weaning. *Zivocisna Vyroba*, **39**, 331-342.

Kumprecht, I. and Zobac, P. (1998) Study of the effect of a combined preparation containing *Enterococcus faecium* M-74 and mannan-oligosaccharides in diets for weanling piglets. *Czech Journal of Animal Science,* **43**, 477-481.

Lewis, D. and Swan, H. (1971) The role of intestinal microflora in animal nutrition. In *Microbes and Biological Productivity, Society for General Microbiology Symposium 21*, pp 149-175. Edited by D.E. Hughes and A.H. Rose, Cambridge University Press, Cambridge.

Lubke, L.L. and Garon, C.F. (1997) The antimicrobial agent mellitin exhibits powerful *in vitro* inhibitory effects on the Lyme disease spirochaete. *Clinical Infectious Diseases*, **25**, S48-S51, Suppl. 1.

Maczulak, A.E., Wolin, M.J. and Miller, T.L. (1989) Increase in colonic methanogens in aging rats. *Applied and Environmental Microbiology,* **55**, 2468-2473.

Manley-Harris, M. and Richards, G.N. (1994) Thermolysis of sucrose for food-products - a sucrose caramel designed to maximize fructose oligosaccharides for beneficial moderation of intestinal bacteria. *Zuckerindustrie*, **119**, 924-928.

Marounek, M. and Rada, V (1995) Susceptibility of poultry lactobacilli to ionophore antibiotics. *Journal of Veterinary Medicine series B - Zentralblatt für Veterinarmedizin Reihe B- Infectious Diseases and Veterinary Public Health,* **42**, 193-196.

Mathew, A.G., Chattin, S.E., Robbins, C.M. and Golden, D.A. (1998) Effects of a direct-fed yeast culture on enteric microbial populations, fermentation acids and performance of weanling pigs. *Journal of Animal Science,* **76**, 2138-2145.

McFarland, S.P. (1998) The influence of dietary starches on the microflora and fermentation pattern of the porcine colon. M.Sc. thesis, SAC/University of Aberdeen.

Mohan, B., Kadirvel, R., Natarajan, A. amd Bhaskaran, M. (1996) Effect of probiotic supplementation on growth, nitrogen utilisation and serum cholesterol in broilers. *British Poultry Science,* **37**, 395-401.

Mor, A. (2000) Peptide-based antibiotics: A potential answer to raging antibiotic resistance. *Drug Development Research*, **50**, 440-447.

Murumatsu, T., Takasu, O., Furuse, M., Takaski, I. and Ikumura, J. (1987) Influence of the gut microflora on protein synthesis in tissues and in the whole body of chicks. *Biochemical Journal*, **246**, 475-479.

Nemcova, R., Laukova, A., Gancarcikova, S. and Kastel, R. (1997) In Vitro studies of porcine lactobacilli for possible probiotic use. *Berliner und Munchener Tierarztliche Wochenschrift,* **110,** 413-417.

Nord, C.E. and Edlund, C. (1990) Impact of antimicrobial agents on human intestinal microflora. *Journal of Chemotherapy,* **2,** 218-237.

Nurmi, E. and Rantala, M. (1973) New aspects of *Salmonella* infection in broiler production. *Nature,* **241,** 210-211.

Oda, Y. and Tatsumi, Y. (1993) New lectins from bulbs of Crocus-sativus. *Biological and Pharmaceutical Bulletin,* **16,** 978-981.

Orban, J.I., Patterson, J.A., Sutton, A.L. and Richards, G.N. (1997) Effect of sucrose thermal oligosaccharide caramel, dietary vitamin-mineral level and brooding temperature on growth and intestinal bacterial populations of broiler chickens. *Poultry Science,* **76,** 482-490.

Orban, J.I., Patterson, J.A., Adeola, O., Sutton, A.L. and Richards, G.N. (1997) Growth performance and intestinal microbial populations of growing pigs fed diets containing sucrose thermal oligosaccharide caramel. *Journal of Animal Science,* **75,** 170-175.

Otvos, L. (2000) Antibacterial peptides isolated from insects. *Journal of Peptide Science,* **6,** 497-511.

Piattoni, F., Demeyer, D.I. and Maertens, L. (1996) *In vitro* study of the age-dependent caecal fermentation pattern and methanogenesis in young rabbits., *Reproduction Nutrition Development,* **36,** 253-261.

Pluske, J.R., Williams, I.H. and Aherne, F.X. (1996a) Maintenance of villous height and crypt depth in piglets by providing continuous nutrition after weaning. *Animal Science,* **62,** 131-144.

Pluske, J.R., Pethick, D.W. and Mullan, B.P. (1998) Differential effects of feeding fermentable carbohydrate to growing pigs on performance, gut size and slaughter characteristics. *Animal Science,* **67,** 147-156.

Pollman, D.S., Danielson, D.M. and Peo, E.R. (1980) Effects of microbial feed additives on performance of starter and growing-finishing pigs. *Journal of Animal Science,* **51,** 577-581.

Postgate, J. (1992) *Microbes and Man, Third Edition.* Cambridge University Press, Cambridge.

Quigley, J.D., Drewry, J.J., Murray, L.M. and Ivey, S.J. (1997) Body weight gain, feed efficiency and fecal scores of dairy calves in response to galactosyl-lactose or antibiotics in mik replacers. *Journal of Dairy Science,* **80,** 1751-1754.

Ravindran, V. and Kornegay, E.T. (1993) Acidification of weaner pig diets: a review. *Journal of the Science of Food and Agriculture,* **62,** 313-322.

Reid, C-A. (1999) Fermentation of resistant starch: implications for colonic health in the monogastric animal. PhD. Thesis, Robert Gordon University, Aberdeen.

Reid, C-A., Hillman, K. and Henderson, C. (1998) The effect of retrogradation, pancreatin digestion and amylose/amylopectin ratio on the fermentation of starch by *Clostridium butyricum* (NCIMB 7423) *Journal of the Science of Food and Agriculture*, **76**, 221-225.

Reid, C-A. and Hillman, K. (1999) The effects of retrogradation and amylose/amylopectin ratio of starches on carbohydrate fermentation and microbial populations in the porcine colon. *Animal Science*, **68**, 503-510.

Rychen, G. and Nunes, C.S. (1993) Effects of a microbial probiotic (*Sporolactobacillus* P44) on postprandial porto-arterial concentration differences of glucose, galactose and amino-nitrogen in the growing pig. *Reproduction Nutrition Development*, **33**, 531-539.

Rychen, G. and Nunes, C.S. (1995) Effects of 3 microbial probiotics on postprandial porto-arterial concentration differences of glucose, galactose and amino-nitrogen in the young pig. *British Journal of Nutrition*, **74**, 19-26.

Sperandio, V., Mellies, J.L., Nguyen, W., Shin, S. and Kaper, J.B. (1999) Quorum sensing controls expression of the type III secretion gene transcription and protein secretion in enterohaemorrhagic and enteropathogenic *Escherichia coli*. *Proceedings of the National Academy of Sciences of the United States of America*, **96**, 15196-15201.

Stewart, C.S., Hillman, K. and Maxwell, F. (1995) Effects of probiotics on the gut microbiology of animals. In *Medical and dental aspects of anaerobes*, pp 109 - 120. Edited by B.I. Duerden, W.G. Wade, J.S. Brazier, A. Eley, B. Wren and M.J. Hudson, Science Reviews Press, Northwood,

Strauss, E. (1999) A Symphony of Bacterial Voices. *Science*, **284**, 1302-1304.

Thomke, S. and Elwinger, K. (1998) Growth promotants in feeding pigs and poultry III. Alternatives to antibiotic growth promotants. *Annales de Zootechnie*, **47**, 245-271.

Tortuero, F., Rioperez, J., Fernandez, E. and Rodriguez, M.L. (1995) Response of piglets to oral administration of lactic acid bacteria. *Journal of Food Protection*, **58**, 1369-1374.

Van der Waaij, D., Berghuis de Vries, J.M. and Lekkerkerk van der Wees, J.E.C (1971) Colonization resistance of the digestive tract in conventional and antibiotic treated mice. *Journal of Hygeine (Cambridge)*, **69**, 405-411.

Vassalo, M., Fialho, E.T., de Oliveira, A.I.G., Teixeira, A.S. and Bertechini, A.G. (1997). Probiotics for piglets from 10 to 30 kg of life weight. *Revista da Sociedade Brasileira de Zootechnica - Journal of the Brazilian Society of Animal Science*, **26**, 131-138.

Vogeli, P., Kuhn, B., Kuhne, R., Obrist, R., Stranzinger, G., Huang, S.C., Hu, Z.L., Hasler-Rapacz, J. and Rapacz, J. (1992) Evidence for linkage between the swine L blood group and the loci specifying the receptors mediating adhesion of K88 *Escherichia coli* pilus antigens. *Animal Genetics*, **23**, 19-29.

Wagner, R.D. and Balish, E. (1998) Potential hazards of probiotic bacteria for immunodeficient patients. *Bulletin de l'Institut Pasteur*, **96**, 165-170.

Wallace, J.S., Stanley, K.N., Currie, J.E., Diggle, P.J. and Jones, K. (1997) Seasonality of thermophilic *Campylobacter* populations in chickens. *Journal of Applied Bacteriology*, **82**, 219-224.

Wilhelm, M. P., Lee, D.T. and Rosenblatt, J.E. (1987) Bacterial interference by anaerobic species isolated from human feces. *European Journal of Clinical Microbiology*, **6**, 266-270.

Yeo, J. and Kim, K.I. (1997) Effect of feeding diets containing an antibiotic, a probiotic, or yucca extract on growth and intestinal urease activity in broiler chicks. *Poultry Science*, **76**, 381-385.

Zani, J.L., daCruz, F.W., dos Santos, A.F. and GilTurnes, C. (1998) Effect of probiotic CenBiot on the control of diarrhoea and feed efficiency in pigs. *Journal of Applied Microbiology*, **84**, 68-71.

Zuanon, J.A.S., Fonseca, J.B., Rostagno, H.S. and Silva, M.D.E. (1998) Effects of growth promoters on broiler chickens performance. *Revista Brasileira de Zootechnica - Brazilian Journal of Animal Science*, **27**, 999-1005.

8

TRACING MODES OF ACTION AND THE ROLES OF PLANT EXTRACTS IN NON-RUMINANTS

C. KAMEL

Technical Director, AXISS France SAS, Archamps, France

Introduction

Global population is steadily increasing. In 25 years time, there will be almost 9 billion inhabitants on Earth (United States Census Bureau, 1998; FAOSTAT, 1998) expecting enough food to meet their daily nutritional needs. This objective can only be met by a 2% increase per year in world food production, which is expected for pigs and poultry in the next 20-25 years (FAOSTAT, 1998).

At the same time, there is a growing consumer consciousness about the necessity of food safety. This is no doubt due in part to serious issues in recent decades, such as the use of hormones and beta-agonists in animal nutrition, the presence of heavy metals in fish, the overuse of sludge, the presence of pesticides in fruit, the dioxins in meat, the use of antibiotics in animal feed, the production of genetically modified organisms, and Bovine Spongiform Encephalopathy.

"Let your food be your first medicine" (Hippocrates, 377 BC) was probably the first time that the link was made between nutrition and well-being. This old aphorism has come back into fashion in human and animal nutrition over the last few years as global markets have returned to a more "natural" source. In the animal feed industry, most diets can provide the diverse blend of nutrients needed for growth, maintenance and overall well-being. In some instances, however, diet formulation alone may not supply adequate amounts of the required nutrients, or the animal may not be in a state to make proper use of it, as is the case with ageing, during pregnancy and lactation, or under stress or disease challenge. During the twentieth century, much work was directed towards elucidating the biochemical structures and physiological roles of many nutritional additives, especially the routine inclusion of in-feed antibiotic growth promoters in pig and poultry diets over the last 50 years. These have proven to be an effective method of both enhancing animal performance and reducing digestive disorders that often presented themselves as scouring, particularly in weaner pigs. At the same time, increased concerns over food safety, environmental contamination and general health risks have made "natural" the norm, promoting the trend toward alternative strategies to manage and feed young animals without reliance on antibiotics. In response to this void in the marketplace, a new generation of alternative products has vied to take their place, many of which present more questions than answers. These have been defined by Rosen (1996) as "microfeedingstuffs used orally in a relatively small amount to improve the intrinsic value of the nutrient mix in the animal diet." Whether labelled under the name of 'adaptogens', 'dietetics', 'neutraceuticals', 'neutricines' or 'multi-functional additives', these are marketed for their benefits to animal nutrition. They include natural plant extracts; much research has focused on the specific beneficial effects of these as feed micro-ingredients, but this is by no means exhaustive. In particular, it remains unclear if a particular action, e.g. antioxidant, antiseptic, immunomodulator, can be associated with a specific molecule; this is complicated even more because one active substance may have multiple actions. In some cases, while searching for the "magic bullet", what results is more of a "dietary machine gun." Nonetheless, the elucidation of well-defined modes of action will provide the scientific basis for establishing the efficacy and safety of these additives to develop a long-term strategy for their use in specific diet formulations and other feed supplements. Only then will the industry and consumer feel confident about the potential of these products and embrace them as part of standard practice, as once they did with the antibiotic growth promoters.

Historical overview

One generation of productivity enhancers includes additives based on natural plant extracts and their intrinsic active principles. The use of plants and their respective extracts dates back some thousands of years to the ancient Egyptians, Chinese, Indians and Greeks. Even Christopher Columbus discovered America while looking for a shortcut to many of the herbs and spices found today in practically everyone's kitchen in one form or another. In addition, many natural extracts have provided the basis for modern medications, such in the case of digoxin from the digitalis plant, ephedrine from the Chinese herb ma huang, and common aspirin from willow. Clearly, much can be learnt from history, but the empirical knowledge of plants has risen to another level, due largely to modern technologies that have led to the systematic isolation and characterisation of active principles contained in these plant sources. As summarised in Table 8.1, many of the current extracts have beneficial multi-functional aspects that derive from their specific bioactive components. However, this represents only a fraction of the phytogenic elements known to man. To date, more than 10,000 species of plants known in the modern

Table 8.1. Summary table representing common plant extracts, their utilised parts, main active substances and reported properties (Richard, 1992; Charalambous, 1994).

Vegetal form	Utilised parts	Main Compounds	Reported properties
Aromatic spices			
Nutmeg	Seed	Sabinene	Digestion stimulant, antidiarrhoeic
Cinnamon	Bark	Cinnamaldehyde	Appetite and digestion stimulant, antiseptic
Clove	Cloves	Eugenol	Appetite and digestion stimulant, antiseptic
Cardamom	Seed	Cineol	Appetite and digestion stimulant
Coriander	Leaf, seed	Linalol	Digestion stimulant
Cumin	Seed	Cuminaldehyde	Digestive, carminative, galactagogue
Anise	Fruit	Anethol	Digestion stimulant, galactagogue
Celery	Fruit, leaf	Phtalides	Appetite and digestion stimulant
Parsley	Leaf	Apiol	Appetite and digestion stimulant, antiseptic
Fenugreek	Seed	Trigonelline	Appetite stimulant
Pungent spices			
Capsicum	Fruit	Capsaicin	Antidiarrhoeic, anti-inflammatory, stimulant, tonic
Pepper	Fruit	Piperine	Digestion stimulant
Horseradish	Root	Allylisothio-cyanate	Appetite stimulant
Mustard	Seed	Allylisothio-cyanate	Digestion stimulant
Ginger	Rhizome	Zingerone	Gastric stimulant
Aromatic herbs and spices			
Garlic	Bulb	Allicin	Digestion stimulant, antiseptic
Rosemary	Leaf	Cineol	Digestion stimulant, antiseptic, antioxidant
Thyme	Whole plant	Thymol	Digestion stimulant, antiseptic, antioxidant
Sage	Leaf	Cineol	Digestion stimulant, antiseptic, carminative
Bay laurel	Leaf	Cineol	Appetite and digestion stimulant, antiseptic
Peppermint	Leaf	Menthol	Appetite and digestion stimulant, antiseptic

world remain undiscovered, and 90% of the molecules that account for many of the actions remain a mystery. Through the integration of current knowledge, these molecules will become easier to find, identify and characterise. Furthermore, information regarding dose-response, metabolism, toxicity and maximum residue levels (MRL) should also be accumulated, with the intention of establishing an international database of information available to feed experts and consumers alike.

Back to basics

RELATING STRUCTURE TO FUNCTION

Plant extracts contain a vast source of different molecules that have intrinsic bioactivities on animal physiology and metabolism. One example is the phenolic compounds, a diverse series of either single (simple phenolic) or multiple (polyphenolic) aromatic rings differing principally in their respective side chains (Figure 8.1). In this sense, the phytogenic active components share something in common with organic acids, which also represent a non-homologous group, varying in the numbers of carboxy groups, hydroxy groups and carbon-carbon double bonds at the molecular level. These structural differences confer different stabilities to polyphenolic compounds over a standard pH range of 3-11. A funded project is ongoing at TEKTRAN (Technology Transfer Automated Retrieval System) of the United States Department of Agriculture Research Service to investigate the family of phytogenic phenolic compounds. In this programme, series of structurally well-defined phenolic compounds are being tested for their stability coefficients by changes in the ultraviolet absorption spectra of the phenolic compounds left standing in different pH buffers for several periods up to 72 hours. The results of this screening have suggested that the stability of the phenolic compounds strongly depended on the structure of the compound, the pH and length of storage buffer time (Friedman and Jurgens, 1999).

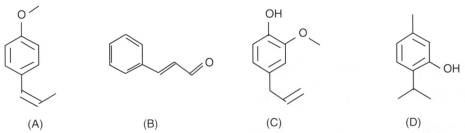

(A)   (B)   (C)   (D)

Figure 8.1 Chemical structures and molecular weights (MW) for common plant active substances. (A) anethol, MW 148.2; (B) cinnamaldehyde; MW 132.2; (C) eugenol, MW 164.2; (D) thymol, MW 150.2.

Principal Components Analysis (PCA), performed at the Department of Microbiology at the University of Auvergne Faculty of Medicine, Clermont-Ferrand, France, has also been used to study the properties of a preliminary screening panel of plant extracts. The pH distribution of this panel (data not shown) ranged from pH 4.3 (cinnamon extract) to pH 6.7 (ginger extract). This analysis continues to support the theory that not all extracts are created equal, but are representative of a group of structurally and chemically distinct molecules.

In addition, previous work has shown that, like organic acids, these chemical properties may correspond structurally to the incidence of polar groups, the number of double bonds, molecular size, and solubility in non-polar matrices. This, in turn, may dictate the activities observed *in vitro*: for example, antiseptic (Hammer *et. al.*, 1999), antifungal (Apisariyakul *et. al.*, 1995); antioxidant (Aeschbach *et. al.*, 1994); as well as *in vivo*: for example, stimulation of feed intake (Petit *et. al.*, 1993), stimulation of digestive enzymes (Platel and Srinivasan, 1996).

NATURAL PLANT EXTRACTS SHOW STRUCTURE-SPECIFIC MIC_{50} SPECTRA

The antiseptic effects of natural plant extracts are incontrovertable. The scientific literature is filled with work supporting the minimum inhibitory concentration (MIC_{50}) and minimum bacteriocidal concentration (MBC_{50}) of these substances. What has remained unclear is the relationship between structure of the active substances found in plant sources and their effects. Preliminary work carried out in collaboration with the University of Auvergne Department of Microbiology has confirmed that different extracts do indeed have different MIC_{50} spectra when screened against a diverse bacterial panel (Tabel 8.2).

Table 8.2. **MIC_{50} values (ppm) for selected plant extracts against a panel of Gram-positive, Gram-negative and anaerobic bacteria (performed in collaboration with the Department of Microbiology, University of Auvergne, Clermont-Ferrand, France). NE: No Effect**

MIC_{50} values	E. coli H7:O157	S. typhimurium	C. coli	C. perfringens
Capsaicin	NE	NE	NE	50
Cinnamaldehyde	1,000	1,000	1,000	1,000
Eugenol	500	500	500	500
Allicin	NE	NE	NE	100
Carvacrol	500	500	500	500
Cineol	50,000	50,000	50,000	50,000

From these studies, romarin showed very little effect against gram-negative micro-organisms. Cinammon, clove and oregano showed moderate and wide-spectra efficacy against the panel. The spice extracts, capsicum and garlic, showed specific efficacy against *Clostridium sp.* alone. These differences in efficacy at the *in vitro* level can be clearly linked to the level of their active substances, with purity of the plant extract playing a critical role (C. Forestier, University of Auvergne: personal communication). While their *in vitro* effects are well documented, the mode of action by which these substances exhibit their effect on bacteria is currently being investigated. On-going work seems to confirm that these substances, like antiseptic agents in general, exert their effects by disrupting the microbe at the cellular membrane. The effect of natural plant extracts on the hydrophobicity of an *E. coli* strain of post-weaning pig origin was determined by the salt aggregation test, a measure of the aggregative (binding) properties of strains of bacteria. The preliminary panel included extracts from St. John's Wort, cassia, cinnamon and thyme, which showed MIC_{50} values at 250 ppm or more on *E. coli sp.* The results shown indicate a strong increase (40-60%) in hydrophobicity of the microbial species by the addition of St. John's Wort and cassia. Cinnamon and thyme showed more moderate effects (Table 8.3). These differences in hydrophobicity correlated strongly with the MIC_{50} values for the extracts investigated (data not shown). Therefore, one of the probably and potentially important action mechanisms of certain plant extracts may be their ability to influence the surface characteristics of microbial cells and thereby their putative virulence properties. This has interesting implications in the animal gut, where the adhesion of microbes to host cells is of fundamental importance in the development of Gram-negative, microbe-induced infections, and which is influenced strongly by the surface hydrophobicity of the microbial cell (Mbwambo et. al., 1996).

Table 8.3. Hydrophobicity of E. coli (suis strain 4596) in the presence of St. John's wort, cassia, cinnamon or thyme. Values expressed as percentage aggregation by the Salt Aggregation Test.

Aggregation	St. John's wort extract	Cassia extract	Cinnamon extract	Thyme extract
E. coli (suis 4596)	58%[c]	42%[b]	23%[a]	28%[a]

Values on the same row with different supercripts are significantly different.

The role of complementary *in vitro* modelling techniques

The feed industry, and nutrition of the young animal in particular, is shifting into a phase of intensified research and development. Few companies, however, have adequate manpower or resources to put into place and follow long-term research programmes. On the other hand, public institutions often have difficulty finding the financial support necessary for project funding, given the dwindling government-sponsored funding programmes available. While this has become the norm in human nutrition and medical research, the possibilities are just beginning in the animal feed industry for a pooling of resources from the public and private sectors to work jointly toward one or more common objective(s).

One example where this may be most easily realised in the immediate future is in the development and validation of complementary *in vitro* systems modelled after various aspects of animal behaviour and physiology. There is a growing need for the development of systems which will be able to complement the information obtained through classical *in vivo* animal testing, while adding a dimension toward understanding small yet significant mechanisms, benefits or constraints which may remain hidden due to internal or external variability. Research and development on new feed-additive products has been, and will continue to be, one area where *in vitro* modelling operates at its best. Probably the greatest advantages of these models lie in qualities that are not so obvious. Trials with most systems are neither time nor labour intensive and cost only a fraction of comparable live animal trials. There are also advantages in the case of animal welfare considerations, as no live animals are directly involved.

These models are tools for innovative feed-additive companies whose goal is to aid nutritionists and feed compounders in the difficult, time-consuming challenge of evaluating products and their combinations within a range of diet formulations and production practices. Using these models, it has been possible to screen a "library" of plant extracts to investigate their effects on many aspects of bioavailability, normal digestion and gut flora mechanics, environmental and animal impact as it pertains to traceability and residues.

IN VITRO STUDIES: GASTRIC ACID SECRETION

It is proposed that acidification of the diet may provide a prophylactic measure similar to that provided by feed antibiotic growth promoters. While antibiotics are designed to inhibit growth in general, the acidifier would cause the beneficial rather than the harmful microorganisms to dominate in the gastro-intestinal tract. Supplementation of the diet for weanling pigs with acidifying agents has been shown, in many cases, to increase live-weight gain and feed conversion ratio, and to reduce the incidence of diarrhoea. However, the performance response to supplementation with dietary acidifiers is often variable. This has led to investigation of the possible modes of action of the acidifying agents but, although several hypotheses have been proposed, the exact mechanisms remain unclear. It is generally considered that feed acidifiers lower gastric pH, resulting in increased gastric proteolytic enzyme activity, and improved gastric emptying. Furthermore, acids entering the duodenum stimulate the exocrine pancreas to secrete bicarbonates. Therefore, it was expected that feed acidifiers would increase exocrine pancreatic secretion. The hypothesis that lowering dietary pH with feed acidifiers reduces gastro-intestinal pH has been tested in several studies. However, only a few have shown that dietary acidification significantly decreased gastric pH.

One way to understand the mechanisms behind different classes of products is by using *in vitro* modelling systems, which allow animal variability to be eliminated.

One such system is the non-ruminant intestinal model, developed by Drs. Robert Haavenar and Mans Minekus at the TNO Institute in Zeist, Netherlands. Briefly, this accurate "mechanical" model provides continuous monitoring of the natural parameters of digestion and its by-products. This can in turn determine the traceability of ingested feed components as well as their possible effects on the animal and their impact on the environment (Kamel, 1999). Certainly any type of digestion modelling has its limits, but this model has provided clear in-roads into the modes of action of various plant extracts and their possible usage in combination with other alternatives such as organic acids. Feeds that included certain plant extracts demonstrated effects in the gastric compartment similar to those observed with specific organic acids (Figure 8.2). Specifically, some extracts exerted an acid-sparing effect when included in a feed whose basic pH nature could compromise production of gastric acid, which can result in poor feed digestibility and potential overgrowth of pathogenic bacteria. Using a weaning porcine gastro-intestinal model, 200g diet samples were added to the gastric compartment and the amount of acid needed to maintain the pH at a level of pH 3.0 during a 6-hour period was recorded. The samples were taken from three groups of a piglet field trial (performance results shown in Table 8.4). The feed samples were based on a diet of wheat and soya, with no enzyme supplement, differing only in the micro-supplementation with either plant extracts at 2 g/kg, or formic acid (85%) at 10 g/kg and a combination of both. The results are summarised in Figure 8.2.

In general, responses confirm previous observations that the addition of certain plant phenols can act as an acid-sparing agent against the highly basic components found in a wheat and barley based weaning diet. Preliminary evidence suggests that the addition of an acid with a low pKa (i.e. formic or lactic acid) intensifies this effect as early as 60 minutes post-digestion. Whether extract and acid act alone or together to reduce gastric pH significantly, and whether this brings about influences on performance parameters (such as feed utilisation), remains to be seen.

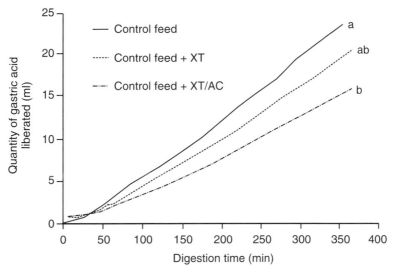

Figure 8.2. Experimental means for gastric acid secretion required to maintain a gastric pH of 3.0 for three different feed formulations in a porcine gastro-intestinal model. The three feeds are: Control (neutral) feed with no additive; control feed with XT (plant extracts based on cinnamaldehyde, thymol and capsaicin; and control feed with XT + AC (formic acid 85%).

In vivo animal studies comparing the performance benefits of plant extracts and organic acids either alone or in combination

For many years, feed additives have been used in young animal diets where digestion problems at weaning lead to scouring and a growth check. The slow, uneven acidification of the stomach contents of young animals plays a major role in the classic over-proliferation *of E. coli*, which then leads to enterotoxaemia in weaned piglets. The nutritional effectiveness of organic acids in piglet and broiler production has resulted in their routine use in young animal diets. However, raw material changes and evidence for "acid-resistant" micro-organisms, such as *E. coli H7:O157*, have led many nutritionists to use acid blends or a combination of acids with other alternatives, such as enzymes or probiotics. Early research into the modes of action of certain plant extracts and phytogenic active substances indicates that a synergism may occur between this class of compounds and organic acids. Development using many of the before-mentioned laboratory and *in vitro* modelling systems have shown promise in optimising the actions of these products alone and in combination with each other. A series of post-weaning pig trials is underway to confirm the promising combinations studied. One such trial was conducted on a commercial farm in Switzerland in pigs weaned at 21 days of age. The results (Table 8.4) show the benefits of inclusion of a precise combination of plant extracts on both live-weight gain (DWG, + 7.3% compared with control) and feed conversion (FCR, + 5.9% compared with control). The addition of formic acid to the plant extracts potentiated the observed effects even more dramatically, as well as having a significant effect on the incidence of diarrhoea. The evidence that this strategy may be useful in reducing stress associated with early-weaning programmes is currently being studied at several levels to determine the underlying mechanisms *in vivo*.

Table 8.4. Means for daily feed intake (DFI), daily live-weight gain (DLWG), feed conversion ratio (FCR) and diarrhoea scores of post-weaning pigs fed from 21 to 70 days on diets either containing XT (plant extracts based on cinnamaldehyde, thymol and capsaicin), FA85 (formic acid 85%) or a combination of XT and FA85.

Parameter	Negative control group Neutral feed	Experimental group 1 Neutral feed + XT	Experimental group 2 Neutral feed + XT/FA85
Number of animals	28	28	28
Start weight (kg)	6.90	6.92	6.90
End weight (kg)	21.21[b]	22.36[ab]	23.32[a]
DLWG (g/d)	292[b]	315[ab]	335[a]
DFI (g/d)	491[b]	503[b]	530[a]
FCR (kg/kg)	1.68[b]	1.58[a]	1.52[a]
Diarrhoea Score	39.0[b]	33.8[ab]	28.5[a]

Values on the same row with different supercripts are significantly different.

From the above trial in weaning pigs, it can be clearly seen that a diet based on wheat and barley gives particularly poor performance results in the absence of exogenous feed enzymes. This is due to the presence of anti-nutrients in wheat (pentosans) and barley (beta-glucans). The addition of beta-glucanases can improve diet digestibility. In addition, wheat and barley exhibit particularly high acid-binding coefficients. Therefore, the addition of plant extracts XT alone, but particularly in combination with organic acid FA85, improves digestibility, as evidenced by improved feed conversion ratios and higher live weights. This follows previous work in a number of published works showing the benefits of specific spice principles in enhancing the activities of non-ruminant digestive enzymes (Platel and Srinivasan, 1996). In addition, the combination also leads to a lower incidence of diarrhoea (27% lower than the control group), probably due to the reduction in volume of digesta, which can often act as a substrate for pathogenic bacterial growth.

ESTABLISHING REPRODUCIBLE METHODS OF DETECTION FOR NEW ALTERNATIVE PRODUCTS

Traceability and residues are probably two of the "hottest" topics of discussion in the modern global feed industry. The need to be able to follow the course of a product and its active ingredients from start to finish is seen as a fundamental requirement. Reliable detection techniques found in most analytical laboratories, such as GC (gas chromatography), LC (liquid chromatography), MS (mass spectrophotometry) and HPLC (high-pressure liquid chromatography) as well as their combinations, are necessary to realise this objective. Early work has shown that the phenolic compounds (Figure 8.2) can be identified and quantified, with detection systems reaching the sensitivity levels of picograms of material depending on the system in place (1 picogram is 10^{-12}g). These analytical techniques will be vital for implementing programmes of feed traceability, as well as establishing minimum residue levels (MRL) in carcass, egg and milk evaluation programmes for these categories of products.

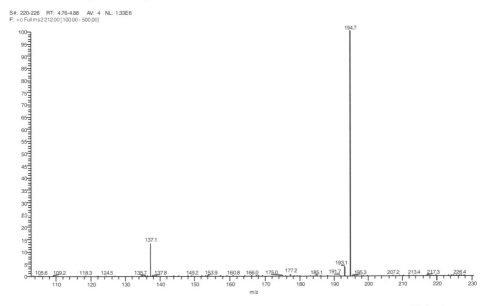

Figure 8.3. Sample extract representing 50 picograms of zingerone, from *Zingiber officinale* Roscoe, analysed by electrospray ionisation in LC-MS.

In vitro intestinal epithelium cell culture systems

The first step in any traceability approach is to determine what level of the compound crosses the intestinal epithelium. If the substance in question and its respective metabolites remain intraluminal and are finally excreted in the urine or faeces, there is little need to check for its deposition in internal depots. In this regard, immortalised epithelium cell systems (e.g. Caco-2 cell line), or certain primary pig intestinal cell lines, will figure as an important tool in tracking these new substances (Figure 8.4). Indeed, there is an extensive database of several thousand scientific publications relating to these cells, making them an extremely well-characterised research and development tool for many areas of scientific research. Their use has been quite extensive in the area of pharmacology, especially since there is a high degree of correlation between the permeability to pharmacological agents and the absorption of these same compounds in live subjects. This demonstrates only one value of the system in predicting potential bioavailability and absorbability of active compounds. In addition, it can be used to screen for detrimental effects of novel compounds following ingestion. The potential of this system in the traceability, safety and toxicity studies in the elucidation of plant extracts and their intrinsic active substances is exciting. Preliminary screening of panels of extracts and their active components has shown that these substances can be traced to different intracellular, as well as transcellular, destinations. The differences seen among phytochemicals may be due to the cellular sorting signals that are found deep in their molecular structure.

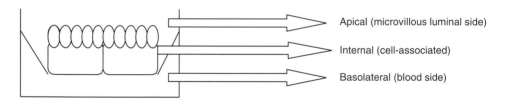

Apical (microvillous luminal side)

Internal (cell-associated)

Basolateral (blood side)

Figure 8.4. Schematic representation of a primary intestinal cell culture system illustrating its two interfaces: apical (luminal) and basolateral (blood). Substances added at the apical side are monitored over time along the cell transport system.

The future of this approach will lie in the full characterisation of these natural active substances on a molecule-by-molecule basis in an effort to associate families of compounds with specific activities and targeting routes. Indeed, research supports the idea that that not all extracts are created equal, but represent a group of structurally and chemically distinct molecules with specific and quantifiable activities, some of which are synergistic in nature. Based on preliminary screening experiments, synergistic combinations of plant active principles have been shown to provide benefits in the field (Table 8.5). Their cost-effective benefits in broilers, for example, can approach the results seen with antibiotic growth promoters still available in the marketplace.

Table 8.5. Summary of university, institute and field trials, carried out in Western Europe, Central Europe and Scandinavia, illustrating the effects on broiler performance of using plant extracts based on capsaicin and polyphenols at a final feed inclusion level of 200 ppm.

Parameter	Control feed No additive	Plant extracts XT 200 ppm
Number of trials	6	6
Average trial duration (days)	37	37
Number of animals	31,756	49,756
Initial weight (g)	39.4	39.4
Final weight (g)	1770	1849
Live-weight gain (g/d)	46.8	48.9
Feed intake (g/d)	81.9	85.1
Feed conversion ratio	1.78	1.74

Individual trial reports available upon request from the author.

One general observation in the majority of these trials is the evidence of an improved uniformity within the experimental groups fed on a diet supplemented with plant extracts. Clearly, one of the problems encountered in current "all-in/all-out" programmes is birds that fall behind the main group. These "laggers" can present significant problems to handlers in additional costs for prolonged finishing, floor utilisation, or extra care and treatment. Many natural plant extracts, due to their beneficial effects on feed efficiency and friendly gut flora populations, appear to optimise performance within the genetic framework of the bird. Therefore, their use may yield more important benefits, leading to a more uniform, cost-effective output.

Current research is taking account of public concern in investigating the advantages from the programmed administration of these phytogenic additives on the animal carcass. Only recently has the industry come to appreciate some of the many factors involved in determining meat quality and composition. One of those most obvious is the area of genetics, where work over the last few years in the United Kingdom has shown that certain exotic genes, such as those pertaining to the Duroc breed, can influence eating characteristics of meat. However, despite the excellent research being carried out by hybrid companies in generating pigs that grow faster and are leaner and more efficient than their predecessors, there have been costs in terms of meat quality. Colour and drip loss have worsened in recent years and the level of marbling fat has deteriorated, with the associated effect on eating quality. While the majority of studies have focused on raw material changes in the diet, supplements such as those based on plant extracts may also contribute to meat colour and drip loss, reduction in intermuscular fat stores, and shifts in fatty acid content and profile. Until now, the benefits from inclusion of plant extracts as a diet supplement have been measured largely on body composition. This includes several independent European university and trial station studies that have shown significant benefits in carcass backfat and lean characteristics, all in the absence of quantifiable residue deposition of these phytogenic substances in meat, fat or soft organ tissue (Table 8.6).

Table 8.6. Trials 1485 and 1602: Effect of inclusion of a micro-additive based on natural plant extracts in wheat and barley diets on slaughter characteristics in 128 growing and finishing pigs.

Parameter	Trial 1485		Trial 1602	
	Control feed *No additive*	*Natural extracts* *XT*	*Control feed* *No additive*	*Natural extracts* *XT*
Carcass weight (kg)	82.0	81.3	83.6	84.2
Dressing percentage (%)	59.9	60.7	54.9	55.7
Backfat thickness P1 (mm)	18.0[b]	16.7[a]	18.9	19.1
Muscle depth (LD, mm)	52.4	52.8	52.2[b]	54.8[a]
Meat quality		*% Distribution*		
European Meat Classification Scheme	S		0b	10a
	E		50b	42a
	U		50	48

Values on the same row with different superscripts are significantly different.

The inclusion of plant extracts as micro-supplements in current formulations for non-ruminant diets appears to yield improvements in certain carcass characteristics. At the moment, however, there are more questions than answers, as the framework within which these promising benefits operate needs to be better defined. In particular, more information must be obtained with respect to the specific class of phytochemicals responsible for these effects, their dose-response, their effects in combination with different commercial feed formulations, as well as the contribution of animal genetics and rearing conditions, in order to better explain the processes involved. Furthermore, the effects on carcass yield must be studied in relation to effects on carcass quality (i.e. meat structure and composition) in order to get a more complete picture.

Conclusions

Plant extracts represent one of a myriad of alternatives in response to the void created by the ban on antibiotic growth promoters in Europe. These substances have been in use since the beginning of recorded history, but little is known about the mechanisms that lead to the benefits seen in both man and animals. Furthermore, the supplier of feed additives will be responsible for a "checklist" of issues surrounding these components, such as: (i) composition identification; (ii) technical and performance efficacy; (iii) toxicity analysis; (iv) feed traceability; (v) residue analysis; and (vi) exposure and handling risks. Although this will be an arduous task for many, it will represent a responsible, informed approach within the feed industry and towards the consumer.

References

Aeschbach R., Löliger, J. Scott, B.D., Murcia, A., Butler, J. Halliwell, B and Aruoma, O.I. (1994). Antioxidant actions of thymol, carvacrol, 6-gingerol, zingerone and hydroxytyrosol. *Food Chemistry and Toxicology,* 32(1), 31-6.

Apisariyakul A., Vanittanakom, N., Buddhasukh, D. (1995). Antifungal activity of turmeric oil   extracted from Curcuma longa (Zingiberaceae). *Journal of Ethnopharmacology*, 49(3), 163-9.

Charalambous, G. (ed) (1994). Spices, herbs and edible fungi. Elsevier, London.

Friedman, M. and Jurgens H. (1999). Effect of pH on the stability of plant polyphenols. *U.S. Agriculture Research Report*, United States Department of Agriculture.

Kamel, C. (1999). Novel tools to test new feed additives. *Feed International,* 20(8), 22-26.

Mbwambo, Z.., Luyengi, L. and Kinghorn, D. (1996). Phytochemicals: A glimpse into their structural and biological variation. *International Journal of Pharmacognosy*, 34(5), 335-343.

Petit, P., Sauvaire, Y., Ponsin, G., Manteghetti, M. Fave, A. and Ribes, G. (1993). Effects of a fenugreek seed extract on feeding behaviour in the rat: Metabolic-endocrine correlates. *Pharmacology Biochemistry and Behavior*, 45, 369-74.

Platel, K. and Srinivasan K. (1996). Influence of dietary spices or their active principles on digestive enzymes of small intestinal mucosa in rats. *International Journal of Food Science and Nutrition*, 47(1), 55-9.

Richard, H. (ed) (1992). Spices and aromats, Lavoisier, Paris.

Türii, M., Türi, E., Koljalg, S., and Mikelsaar, M. (1997). Influence of aqueous extracts of medicinal plants on surface hydrophobicity of *Escherichia coli* strains of different origin. *APMIS,* 105, 956-62.

9

FUTURE DEVELOPMENTS IN ORGANIC FARMING – IMPLICATIONS FOR THE ANIMAL FEED INDUSTRY

NICOLAS LAMPKIN
Institute of Rural Studies, University of Wales, Aberystwyth

Introduction

Organic farming is increasingly recognised by consumers, farmers, environmentalists and policy-makers as one of a number of possible models for environmental, social and financial sustainability in agriculture. It has taken a long time to get this far, since roots of organic farming can be traced back more than 100 years and certified organic production in the UK dates back 25-30 years. Yet little more than 3% of agriculture in Europe is organic, and much less than that in other parts of the world. Can it expand to become a major part of agriculture in the next few years and what implications does that have for the agricultural supply industry, including feed compounders?

The growth of organic farming in the EU and the UK

Recent years have seen very rapid growth in organic farming (for detailed statistics, see Foster and Lampkin, 1999, 2000). In 1985, certified and policy-supported organic production accounted for just 100,000 ha in the EU, or less than 0.1% of the total agricultural area. By the end of 1999, this figure had increased to nearly 3.5 million ha, just over 2.5% of the total agricultural area. It is likely that 4.0 million ha is managed organically at the end of 2000, representing a 40-fold increase in 15 years (Figure 9.1). These figures hide great variability within and between countries. Several countries have now achieved 5-10% of their agricultural area managed organically (Figure 9.2), and in some cases more than 30% on a regional basis. Countries like Austria, Italy, Sweden and Switzerland, and recently the UK, have seen the fastest rates of growth. In the UK, the organic land area grew from 6,000 to 50,000 ha between 1985 and 1996, but has increased dramatically to

472,515 ha in 2000. The number of farms has also increased from 865 in 1996 to 3,182 in 2000.

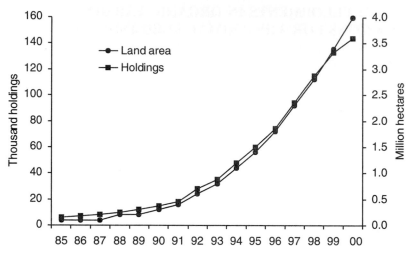

Figure 9.1. European Union organic and in-conversion holdings and land area.

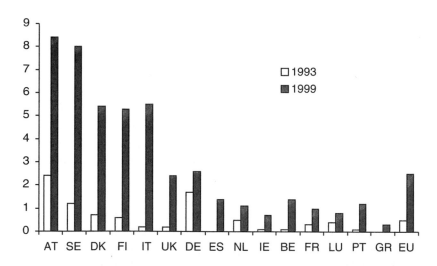

Figure 9.2. European Union organic and in-conversion land area as a % of total utilisable agricultural area in each country.

Alongside the increase in the supply base, the market for organic produce has also grown, but statistics on the overall size of the market for organic produce in Europe are still very limited. Some estimates have suggested that the retail sales value of the European market for organic food was of the order of 5-7 billion Euro in 1997 and possibly 7-9 billion Euro in 1999. The UK share of this is approximately 10%, with the UK organic market forecast to grow from £260 million retail sales value in 1997 to £750 million in 2000 and possibly exceeding £1 billion in 2001 (Soil Association, 1999). However, the rate of land conversion in the UK, despite recent growth, is not keeping pace with demand (Figure 9.3), leading to increased reliance on imports, currently estimated at 70-80% of retail sales.

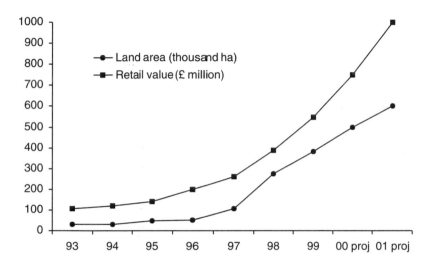

Figure 9.3. UK organic and in-conversion land area and market retail sales value (projections for 2000 and 2001).

More than 80% of the expansion in the EU land area has taken place in the last eight years, since the implementation in 1993 of EC Regulation 2092/91 defining organic crop production (now reinforced by the new EU Regulation 1804/1999 on organic livestock production), and the widespread application of policies to support conversion to and continued organic farming as part of the agri-environment programme (EC Reg. 2078/92), now continuing under the Rural Development Regulation. The organic regulations have provided a secure basis for the agri-food sector to respond to the rapidly increasing demand for organic food across Europe. The agri-environment and rural development regulations have provided the financial basis to overcome perceived and real barriers to conversion. Accompanied by high consumer demand, in part due to BSE, GMO and other food scares, as well as by the crisis in conventional agriculture (particularly in the UK) farmers have been keen to look at alternative options such as organic farming.

However, uptake has been significantly skewed towards grassland-based livestock farms, and there has yet to be a corresponding conversion by arable producers, leading to a significant imbalance in the demand for and availability of livestock feed. This has meant that prices of organic cereals for livestock feed have often been at levels similar to cereals for human consumption; in the case of feed wheat and field beans exceeding £200/t, which is some 2.5-3 times the price of conventional feeds; the price for "in-conversion" products is approximately twice that of conventional equivalents.

Potential for more widespread conversion

Although growth trends in individual countries have varied considerably, with periods of rapid expansion followed by periods of consolidation and occasionally decline, overall growth in Europe has been consistently around 25% per year for the last ten years, i.e. exponential growth. If these growth rates are projected forward to 2010, this gives some indication of the potential significance of organic farming within a relatively short period. Continued 25% growth each year would imply a 10% share by 2005 and nearly 30% by 2010. Accepting that this rate of growth cannot be sustained indefinitely, a slower rate of growth of 15% each year would still result in 5% of EU agriculture by 2005 and 10% by 2010. In the UK, current growth rates are much faster, but an average 25% a year growth from the 2% base at the end of 1999 would see the UK keeping up with the EU average (Figure 9.4).

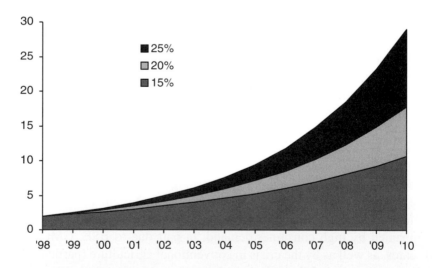

Figure 9.4. Projected growth for organic farmland in the European Union

At the EU level, 10%, whether achieved by 2005 or 2010, may still sound like a small proportion of the total, but it is very significant in absolute terms (Table 9.1). It represents nearly 15 million ha and possibly 600,000 farms (this number is likely to be lower given the rapid rate of structural change in agriculture at present) compared with the current total of 140,000 organic holdings. 10% is close to the UK agricultural land area (16 million ha) and more than double the total number of UK holdings in 1993. This level of growth has significant implications for the provision of training, advice and other information to farmers, as well as for the development of inspection and certification procedures. It also has implications for the development of the market for organic food, as it progresses from niche to mainstream status, with a likely retail sales value in 2005 of 25-35 billion Euro at the EU level.

Table 9.1. Projected growth for organic farming and food sales in the European Union (upper estimate)

	1998	2005	2010
Share of Utilisable Agricultural Area (%)	2%	10%	30%
Holdings (thousand)	120	600	1800
Land area (million ha)	3	15	45
Retail sales value (billion EURO)	5-7	25-35	75-100

Source: Lampkin, University of Wales

Pre-conditions for widespread conversion

Projections into the future based on past performance are not sufficient to realise the potential of organic farming. There is no guarantee that the rates of growth seen in the past will continue, and the normal expectation would be for rates of growth to decline eventually. A better understanding of the factors lying behind the growth of organic farming, and in particular the differences between countries, is needed.

In many respects, the development of organic farming has parallels to the traditional adoption-diffusion model for the adoption of innovations. Over time, the individualistic and socially-isolated innovators or pioneers are followed by the early adopters typified as community opinion leaders, to be followed in turn by the majority of farmers. In many countries, including the UK, this shift can be clearly seen. However, the rate at which this change takes place depends on the complexity of the innovation, and the adoption of organic farming is clearly a complex innovation (Padel, 2001).

The adoption-diffusion model does not seem to explain why the development of organic farming may be characterised by periods of stagnation followed by very rapid growth, as we have seen in the UK. A possible explanation for this is that farmers need to perceive the *need* to change before significant change will take place. A period of financial prosperity, as UK farmers experienced between 1992 and 1995 due to the CAP Reform package combined with the low value of the pound, was clearly not the basis for change. The reversal of circumstances since 1996, with the BSE crisis, the high value of the pound, and falling prices and agricultural support levels, has changed this perception dramatically. Similarly unsettling circumstances have arisen in other countries, for example in eastern Germany following re-unification, and in Austria on accession to the EU, leading to large increases in the number of farms converting.

The perception of the need for change needs to be accompanied by a conviction that organic farming is a suitable alternative. This requires a high degree of confidence-building because of the perceived financial, social and psychological barriers to conversion. It is not simply a case of 'more profits = more farmers' as many might argue. A preliminary assessment of this issue indicates that four key factors are involved:

- policy signals from government and other policy-related institutions;
- market signals from the food industry;
- access to information;
- the removal of institutional blockages or antagonisms.

It seems clear that each of these has been problematic in the UK at some point in the last decade. However, the changes that have taken place in all four areas over the last two years mean that the UK is now poised for substantial growth. Assuming that this analysis is correct, it provides a new basis for future policy development to encourage organic farming, particularly in the context of Agenda 2000, with a focus on integrated action plans rather than single measures like the organic support schemes under the agri-environment programme (Lampkin *et al.*, 1999).

Implications of widespread conversion

An expansion of organic farming to 10% of western European agriculture by 2005 or even 30% by 2010 has significant implications for the agricultural industry with respect to the likely impact on food production, surpluses and global food security, rural employment and incomes, international trade and consumption patterns. These issues have not even begun to be analysed seriously – the limited studies that have been done to date suffer from serious problems with underlying assumptions, availability of data, and the limited range of factors analysed.

For example, it is known that crops yields are 20-40% lower in organic systems compared with conventional systems in western Europe (Lampkin and Padel, 1994;

Lampkin and Measures 2001). Rotational constraints prevent crops being grown so frequently on arable farms, so that the overall yield reduction for cereal crops may be 40-50%. The production of grain legumes is likely to increase substantially, while some studies suggest that vegetable crops will stay stable or increase slightly, and other crops such as oilseeds and sugar beet will decline by as much as 75%. However, most of these assumptions reflect current demand patterns, which place more emphasis on horticultural crops and less emphasis on crops for processing such as oilseeds and sugar beet. The extent to which these crops are produced organically will depend on the market demand for them and this is still very difficult to predict.

Future livestock production levels are equally difficult to determine. It is reasonable to suppose that the production of ruminant livestock might decline by 10-30%, and that pig and poultry production levels might be substantially lower as a consequence of reverting to land-based, extensive systems.

A further concern about the widespread adoption of organic farming is the potential for erosion of premium prices as supply increases, leading to reductions in farm incomes and reduced attractiveness of the organic option. A number of factors that imply this may not happen in the near future need to be considered: expansion of demand as more outlets stock organic products, improvements in the efficiency of processing, marketing and distribution through economies of scale, and improvements in technical efficiency on the part of organic producers. Given that nearly 70% of organic farmland in the EU is grassland, the projected demand for livestock feed is likely to contribute to sustaining demand for cereals and grain legumes, and therefore prices in the arable sector, for some time to come.

Implications for the feed industry

To take advantage of the potential for growth in the organic sector, the livestock feed industry needs to be aware of the following issues:

a) the need to adhere to EU Regulation 1804/1999 (as implemented by UKROFS in the UK (UKROFS, 2000)), which contains lists of permitted feedstuffs and additives and prohibits or restricts the use of GMOs, amino acids, and in-feed medication. Although the majority of feed fed must be organically produced, there is a small allowance for conventionally produced feeds that are difficult to obtain organically in the short to medium term. In-conversion ingredients at also permitted and some cereals and pulses are available in the UK are lower prices than fully organic grains.

b) problems of price and demand – high prices for organic ingredients due to high demand and lack of supply are leading dairy producers in particular to reduce concentrate use and rely more heavily on milk from forage. However, high rates of conversion of dairy farms, and increasing demand for organic feed for pig and in particular poultry production, may compensate for reduced demand from individual dairy producers.

c)   problems of availability – production of raw materials in the UK is still very limited and imports are currently being obtained from as far afield as China. The feed industry may need to follow the example of other sectors and be pro-active in encouraging arable farmers to convert to meet requirements. This could take the form of partnerships or contracts with converting producers, providing a higher price for first-year and in-conversion products, even though some of these might have to be utilised in conventional feeds. A strategy based solely on importing at lowest cost will not succeed in developing domestic supply and may lead to problems if demand elsewhere catches up with the levels experienced in the UK.

Conclusions

The organic sector is growing rapidly and has high potential for future growth, to become a small but not insignificant part of the agricultural industry in the next decade. The organic sector is a global phenomenon that is here to stay and therefore investing in this area should bring worthwhile returns.

As in food processing and retailing, the feed industry will need to respond to the demand and contribute to the development of supply. However, the organic sector is not a short-term option – significant commitment is needed to build a base for growth. Demand cannot be met overnight and the integrity and consumer trust built up by the sector needs to be respected.

References

Foster, C. and N. H. Lampkin (1999, 2000) *European organic production statistics 1993-1996.* Organic farming in Europe: Economics and Policy, Volume 3, Universität Hohenheim, Stuttgart-Hohenheim. (the 1993-1998 report, and 1985-2000 land area and holdings data, can be found at http://www.organic.aber.ac.uk).

Lampkin, N. H., Foster, C., Padel, S. and P. Midmore (1999) *The policy and regulatory environment for organic farming in Europe.* Organic farming in Europe: Economics and Policy, Volume 1 and 2, Universität Hohenheim, Stuttgart-Hohenheim.

Lampkin, N. H. and M. Measures (eds.) (2001) *2001* Organic *Farm Management Handbook.* (4th edition) Welsh Institute of Rural Studies, University of Wales, Aberystwyth.

Lampkin, N. H. and S. Padel (eds.) (1994) *The economics of organic farming - an international perspective.* CAB International, Wallingford.

Padel, S. (2001) Adoption of organic farming – a typical example for the diffusion of an innovation. *Sociologia Ruralis.* **41**(1): 40-61.

Soil Association (1999) *1999 Organic Food and Farming Report.* Soil Association, Bristol.

UKROFS (2000) *UKROFS standards for organic food production.* UK Register of Organic Food Standards, MAFF, London.

10

FEEDING ANIMALS ORGANICALLY - THE PRACTICALITIES OF SUPPLYING ORGANIC ANIMAL FEED

S. Wilson
BOCM PAULS LTD, P O Box 39, 47 Key St, Ipswich IP4 1BX

Introduction

On the 24th August 2000, Council Regulation (EC) No. 1804/1999 came into force within the UK. This legislation, which supplements earlier legislation (Regulation (EEC) No 2092/91), sets out standards concerning the production of organic agricultural products. This includes standards regarding the production of organic feedingstuffs to be used for organic livestock production.

This chapter will discuss the key aspects of the legislation relating to feed and highlight the practical difficulties that result from trying to formulate a diet to meet the standards. Additional costs will occur for the feed compounder resulting from the implementation of the legislation and these will be discussed in relation to the feed itself and also the operational logistics.

Some personal views as to what are the important areas for the future are also given.

Aspects of the legislation relating to feed manufacture

It is the job of the commercial nutritionist to interpret the legislation and to produce a diet that not only satisfies the regulations but also satisfies nutritional needs of animals and the customer's aspirations for cost effectiveness. This role, which is challenging at the best of times, becomes exasperating when it comes to organic production.

The legislation that affects the production of organic compound feeds is in Council Regulation (EC) No. 1804/1999 "on organic production of agricultural products and indications referring thereto on agricultural products and foodstuffs to include livestock production". This legislation was laid before Brussels on the 19th July 1999 with the requirement to be implemented into national legislation by

24th August 2000. This legislation supplemented earlier legislation, Regulation (EEC) No. 2092/91 which came in force on 22nd July 1999 and covered unprocessed agricultural crop products and products for human consumption consisting of ingredients of plant origin. It also laid down the principle for the inspection and labelling of such products.

Within the UK, the United Kingdom Register of Organic Food Standards (UKROFS) has the responsibility of implementing the European standard.

Within the legislation there is a whole section dedicated to feed and it is recommended that anyone contemplating the production of an organic diet should read the section in some detail; however the key issues that it covers are:

- Livestock must be fed organically-produced feedingstuffs; however, due to limited supply of organically-produced material, a derogation exists which permits the use of conventional materials up to a limit of 10% for herbivores and 20% for other species. The figures are to be calculated annually as a proportion of the dry matter of feedingstuffs from agricultural origin. The maximum permitted on any one day is 25%.
- Up to 30% of the feed may comprise "in-conversion" feedingstuffs. When the in-conversion feedingstuffs come from a unit belonging to the livestock producer the proportion can be increased to 60%.
- Herbivores must receive at least 60% of the daily dry matter as roughage, fresh or dried fodder or silage. This may be reduced to 50% for animals in dairy production for a maximum period of three months in early lactation.
- Fattening poultry must be fed a diet containing at least 65% cereals.
- Raw materials produced with the use of chemical solvents or with genetically modified organisms or products derived therefrom, are not permitted in organic feedingstuffs.

It is worth at this stage discussing in a little more detail the practical considerations covering the exclusion of materials produced with the use of genetically modified organisms. The main problem here is one of definition. The legislation talks of the prohibition of materials produced with the use of "genetically modified organisms or products derived therefrom". The problem here is how far back down the production chain is it reasonable to check to see if GM technology has been used. It is well known that there are sources of macro-ingredients on the market, such as maize and soya, where the material has been produced directly using genetic modification (GM) technology; these materials are clearly excluded from inclusion into organic diets. The situation becomes a little less clear when considering materials manufactured by the use of genetically-modified organisms.

Both vitamin B_2 and vitamin B_{12} are generally manufactured via a fermentation process and in the case of vitamin B_{12} the organism predominantly used in the process is a genetically modified strain. Such vitamins are therefore produced using genetic modification technology and are therefore excluded from organic diets. Unless alternative supplies can be sourced which do not use such technology then the organic diet must rely solely on the vitamin B_{12} supplied via the raw materials.

However, the situation is not even completely clear regarding those vitamins that are not manufactured using GM technology. Many of the materials are beaded and coated in starch; if the starch was produced from maize, then in theory it must be established that the maize was of a non-GM variety before the material can be used.

There is an urgent need here to have clarification on the extent of exclusion of GM technology since, with the ever widening use of such technology, the day will rapidly come when it will be impossible to produce a balanced diet which is completely free of all contact with genetically modified material.

• The availability of raw materials to be used in organic diets is limited by the regulations to those materials listed in Tables 10.1-6. The legislation also states that by no later than 24th August 2003 those materials listed in these tables will be reviewed with the aim of removing, in particular, conventional feed materials of agricultural origin produced organically in sufficient quantity in the community.

Table 10.1 Conventional Feed Materials of Plant Origin Permitted For Use In Organic Diets (Council Directive (EC) No. 1804/1999)

1.1	Cereals, grains, their products and by products
1.2	Oil seeds, oil fruits, their products and by products
1.3	Legume seeds, their products and by products
1.4	Tuber roots, their products and by products
1.5	Other seeds and fruits, their products and by products
1.6	Forages and roughages
1.7	Other plants, their products and by products

NB. Further clarification of the materials included within each category is presented within the published reference.

Table 10.2 Feed Materials of Animal Origin (Conventional or Organic) Permitted For Use In Organic Diets (Council Directive (EC) No 1804/1999)

2.1 Milk and milk products
2.2 Fish, other marine animals, their products and by products

NB. Further clarification of the materials included within each category is presented within the published reference.

Table 10.3 Feed Materials of Mineral Origin Permitted For Use In Organic Diets (Council Directive (EC) No 1804/1999)

3.1 Sodium
- Unrefined sea salt
- Coarse rock salt
- Sodium sulphate
- Sodium carbonate
- Sodium bicarbonate
- Sodium chloride

3.2 Calcium
- Lithotamnion and Maerl
- Shells of aquatic mammals
- Calcium carbonate
- Calcium lactate
- Calcium gluconate

3.3 Phosphorus
- Low dicalcium phosphate precipitate
- Defluorinated dicalcium phosphate
- Defluorinated monocalcium phosphate

3.4 Magnesium
- Anhydrous magnesium
- Magnesium sulphate
- Magnesium chloride
- Magnesium carbonate

3.5 Sulphur
- Sodium sulphate

NB: Further clarification of the materials included within each category is presented within the published reference.

Table 10.4 Feed Additives Permitted For Use In Organic Diets (Council Directive (EC) No 1804/1999)

4.1 Trace elements
- Iron
- Iodine
- Cobalt
- Copper
- Manganese
- Zinc
- Molybdenum
- Selenium

Table 10.4 Contd.

4.2 Vitamins - Those permitted under Directive 70/524/EEC
4.3 Enzymes - Those permitted under Directive 70/524/EEC
4.4 Micro organisms - Those permitted under Directive 70/524/EEC
4.5 Preservatives
• Formic acid
• Acetic acid
• Lactic acid
• Propionic acid
  NB: For silage only
4.6 Binders
• Colloidal silica
• Kieselgur
• Sepiolite
• Bentonite
• Kaolinitic clay
• Vermiculite
• Perlite

NB. Further clarification of the materials included within each category is presented within the published reference.

Table 10.5 Certain Products Used In Animal Nutrition

There are no permitted products listed within this category.

Table 10.6 Processing Aids Used In Feedingstuffs

6.1 Processing aids for silage
• sea salt
• coarse rock salt
• enzymes
• yeasts
• whey
• sugar
• sugar beet pulp
• cereal flour
• molasses
• lactic, acetic, formic and propionic bacteria

There are also sections within the legislation that, whilst not referring directly to the diet, do have a bearing on the type of diet that can be produced.

Such sections state:

- Feed is intended to ensure quality production rather than maximising production, while meeting the nutritional requirements of the livestock at various stages of their development.
- Disease prevention in organic livestock production shall be based on the following principles:
  The use of high quality feed, together with regular exercise and access to pasturage, having the effect of encouraging the natural immunological defence of the animal
- Minimum slaughter ages for poultry are stated
- With regard to selection of breeds or strains, preference is to be given to indigenous breeds or strains.
- Livestock will have access to range or pasture.

Whilst the specific regulations do not place a quantifiable restriction on the diet, they must influence the decision of nutritionists in choice of diet specifications.

As with any other section of the livestock industry, the organic sector is interested in achieving the most cost effective level of production and hence the diet specification needs to reflect this. However, as already noted, restrictions are placed on the commercial nutritionist through the breeds being used, the environmental conditions under which they are kept, the level of output expected, and, possibly most importantly, the range of raw materials available. To illustrate this latter point, the lists of permitted ingredients will be considered further.

Table 10.1 lists those conventional materials of plant origin that can be used in organic diets as part of the conventional component. Category 2 within the table lists oil seeds, oil fruits and their products and by-products; within the full legislation the categories are further expanded to list those materials included within each category. However it regards the meal as the product and the hulls, if any, as the by-product; there is no mention of the liquid oil. Therefore liquid vegetable oils from conventional sources are not permitted. Liquid oils from organic sources would be allowed, but their availability and cost would tend to exclude their use in animal feed.

The obvious disadvantage of not being able to use a liquid oil is that the energy level of the diet has to be set at a value that can be achieved from either non-oil energy sources or oil as a component of other materials, e.g. full fat soya. The other potential negative aspect of the lack of liquid oil is that the physical quality of the diet may be inferior.

The vitamins that can be included in organic diets are listed on Table 10.4 as those authorised under Directive 70/524/EEC; however this is qualified by stating that the vitamins should be preferably derived from those naturally occurring in raw materials and, in the case of ruminants, that this must be the case. This inability to use synthetic vitamins in diets for ruminants has caused several practical problems and appears that the veterinary profession already has to prescribe vitamin supplements to counteract nutritional deficiencies in certain instances.

The requirement for non-GM products also leads to the exclusion of certain vitamins, as already mentioned.

Within the European Community amino acids permitted for use in conventional animal feedingstuffs are listed in Council Directive 82/471/EEC "concerning certain products used in animal nutrition". Within the organic legislation, whilst Table 10.5 is titled "Certain products used in animal nutrition", there are no products listed within that category. Therefore proteins obtained from yeast, non-protein nitrogen compounds and amino acids are excluded from organic diets.

It is obviously possible to formulate a diet without the use of synthetic amino acids; however the amino acid levels achieved are certainly sub-optimal for cost effective production in pig and poultry operations.

Diet specification

The nutrient specification of any diet is not only influenced by the level of production desired and the genotype of the animal being fed, but also, on a practical level, the range of raw materials available for use.

Limitations on the materials that can be included in organic diets have already been described, i.e. no solvent-extracted materials, no materials produced using GM technology, no synthetic amino acids, no non-organic liquid oils and fats.

It is also clearly stated in the legislation that the diets should be designed for "quality production" rather than "maximising production". The meaning of this statement is difficult to quantify, although the intention is clear.

The legislation states with regard to "origin of the animals" that preference is to be given to indigenous breeds and strains. These types of animals would generally have a lower nutrient requirement as a result of their lower genetic potential.

It is therefore obvious that in order to comply with the new legislation certain compromises have to be made with regard to the diet formulation, both in terms of nutrient content and also physical quality. There is also a significant cost increase.

In order to assess the cost implications of the new legislation it is beneficial to carry out a stepwise formulation exercise so that the various aspects can be individually costed. For the purpose of the exercise a diet for laying hens has been used as an example (Table 10.7). Starting with a conventional diet, with the full range of materials on offer, an initial cost of 100 units is established. If maximum constraints on the usage of conventional and in-conversion materials are imposed, the cost increase to 165 units, but also the lysine level moves off its minimum constraint of 7.6 g/kg to a new least-cost value of 8 g/kg. Further stepwise changes involving the removal of GM sources, solvent extracted materials, synthetic amino acids and liquid vegetable oil show not only subsequent cost changes, but also associated nutrient changes (lower ME and lower methionine) in order to make the diets feasible. The final solution (version 5) meets all the requirements of the organic regulations, but has had to have its nutrient constraints amended so that it will formulate. The cost increase (100 units to 188 units) is therefore a combination of legislative changes and nutrient changes. In order to quantify the cost impact of the legislation alone, the starting point (version 1) can be adjusted so that it includes the necessary nutrient changes (version 6). The true cost impact of the legislation alone is therefore the difference between versions 5 and 6 (188 units versus 95 units), which equates to a doubling in cost.

Table 10.7 Cost implications of the new organic regulation

	Formulation version					
	1 →	2 →	3 →	4 →	5 →	6 →
Diet Cost	100	165	193	191	188	95
Crude Protein (g/kg)	165	165	165	165	169	165
ME-P (MJ/kg)	11.3	11.3	11.1	11.1	10.5	10.5
Lysine (g/kg)	7.6	8	8	8.2	8.4	7.6
Methionine (g/kg)	4	4	4	3.2	3.2	3.2
In Conversion (%)	0	30	30	30	30	0
Organic (%)	0	50	50	50	50	0
Conventional (%)	100	20	20	20	20	100
GM Sources	+	+	-	-	-	+
Solvent Extracted	+	+	-	-	-	+
Synthetic Amino Acids	+	+	+	-	-	+
Liquid Vegetable Oil	+	+	+	+	-	+

It is evident therefore from the data in Table 10.7 that the impact of the full implementation of the new regulations is an approximate doubling of the ration costs on the basis of the same specification. However, with no adjustment for specification, a 90% increase in costs is more likely. On a fairly conservative basis, this will equate to an additional feed cost of approximately 30 pence/dozen eggs (UK prices).

As already seen, the exclusion of synthetic amino acids and non-organic liquid vegetable oils will have a dramatic effect on the specification of the diet to be formulated. As a consequence of this, the relative values of different materials are likely to change (Table 10.8).

Table 10.8 Change in relative value of different materials

	Conventional Diet	*Organic Diet*
Fish meal	371	640
High Oil Sunflower	155	293
Wheatfeed	95	3
Wheat	100	88

(All values are relative to wheat in a conventional diet)

It is also important to note that, as a direct consequence of the removal of the synthetic amino acids, diets with higher crude protein levels will have to be produced, and, more importantly, with an unbalanced amino acid content. As everyone is well aware, this can result in wet droppings and increased production of waste nitrogen, all at a time when both the organic regulations and Integrated Pollution Prevention Controls (IPPC) are trying to control the production of nitrogenous waste.

Whilst UKROFS have the responsibility of implementing the European legislation into the UK, individual organic producers must be registered with one of the approved organic certification bodies. Each of these bodies has the right to implement standards that are stricter than those covered by the European legislation, and some do set higher requirements. From the compounder's point of view, this can have major practical implications.

If two organic producers within the trading area of a feed compounder are registered with different certification bodies then, in theory, the feed compounder may have to produce two different ranges of diets to satisfy the specific requirements of the individual certification bodies. This is obviously commercially impractical, but is a serious factor to bear in mind when setting out to produce a range of organic diets.

Aspects affecting operational logistics

So far all of the cost considerations have been centred on the diet; however it is important to note that the production and distribution of organic diets incur additional costs over and above those directly associated with the feed.

Mill accreditation - in order for any feed mill to be allowed to manufacture organic diets it must first become registered with one of the approved sector bodies. The mills will then have to be audited annually for which there is a charge. These charges will vary depending on the individual body, but could vary from a few hundred pounds per year up to several thousands.

Production constraints - Due to limitations placed on the manufacture of organic diets within a conventional feed mill, additional production costs are likely to occur. These are likely to include the need to flush the plant prior to manufacturing organic feed and dedicated finished product storage.

It is difficult to put an exact cost on such practices, however it is likely to be in the order of 10-20 pence per tonne (0.1-0.2%).

Transport utilisation - In the modern commercial environment of animal feed, transport costs now exceed production costs. Therefore any additional constraints are likely to have an effect on an already significant cost.

Due to the size of production units within the organic sector, it is likely that delivery sizes are going to be less than the commercial average; this, along with the need to keep organic feed well segregated from conventional feed during delivery, will place additional constraints on logistics, and hence increase the cost of delivery. Again, it is difficult to put absolute value on the additional costs, but a figure of one pound (£1) per tonne (0.6-0.7%) would not be excessive.

Bin availability - Within any feed mill, bins for storing raw materials are always at a premium and a reduction in their number, and hence reduced raw material availability, will have a significant detrimental cost effect on all diets manufactured. The introduction of organic diets into a mixed feed mill will have the effect of reducing bin availability for the non-organic diets, due to the introduction of several new organic raw materials. These materials are unlikely to feature in non-organic diets due to cost and there is, therefore, an increase in the cost of non-organic diets due to the introduction of organic feeds, which again should be considered as part of the evaluation.

Finished Product Labelling - Whilst not directly adding to the actual cost of the diet, the additional labelling requirements of the organic diet is a practical consideration that needs to be addressed.

There is a need to state that the feed is organic, the proportions of organic, conventional and in-conversion materials and a breakdown of materials by type.

Current feed labelling is controlled by the Feedingstuffs Regulations, which do not permit the use of the adjective "organic" within the statutory statement. However, there is scope to include additional detail on the feed label, providing it is clearly separated from the statutory statements and can be substantiated.

Future developments

Looking to the future, it is evident that the new legislation will impose significant additional costs on the organic sector. In order for the UK industry to be able to compete, it is imperative that it works towards simplifying the current legislation as much as possible and tries to modify those aspects which add significant additional costs to the diet.

Whilst it is accepted that the individual organic-sector bodies have the right to impose stricter standards than the UKROFS guidelines, it is suggested that, when it comes to feed aspects, the requirements are standardised so that a feed mill does not have to incur additional costs in producing different diets for the different sector bodies.

Much has been said about the level playing fields of Europe, and there is a very real danger that, as the costs of organic production increase within the UK, there will be increased imports of produce from abroad. The UK producers do not fear competition if everyone is playing by the same rules; however, unfair competition, either arising from higher standards being imposed in the UK or different interpretation being applied elsewhere, makes it very difficult to establish a thriving industry.

Finally, all must support constructive lobbying of both the UK Ministry and Brussels to improve the current legislation.

Article 14 of the legislation sets out guidelines for its modification; however, any such process will inevitably take time and would require a Europe-wide approach. In this respect, support should be given to the work that UKASTA is doing, in conjunction with FEFAC, to raise many of the conflicting aspects of the legislation at a European level with a view to getting them amended. Nearer to home, there is ongoing dialogue between UKASTA, UKROFS and various industry working groups to discuss the problem and to try and find a working solution. Due to the welfare and health implication, one aspect currently being considered is to allow veterinary surgeons to authorise the use of synthetic amino acids, and presumably this could be extended to the use of synthetic vitamins for ruminants. Clarification is currently being sought to see if this is a feasible solution.

Most feed compounders want to see a thriving and expanding organic sector as they see it as a new market opportunity; however, as with any new venture the advantages and disadvantages of producing a new range of feed must be weighed up very carefully and the many practical difficulties considered.

If eventually the legislation can be amended to make the production of cost-effective organic diets more easily achievable, this will be to the benefit of the industry as a whole.

References

Council Directive (EC) No 1804/1999 supplementing Regulation (EEC) No 2092/91 on organic production of agricultural products and indications referring thereto on agricultural products and foodstuffs to include livestock production (1999). (OJ No L222).

11

INTERACTIONS BETWEEN THE IMMUNE SYSTEM, NUTRITION, AND PRODUCTIVITY OF ANIMALS

ELIZABETH A. KOUTSOS AND KIRK C. KLASING
Department of Animal Science, 1 Shields Ave., University of California, Davis, CA 95616

Introduction

Many important interactions exist between the nutrition of an animal and the physiological processes involved in its immune function, growth and reproduction (Figure 11.1). Understanding these interactions enables the nutritionist to contribute to the welfare of animals, reduce production costs and protect consumers from food-borne pathogens. This chapter will focus on the underlying mechanisms and practical consequences of these interactions. Much of the mechanistic information was developed in rodent models; however production animal species, especially poultry and pigs, are often used to elucidate the quantitative and practical applications of interactions between nutrition and immunity.

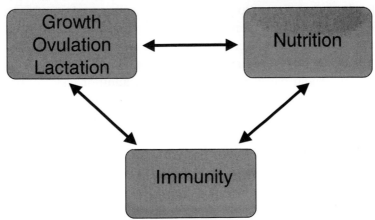

Figure 11.1 Nutrition interacts with the physiological processes involved in immune function, growth and reproduction. For example, nutrition modulates the immune system and immune responses modulate nutritional needs.

The systemic immune response – overview

The immune system functions to recognize the presence of macromolecules of non-self origin. Lymphocytes primarily recognize protein antigens through their cell surface receptors (e.g. membrane bound immunoglobulin), while phagocytes, such as macrophages (mf) and heterophils/neutrophils, recognize lipid and carbohydrate antigens through their surface pattern receptors. Antigen recognition results in leukocyte activation, cytokine secretion, and the engagement of effector mechanisms directed at removal of the pathogen.

The pro-inflammatory cytokines, including interleukin-1 (IL-1), IL-6, IL-8, tumour necrosis factor-α (TNFα), interferon-α (INF-α) and INF-γ act as communication molecules between immune cells and other cells of the body. Secretion of these cytokines occurs rapidly after an infectious challenge and they act locally to regulate the immune response. The pro-inflammatory cytokines reach systemically relevant levels within a few hours of a challenge. Quantitatively, their production is regulated by a complex network of co-stimulatory and feedback loops that respond to various stimuli. Among the mediators that counterbalance the pro-inflammatory cytokines are glucocorticoids, transforming growth factor-ß, receptor antagonists (e.g. IL-1_{ra}), soluble receptors, and acute phase proteins. In addition, some dietary factors can regulate the production of pro-inflammatory cytokines and anti-inflammatory regulators (see later).

The systemic effects of pro-inflammatory cytokines are important for coordination of the various biochemical, cellular and endocrine responses that enhance immune function and direct substrates to the immune system and other protective pathways (Table 11.1). In a homeorhetic response driven by elevated systemic levels of pro-inflammatory cytokines, nutrients are redirected from anabolic pathways related to growth, skeletal muscle accretion or reproduction towards pathways that bolster defence against pathogens. For example, nutrient repartitioning results in an increased rate of protein turnover, which leads to increased body temperature and basal metabolic rate. In addition, accelerated protein degradation may augment host defence by increasing proteosome activity and therefore enhancing expression of peptide fragments from intracellular proteins on MHC class II molecules. This process permits greater detection of intracellular pathogens and is illustrative of the heightened vigilance and activity of the immune system induced by a generalized stress state. In some tissues, such as skeletal muscle, increased proteolysis is not matched by augmented protein synthesis causing decreased tissue accretion that is manifested as impaired growth in young animals or negative nitrogen balance in mature animals. Nutrient repartitioning occurs not only with protein and amino acids, but also with virtually every other essential nutrient (Table 11.2).

Table 11.1 Effects of pro-inflammatory cytokines on metabolism and immune function. Adapted from (Chang and Bistrian, 1998, Klasing, 1988, Ling, Schwartz and Bistrian, 1997).

Cytokine	Metabolic effects	Immune effects

TNFα	↑ weight loss
	↑ catabolism in skeletal muscle
	↑ lipid release by adipocytes
	↑ osteoclast activity
	↑ collagenase gene transcription
	↑ resting energy expenditure
	↑ body temperature
	↑ hepatic cholesterol and
	   triglyceride synthesis
	↓ voluntary food intake
	↓ fatty acid uptake by adipocytes
		↑ phagocytosis by neutrophils
		↑ B cell growth/differentiation
		↑ proliferation of fibroblasts
		↑ cytokine production by mφ,
		  endothelial cells

IL-1	↑ collagen synthesis
	↑ bone resorption
	↑ weight loss
	↑ glucose oxidation
	↑ gluconeogenesis
	↓ voluntary food intake
	↑ body temperature
	↓ fatty acid uptake by tissues
	↓ thyroxin release
		↑ intestinal mucous production
		↑ acute phase protein (APP)
		   production
		↑ growth of fibroblasts, keratinocytes,
		  glial cells
		↑ expression of adherence molecules
		↑ proliferation of vascular smooth
		  muscle
		↑ lymphocyte chemotaxis
		↑ mediator production (IL-2,
		   thromboxane, PGE2, histamine)

IL-6	↑ body temperature
	↑ metallothionein production
	   by hepatocytes
	↓ voluntary food intake
		↑ APP production
		↑ IL-2 production
		↑ growth/differentiation of bone
		   marrow stem cells, macrophages,
		   neurons

IFN-γ	↓ growth of normal/neoplastic cells	↑ cytotoxicity of macrophages ↑ antigen presentation ↑ IL-2, IL-2R production

Mechanisms of the pro-inflammatory cytokines

Pro-inflammatory cytokines exert their wide-ranging actions on metabolism and tissue accretion by a multifactorial mechanism. Most cells have receptors for at least some of the pro-inflammatory cytokines and can respond directly to these mediators when their levels are sufficiently high in blood and interstitial fluids. Additionally, pro-inflammatory cytokines can stimulate neurons that innervate immune tissues such as lymph nodes, spleen, bone marrow or liver permitting communication with areas of the brain that regulate important processes such as body temperature, appetite, and behaviour (Downing and Miyan, 2000). Stimulation of local neural networks modifies vascular tone in specific tissues and, in some cases, decreases gastrointestinal motility, which affects feed intake.   Pro-inflammatory cytokines can also exert their action via the endocrine system, either by modifying the release of hormones, affecting hormone receptor numbers, or impinging upon second messenger pathways. For example, the anabolic drive induced by IGF-1 and insulin is blunted by pro-inflammatory cytokines (Elsasser, Caperna and Rumsey, 1995).

Each pro-inflammatory cytokine has its own set of unique activities as well as a large number of actions that overlap synergistically (Paludan, 2000) with other cytokines (Table 11.1). The specific milieu of pro-inflammatory cytokines released during an infectious challenge depends upon the type of challenging pathogen and its portal of entry.   For example, bacterial pathogens tend to stimulate production of IL-1, IL-6, and TNF-α, while viral pathogens tend to stimulate the production of interferons (Fossum, 1998).   Variation in the amounts and types of pro-inflammatory cytokines partially explains the wide range of specific symptoms expressed in animals with different disease challenges.

Effect of an immune response on animal productivity

An immune response can affect production by several mechanisms, including altering metabolism, altering hormone production, and/or inducing pathology (Klasing and Johnstone, 1991). Sick animals grow slower and less efficiently, miss ovulations, or produce less milk. In the past few decades, it has become clear that a large component of this phenomenon is due to the immune response *per se*, while the pathological consequences of the invading organism are not required for impaired productivity. For this reason, farm management systems are geared toward minimizing the incidence of infectious diseases through practices such as vaccination and rigorous biosecurity programmes.   Feeding growth-promoting levels of antibiotics diminishes challenges from opportunistic and commensal bacteria and consequently improves rates and efficiencies of production. With consumer and legislative pressure to minimize the use of antibiotic growth promoters, an understanding of the connections between immune responses and animal productivity becomes critical.

The immune response causes an overall change in the metabolism and behaviour of an animal. One of the most devastating impacts of this metabolic perturbance on animal performance, attributed to the effects of the pro-inflammatory cytokines, is decreased food intake (Johnson, 1998). In growing broiler chickens, the anorexia that accompanies an inflammatory response to LPS or heat-killed *Staphylococcus aureus* accounts for about 80 % of the decrease in growth rate (Klasing, Laurin, Peng and Fry, 1987). Clearly the immune response can be quite detrimental in production systems where weight gain is critical for profitability.

By altering the hormonal profile of an animal, an immune response can effectively reduce growth rates and reproductive performance (McCann, Kimura, Walczewska, Karanth, Retori and Yu, 1998). Specific binding sites for TNFa have been demonstrated in bovine pituitary, and administration of this cytokine results in a decrease in responsiveness to growth hormone releasing hormone (GHRH) and thyrotropin releasing hormone (TRH) (Elsasser, Kahl, Steele and Rumsey, 1997). GHRH, previously known as somatotropin releasing factor, is involved in the release of growth hormone and prolactin, while TRH stimulates the release of thyrotropin stimulating hormone, thus regulating metabolic rate, growth, and all thyroid functions. In addition to these hormones, as previously stated, the high growth rates of production animals, driven by IGF-1, are reduced by pro-inflammatory cytokine production (Elsasser, *et al.*, 1995). Obviously, reduced responsiveness to hormones can have a dramatic impact on performance.

The immune response can also inflict damage to host tissue, and this pathology is often necessary to eliminate pathogens. For example, during the innate immune response, macrophages are activated to destroy invading pathogens. The macrophage responds to cytokines and other effector molecules by producing antimicrobial factors, including reactive oxygen species and proteases. These mediators effectively destroy invading pathogens. However, host cells are also susceptible to these mediators, and therefore host pathology inevitably results as a consequence of disease. Nutritional strategies can modulate the production of cytokines and effector molecules and thus affect the level of macrophage activation and tissue pathology (Table 11.4). However, the elimination of disease is the only mechanism by which this pathology can be completely prevented.

Genetic implications

Given that many immune responses are associated with decreased growth and productivity, intense genetic selection for one of these processes should result in losses in the other. This negative correlation between immunity and productivity has been observed in a variety of experimental models. For example, in poultry, several long-term selection experiments based on immune response criteria have resulted in genetic lines with widely different immune responsiveness and, consequently, susceptibility to infectious disease. Selection for a high antibody response to sheep red blood cells (SRBC) over 24 generations results in a 15 % decrease in growth rate (Yang, Larsen, Dunnington, Geraert, Picard and Siegel, 2000). Conversely, selecting chickens or turkeys for high growth rates has resulted in impaired immunocompetence, disease resistance and alter pro-inflammatory cytokine release (Bayyari, Huff, Rath, Balog, Newberry, Villines, Skeeles, Anthony and Nestor, 1997, Li, Nestor, Saif, Bacon and Anderson, 1999, Mauldin, Siegel and Gross, 1978, Qureshi and Havenstein, 1994).   For this reason, it is likely that genetic selection for performance characteristics has resulted in an increased susceptibility of today's poultry and livestock to infectious diseases compared with their ancestors. This situation heightens the need for nutritional approaches to optimise immunocompetence.

Effect of an immune response on nutrition

Most research on nutrient requirements of animals is conducted in Federally-approved facilities that have levels of sanitation and pathogen exclusion that cannot be accomplished in real world, field conditions. Consequently, the minimal nutrient requirements summarized in the various National Research Council (NRC) publications may not be relevant to on-farm conditions where there is a prevalence of low-level non-clinical diseases and pathogenic challenges are frequent. Clearly it is of practical importance to understand the impact of these challenges on nutrient requirements.

The immune system must compete with growth and reproductive processes for nutrients, and is a component of the maintenance costs of an animal. When the immune system is engaged in defence against a pathogen, its nutrient demands are increased. Those nutrients are used as substrates for the clonal proliferation of responding lymphocytes, the production of antibodies and the hepatic secretion of acute phase proteins (Figure 11.2). Among these processes, the production of acute phase proteins quantitatively appears to be the most nutritionally demanding process (Klasing and Calvert, 2000). Consequently, the development, maintenance and use of the immune system require nutrients that must ultimately originate from the diet. Understanding the impact of disease challenges on nutrient requirements is complicated by the fact that each disease strain inflicts its own unique pathologies that have differing consequences for nutrient requirements. Those organisms that cause enteric pathology might be expected to have a large impact on digestion and absorption of nutrients, while those that inflict liver or

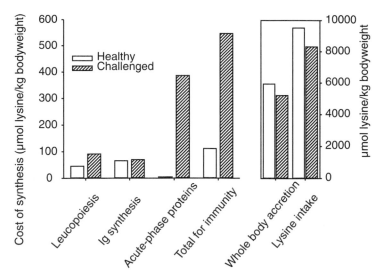

Figure 11.2 Use of lysine for leucopoiesis, immunoglobulin (Ig) synthesis, acute-phase protein synthesis and whole-body protein accretion in 14-day broiler chicks. Chicks were in either a healthy state or challenged with lipopolysaccharide from *Salmonella typhimurium*. The proportion of lysine intake used for immunocompetence was 1.2 and 6.7 % during health and challenged states, respectively. See Klasing and Calvert (2000) for details.

kidney damage disproportionately impact nutrient metabolism. For this reason, it is necessary to divide the impact of infection on nutrient requirements into two components: those effects due to specific pathology and those that are due to the response of the immune system to the pathogen. Presumably, this latter component is more consistent across the multitude of different diseases and can be studied in experimental model systems.

The immune response to a pathogen can affect the nutrition of animals by altering digestion, absorption, metabolism, and excretion of nutrients (Table 11.2), thus affecting substrate availability and dietary requirements. Components of the immune response may increase nutrient use, such as the anabolic components of the immune response (e.g. clonal proliferation of lymphocytes, recruitment of new myeloid cells, and antibody synthesis) as well as acute phase protein production by the liver. Conversely, altered metabolism of a specific nutrient may be a consequence of the immune response, rather than an integral component of that response. For example, retinol is transported in the blood bound to retinol binding protein (RBP). During an immune response, the synthesis of RBP is dramatically reduced, resulting in decreased retinol in the blood (Rosales, Ritter, Zolfaghari, Smith and Ross, 1996). In this situation, transport of a nutrient is altered as a consequence of the acute phase response. In addition, an immune response can increase the need for some substrates. For example, an energetic cost of fever

Table 11.2. Effect of an immune response on nutrient metabolism.

Nutrient	Immune response	Effect	Reference
Protein	Inflammation, infection, trauma	↑ skeletal muscle catabolism ↑ hepatic protein synthesis	(Beisel, 1977)
Cysteine	Acute inflammation	↑ protein production ↑ glutathione production	(Malmezat, Breuille, Pouyet, Mirand and Obled, 1998)
Glucose	Inflammation, infection, trauma	Increased turnover	(Beisel, 1977)
Lipids	Inflammation, infection	↑ hepatic triglyceride, cholesterol synthesis ↓ tissue fatty acid uptake	(Beisel, 1977)
Zinc	Inflammation, infection, trauma	↑ metallothionein ↓ plasma Zn	(Klasing, 1984)
Iron	Inflammation, infection, trauma	↓ dietary Fe absorption ↑ ferritin, lactoferrin ↓ plasma Fe	(Weinberg, 1999), (Butler, Curtis and Watford, 1973)
Vitamin A	Acute Inflammation	↓ RBP synthesis ↓ plasma retinol	(Rosales, *et al.*, 1996)
Carotenoids	Coccidiosis	↓ absorption, pigmentation	(Tyczkowski and Hamilton, 1991)

has been proposed, due to the cost of increasing body temperature (eg. shivering), the increased rates of biochemical reactions due to increased body temperature, and increased cellular utilization of ATP due to futile cycling (Baracos, Whitmore and Gale, 1987).

While the immune response may alter nutrient metabolism, lower rates of productivity during disease results in decreased nutrient needs. For example, in growing pigs and chickens, the overall requirement for lysine decreases during a simulated infectious challenge (Klasing and Barnes, 1988; Webel, Johnson, and Baker, 1998a; Williams, Stahly and Zimmerman, 1997a, b; Williams, Stahly and Zimmerman, 1997b). Presumably this is because lower rates of growth and muscle accretion decrease amino acid requirements much more than are necessary to compensate for the anabolic processes of the immune response.

Finally, some nutrients are metabolised in greater amounts during an immune response, which may increase dietary requirements during a disease state. Many effector mechanisms used by the immune system generate large amounts of reactive oxygen intermediates (ROI). Antioxidants, such as ascorbic acid and vitamin E, protect macromolecules against damage from these ROI. Evidence is accumulating that the requirements for antioxidant nutrients are greater during an inflammatory response than levels required by healthy animals (Webel, Mahan, Johnson and Baker, 1998; Yoshida, Yoshikawa, Manabe, Terasawa, Kondo, Noguchi and Niki, 1999). However, as previously stated, the level of nutrient intake is decreased during an immune response. Therefore, it may be necessary to increase nutrient intake prior to a disease challenge or vaccination.

In addition to conventional nutrients, other dietary factors may also be affected by an immune response. For example, during aflatoxicosis or coccidiosis in the chicken, absorption of dietary carotenoids, responsible for pigmentation of meat and eggs, is significantly decreased (Tyczkowski and Hamilton, 1987, 1991). While carotenoids are not considered essential nutrients, pigmentation of poultry products is often critical for consumer acceptance.

Modulating the immune response via nutrition

The immune response can be modulated by nutrition in several ways. First, the nutritional status of the developing embryo has profound effects on the development of its immune system. Second, nutrients provide the substrates for proliferation of immune cells and effector molecules. On this basis, the nutritional status of an animal can dramatically affect the degree of immune cell proliferation and/or effector molecule production. Third, just as production animals require dietary nutrients, invading pathogens rely on the host for supply of many nutrients. Therefore, nutrient deficiency or excess can affect the ability of a pathogen to replicate *in vivo*. In addition, nutrients can be effective as substrates for inter- and intra-cellular communication, and their availability may be critical for an effective immune response. Finally, the pattern of nutrient intake may affect the endocrine system, which then modulates magnitude, type and/or duration of the immune response.

The time period during which immune cells differentiate and populate the tissues of the body is relatively long compared to other developmental processes and extends well beyond birth. In fact, the high rate of cell proliferation in combination with the complexity of regulatory control over the developmental events is unmatched in any other tissue of the neonate. These developmental processes require a complex series of highly regulated cellular differentiation steps followed by deletion of self-reacting clones of lymphocytes that might cause autoimmunity. For this reason, a chronic deficiency of virtually any required nutrient during the period of immune system development has negative impacts on immunocompetence. Some examples of the deleterious effects of specific nutrient deficiencies on the development of the immune system are shown in Table 11.3. In general, those

Table 11.3. Effect of chronic nutrient deficiency on immune system development.

Nutrient	Species	Effect	Reference
Vitamin A	Rat	↓ body weight ↓ splenocyte # ↓ development/differentiation of all leukocytes	(Nauss, Chew-Chin, Ambrogi and Newberne, 1985) (Semba, 1998)
Vitamin D	Chicken	↓ thymus weight	(Alsam, Garlich and Qureshi, 1998)
Vitamin E	Chicken	↓ CD4-CD8+ thymocytes	(Erf, Bottje, Bersi, Headrick and Fritts, 1998)
Magnesium	Rat	↑ leukocyte #, 1° PMN cells ↓ thymus weight ↑ thymocyte apoptosis	(Malpuech-Brugere, Nowacki, Gueux, Kuryszko, Rock, Rayssiguier and Mazur, 1999)
Zinc	Mice	↓ lymphoid organ development ↓ splenic RBC, leukocyte #	(Beach, Gershwin, Makishima and Hurley, 1980)
Copper	Mice	↓ thymus weight ↓ Cu-Zn SOD activity ↓ ceruloplasmin activity ↓ Ab response to SRBC ↑ spleen weight ↑ anaemia	(Prohaska and Lukasewycz, 1989)

required nutrients that function in regulating cell differentiation (e.g. vitamins A, D) are particularly detrimental to the development of immunocompetence.

Nutrition affects not only the development of the immune system, but also the immune response to disease. Many nutrients modulate the immune system by mechanisms related to the provision of substrates (Table 11.4). The resting immune system is probably not very nutritionally demanding, as it contains some of the most inactive cells of the body (e.g. resting lymphocytes). However, activation of the immune system results in some of the most metabolically active and rapidly proliferating cells of the body. This sudden burst of activity is tightly regulated and critical decisions are made at this time regarding the threshold for a response, the magnitude of the response, and the types of cells recruited for the response. Nutrients are important in supplying the substrates needed for this rapid rate of cell proliferation and the secretion of effector molecules. In fact, the pro-

Table11.4. Effect of nutrient deficiency or excess on immune function.

Nutrient	Species	Effect	Reference
Protein Deficiency	Rat	↓ APP production ↓ NO synthesis	(Grimble and Grimble, 1998) (Wu, Flynn, Flynn, Jolly and Davis, 1999)
	Pig	↓ RBC glutathione production	(Jahoor, Wykes, Reeds, Henry, Del Rosario and Frazer, 1995)
Arginine Deficiency	Rat	↓ NO synthesis	(Wu, *et al.*, 1999)
Vitamin A Deficiency	Rat	↓ body weight ↓ splenocyte mitogenesis to ConA ↓ Natural Killer cell #, cytotoxicity ↓ neutrophil chemotaxis, adhesion, phagocytosis, production of oxidative species ↓ goblet cell number ↓ spleen, thymus weight	(Nauss, *et al.*, 1985) (Zhao, Murasko and Ross, 1994), (Twining, Schulte, Wilson, Fish and Moulder, 1997) (Semba, 1998) (Krishnan, Bhuyan, Talwar and Ramalingaswami, 1974)
	Chicken	↓ body weight ↓ IgA in bile ↓ T lymphocyte proliferation in vitro ↓ antibody response to BSA ↑ morbidity, mortality to *E. coli* ↓ antibody response to *E. coli*	(Rombout, van Rens, Sitjsma, van der Weide and West, 1992) (Friedman, 1989) (Friedman, Meidovsky, Leitner and Sklan, 1991)
Vitamin A Excess	Chicken in vitro	↓ T lymphocyte proliferation ↓ antibody response to BSA ↑ morbidity, mortality to *E. coli*	(Friedman, 1989) (Friedman, *et al.*, 1991)
Vitamin D Deficiency	Chicken	↓ growth, feed consumption ↓ DTH response ↓ thymus weight	(Alsam, *et al.*, 1998)
Vitamin E Deficiency	Chicken	↓ CD4-CD8+ thymocytes	(Erf, *et al.*, 1998)
Vitamin E Excess	Chicken	↓ antibody response to NDV	(Friedman, Bartov and Sklan, 1998)
	Turkey	↓ antibody response to NDV	(Friedman, *et al.*, 1998)
	Rat	↑ splenocyte, alveolar mφ # ↑ splenocyte mitogenesis ↑ NK cell activity ↑ phagocytosis by alveolar mφ	(Moriguchi, Kobayashi and Kishino, 1990)

Table 11.4 contd.

Nutrient	Species	Effect	Reference
Pyridoxine Deficiency	Pig	↓ hematocrit ↓ Ab titre to *S. pullorum*	(Harmon, Miller, Hoefer, Ullrey and Luecke, 1963)
Pantothenic Acid Deficiency	Pig	↓ Ab titre to *S. pullorum*	(Harmon, *et al.*, 1963)
Riboflavin Deficiency	Pig	↓ Ab titre to *S. pullorum*	(Harmon, *et al.*, 1963)
Magnesium Deficiency	Rat	↑ plasma IL-6 ↑ thymus TBAR	(Malpuech-Brugere, *et al.*, 1999)
Zinc Deficiency	Mice	↓ plaque forming ability to SRBC ↓ IgA, IgG, IgM	(Beach, *et al.*, 1980)
Iron Excess	Human	↑ lipid peroxidation ↑ DNA strand damage ↑ microbial replication	(Weinberg, 1999)
Iron Deficiency	Rat	↓ IL-1 activity ↓ sIgA and IgM containing mucosal cells	(Helyar and Sherman, 1987) (Perkkio, Jansson, Dallman, Siimes and Savilahti, 1987)

inflammatory response orchestrates metabolic changes, such as skeletal muscle catabolism, which provide a source of nutrients for these processes even when dietary intake is insufficient.

Just as the immune system is capable of repartitioning nutrients for cell proliferation and effector molecule production, repartitioning of some nutrients occurs as a defence mechanism against invading pathogens. Some nutrients serve as substrates for pathogens and support their proliferation and virulence. For example, iron is the limiting nutrient in the plasma and extracellular fluids of most animals. The pro-inflammatory cytokines released during a pathogen challenge mediate a shift in iron from transferrin in the extracellular fluids to ferritin located in intracellular locations (Weinberg, 1998). This change decreases the amount of iron available to pathogens and decreases their virulence. In neonatal pigs, excessive amounts of iron, provided through the diet or via injection, enhance the proliferation of several types of pathogen (Kadis, Udeze, Polanco and Dreesen, 1984, Knight, Klasing and Forsyth, 1983).   In addition to increased pathogenicity due to nutrient excess, a nutrient deficiency may also directly affect pathogens. In mice, a selenium deficiency has been demonstrated to provide a host environment in which a non-virulent strain of coxsackievirus may mutate and become virulent (Beck, 2000).

The mechanism for this change in the viral genome seems to be related to antioxidant status, as vitamin E-deficiency resulted in similar observations.

In addition to the aforementioned roles, many nutrients are directly involved in, or regulate, the intracellular and intercellular communication of leukocytes (Grimble, 1998). For example, the type of dietary polyunsaturated fatty acids (PUFA) consumed by an animal is reflected in the PUFA content of cellular membranes. As PUFAs are precursors for eicosanoid synthesis, including prostaglandins and leukotrienes, the PUFA content of cellular membranes affects the types and amounts of eicosanoids released during an immune response. This change in regulatory environment is reflected in the type of immune response that predominates during an infectious challenge and in the outcome of the infection (Harbige, 1998). Specifically, feeding n-3 PUFAs of marine origin, primarily eicosapentaenoic acid (EPA; 20:5 n-3) and docosahexaenoic acid (DHA; 22:6 n-3), results in decreased production of PGE_2. Supplementation of EPA results in increased production of PGE_3, a less biologically active molecule than PGE_2, while enrichment with DHA causes suppression of PGE_2 formation via cyclooxygenase inhibition (Chapkin, McMurray and Jolly, 2000). In addition, n-3 PUFAs from marine sources have been shown to reduce IL-12 and IFN-γ release causing a shift from T-helper 1 (T_h1) towards T_h2 responses (Fritsche, Shahbazian, Feng and Berg, 1997). This modulation of the immune response causes increased susceptibility to infections that are controlled by a strong T_h1 response, such as *Listeria*, but an increase in resistance controlled by a strong T_h2 response, such as E. coli sepsis (Fritsche, Anderson and Feng, 2000). In chickens, feeding diets supplemented with n-3 fatty acids significantly decreased the incidence of septicemia by 25 %, but resulted in a 24 % increase in the incidence of tumours (Klasing and Leshchinsky, 2000). Other nutrients that modulate the communication networks within the immune system include vitamins such as A, D, and E, minerals such as iron, and volatile fatty acids like butyrate (Klasing and Leshchinsky, 2000, Säemann, Böhmig, Burtscher, Parolini, Diakos, Stöckl, Hörl and Zlabinger, 2000, Weinberg, 2000)

Antioxidant nutrients, such as vitamin E also modulate an immune response. Reactive oxygen species are responsible for activating transcription factors, such as NF-KB, which are involved in the regulation of viral gene expression, cytokine synthesis, and the acute inflammatory reactions (Lander, 1997, Sen and Packer, 1996). Vitamin E, and other antioxidants, can inhibit the activation of these transcription factors, thus affecting the magnitude of the immune response. In addition to this role, vitamin E can affect arachidonic acid metabolism (Qureshi and Gore, 1997), thereby affecting the immune response in a similar manner to PUFAs.

Finally, the amount of diet fed, the pattern of feeding (e.g. single meal versus *ad libitum*) and the ratio between protein and energy in the diet influence the hormonal milieu. The hormonal profile in turn modulates the immune response via leukocytes, which have receptors for hormones produced by the classical endocrine system, including glucocorticoids, insulin, glucagons, growth hormone and thyroxin (Berczi, Chow and Sabbadini, 1998). For example, feeding protocols that result in periods of over-consumption of food impair both cell mediated and humoral indices of immunity (Klasing, 1988). Often, feeding regimens that restrict feed intake below voluntary levels result in better immunocompetence and enhanced disease resistance (BoaAmponsem, Yang, Praharaj, Dunnington, Gross and Siegel, 1997, Praharaj, Gross, Dunnington, Nir and Siegel, 1996, von Borell, Morris, Hurnik, Mallard and Buhr, 1992).

Conclusions

The immune system functions to recognise and destroy non-self molecules within the body. To maintain this function, leukocytes must proliferate, maintain receptors for recognition of foreign molecules, produce cytokines to regulate the response, and produce antibodies and other effector molecules. These functions require nutrients as substrates, without which the immune response to disease may be insufficient. In addition to the nutrient requirements of the immune system to mount an effective immune response, an immune response will affect nutrient partitioning. In general, an immune stress results in decreased nutrient intake, increased nutrient excretion, and repartitioning of nutrients away from production towards immune-related functions. In addition, production of effector molecules, such as reactive oxygen intermediates, may increase requirements for certain nutrients, such as antioxidants.

The immune response can be manipulated by nutrition. Nutritional status during embryonic development may affect immunocompetence for the life of the animal, while post-natal nutritional status may affect cell proliferation, responsiveness, phagocytic capacity, and effector molecule production. A nutrient deficiency or excess may result in increased susceptibility to disease due to these modifications. In addition, nutritional deficiency or excess may enhance the virulence or pathogenicity of certain organisms. Finally, many nutrients, specifically fatty acids and antioxidants, may modulate the immune response depending on their rate of dietary inclusion. For this reason, selection of dietary ingredients must take into account ratios of fatty acids and overall antioxidant levels. With sufficient knowledge of these interactions between nutrition and the immune response, the producer may optimise productivity and welfare while minimising disease concerns and the costs of over- or underfeeding animals.

References

Alsam, S. M., Garlich, J. D. and Qureshi, M. A. (1998) *Poultry Science,* **77,** 842-849

Baracos, V. E., Whitmore, W. T. and Gale, R. (1987) *Can. J. Physiol. Pharmacol.,* **65,** 1248-1254

Bayyari, G. R., Huff, W. E., Rath, N. C., Balog, J. M., Newberry, L. A., Villines, J. D., Skeeles, J. K., Anthony, N. B. and Nestor, K. E. (1997) *Poultry Science,* **76,** 289-296

Beach, R. S., Gershwin, M. E., Makishima, R. K. and Hurley, L. S. (1980) *Journal of Nutrition,* **110,** 805-815

Beck, M. A. (2000) *American Journal of Clinical Nutrition,* **71,** 1676S-1679S

Beisel, W. R. (1977.) in *Advances in Nutritional Research,* vol. 1 (Draper, H. H., ed.), pp. 125-133, Plenum, New York

Berczi, I., Chow, D. A. and Sabbadini, E. R. (1998) *Domestic Animal Endocrinology,* **15,** 273-81

BoaAmponsem, K., Yang, A., Praharaj, N. K., Dunnington, E. A., Gross, W. B. and Siegel, P. B. (1997) *Journal of Applied Poultry Research,* **6,** 123-127

Butler, E. J., Curtis, M. J. and Watford, M. (1973) *Research in Veterinary Science,* **15,** 267-9

Chang, H. R. and Bistrian, B. (1998) *Journal of Parenteral and Enteral Nutrition,* **22,** 156-166

Chapkin, R. S., McMurray, D. N. and Jolly, C. A. (2000) in *Nutrition and Immunology: Principles and Practice* (Gershwin, M. E., German, J. B. and Keen, C. L., eds.), pp. 121-134, Humana Press Inc., Totowa, NJ

Downing, J. E. G. and Miyan, J. A. (2000) *Immunology Today,* **21,** 281-289

Elsasser, T. H., Caperna, T. J. and Rumsey, T. S. (1995) *Journal of Endocrinology,* **144,** 109-117

Elsasser, T. H., Kahl, S., Steele, N. C. and Rumsey, T. S. (1997) *Comparative Biochemistry and Physiology,* **116A,** 209-221

Erf, G. F., Bottje, W. G., Bersi, T. K., Headrick, M. D. and Fritts, C. A. (1998) *Poultry Science,* **77,** 529-37

Fossum, C. (1998) *Domestic Animal Endocrinology,* **15,** 439-44

Friedman, A., Bartov, I. and Sklan, D. (1998) *Poultry Science,* **77,** 956-62

Friedman, A., Meidovsky, A., Leitner, G. and Sklan, D. (1991) *Journal of Nutrition,* **121,** 395-400

Friedman, A., Sklan, D. (1989) *Journal of Nutrition,* **119,** 790-795

Fritsche, K. L., Anderson, M. and Feng, C. (2000) *Journal of Infectious Diseases,* **182 Suppl 1,** S54-61

Fritsche, K. L., Shahbazian, L. M., Feng, C. and Berg, J. N. (1997) *Clinical Science,* **92,** 95-101

Grimble, R. F. (1998) *Nutrition,* **14**, 634-40

Grimble, R. F. and Grimble, G. K. (1998) *Nutrition,* **14**, 605-610

Harbige, L. S. (1998) *Proceedings of the Nutrition Society,* **57**, 555-62

Harmon, B. G., Miller, E. R., Hoefer, J. A., Ullrey, D. E. and Luecke, R. W. (1963) *Journal of Nutrition,* **79**, 269-275

Helyar, L. and Sherman, A. R. (1987) *American Journal of Clinical Nutrition,* **46**, 346-352

Jahoor, F., Wykes, L. J., Reeds, P. J., Henry, J. F., Del Rosario, M. P. and Frazer, M. E. (1995) *Journal of Nutrition,* **125**, 1462-1472

Johnson, R. W. (1998) *Domestic Animal Endocrinology,* **15**, 309-19

Kadis, S., Udeze, F. A., Polanco, J. and Dreesen, D. W. (1984) *American Journal of Veterinary Research,* **45**, 255-9

Klasing, K. C. (1984) *American Journal of Physiology,* **247**, R901-4

Klasing, K. C. (1988) *Poultry Science* **67**, 626-34

Klasing, K. C. (1988) *Journal of Nutrition,* **118**, 1436-1446

Klasing, K. C. and Barnes, D. M. (1988) *Journal of Nutrition,* **118**, 1158-64

Klasing, K. C. and Calvert, C. C. (2000) in *Proceedings of the VIIIth International Symposium on Protein Metabolism and Nutrition* (Lobley, G. E., White, A. and MacRae, J. C., eds.), pp. 253-264, Wageningen Press, Wageningen

Klasing, K. C. and Johnstone, B. J. (1991) *Poultry Science,* **70**, 1781-1789

Klasing, K. C., Laurin, D. E., Peng, R. K. and Fry, D. M. (1987) *Journal of Nutrition,* **117**, 1629-37

Klasing, K. C. and Leshchinsky, T. V. (2000) in *Nutrition and Immunology: Principles and Practice* (Gershwin, M. E., German, J. B. and Keen, C. L., eds.), pp. 363-373, Humana Press, Inc., Totowa, NJ

Knight, C. D., Klasing, K. C. and Forsyth, D. M. (1983) *Journal of Animal Science,* **57**, 387-95

Krishnan, S., Bhuyan, U. N., Talwar, G. P. and Ramalingaswami, V. (1974) *Immunology,* **27**, 383-393

Lander, H. M. (1997) *FASEB Journal,* **11**, 118-124

Li, Z., Nestor, K. E., Saif, Y. M., Bacon, W. L. and Anderson, J. W. (1999) *Poultry Science,* **78**, 1532-1535

Ling, P. R., Schwartz, J. H. and Bistrian, B. R. (1997) *American Journal of Physiology,* **272**, E333-339

Malmezat, T., Breuille, D., Pouyet, C., Mirand, P. P. and Obled, C. (1998) *Journal of Nutrition,* **128**, 97-105

Malpuech-Brugere, C., Nowacki, W., Gueux, E., Kuryszko, J., Rock, E., Rayssiguier, Y. and Mazur, A. (1999) *British Journal of Nutrition,* **81**, 405-411

Mauldin, J. M., Siegel, P. B. and Gross, W. B. (1978) *Poultry Science,* **57**, 1488-1492

McCann, S. M., Kimura, M., Walczewska, A., Karanth, S., Rettori, V. and Yu, W. H. (1998) *Domestic Animal Endocrinology,* **15**, 333-44

Moriguchi, S., Kobayashi, N. and Kishino, Y. (1990) *Journal of Nutrition,* **120**, 1096-1102

Nauss, K. M., Chew-Chin, P., Ambrogi, L. and Newberne, P. M. (1985) *Journal of Nutrition,* **115**, 909-918

Paludan, S. R. (2000) *Journal of Leukocyte Biology,* **67**, 18-25

Perkkio, M. V., Jansson, L. T., Dallman, P. R., Siimes, M. A. and Savilahti, E. (1987) *American Journal of Clinical Nutrition,* **46**, 341-345

Praharaj, N. K., Gross, W. B., Dunnington, E. A., Nir, I. and Siegel, P. B. (1996) *British Poultry Science,* **37**, 779-786

Prohaska, J. R. and Lukasewycz, O. A. (1989) *Journal of Nutrition,* **119**, 922-931

Qureshi, M. A. and Gore, A. B. (1997) *Immunopharmacology and Immunotoxicology,* **19**, 473-87

Qureshi, M. A. and Havenstein, G. B. (1994) *Poultry Science,* **73**, 1805-1812

Rombout, J. H. W. M., van Rens, B. T. T. M., Sitjsma, S. R., van der Weide, M. C. and West, C. E. (1992) *Veterinary Immunology and Immunopathology,* **31**, 155-166

Rosales, F. J., Ritter, S. J., Zolfaghari, R., Smith, J. E. and Ross, A. C. (1996) *Journal of Lipid Research,* **37**, 962-71

Säemann, M. D., Böhmig G. A. , Ö. C. H., Burtscher, H., Parolini, O., Diakos, C., Stöckl, J., Hörl, W. H. and Zlabinger, G. J. (2000) *FASEB Journal,* **10**, 1096

Semba, R. D. (1998) *Nutrition Reviews,* **56**, S38-S48

Sen, C. K. and Packer, L. (1996) *FASEB Journal,* **10**, 709-720

Twining, S. S., Schulte, D. P., Wilson, P. M., Fish, B. and Moulder, J. (1997) *Journal of Nutrition,* **127**, 558-565

Tyczkowski, J. K. and Hamilton, P. B. (1991) *Poultry Science,* **70**, 2074-2081

Tyczkowski, J. Z. and Hamilton, P. B. (1987) *Poultry Science,* **66**, 2011-2016

von Borell, E., Morris, J. R., Hurnik, J. F., Mallard, B. A. and Buhr, M. M. (1992) *Journal of Animal Science,* **70**, 2714-21

Webel, D. M., Johnson, R. W. and Baker, D. H. (1998a) *Journal of Nutrition,* **128**, 1760-6

Webel, D. M., Mahan, D. C., Johnson, R. W. and Baker, D. H. (1998b) *Journal of Nutrition,* **128**, 1657-60

Weinberg, E. D. (1998) *Cancer Investigation,* **16**, 291-292

Weinberg, E. D. (1999) *Emerging Infectious Diseases,* **5**, 346-352

Weinberg, E. D. (2000) *Microbes and Infection,* **2**, 85-89

Williams, N. H., Stahly, T. S. and Zimmerman, D. R. (1997a) *Journal of Animal Science,* **75**, 2463-71

Williams, N. H., Stahly, T. S. and Zimmerman, D. R. (1997b) *Journal of Animal Science,* **75**, 2481-96

Wu, G., Flynn, N. E., Flynn, S. P., Jolly, C. A. and Davis, P. K. (1999) *Journal of Nutrition,* **129**, 1347-1354

Yang, N., Larsen, C. T., Dunnington, E. A., Geraert, P. A., Picard, M. and Siegel, P. B. (2000) *Poultry Science,* **79**, 799-803

Yoshida, N., Yoshikawa, T., Manabe, H., Terasawa, Y., Kondo, M., Noguchi, N. and Niki, E. (1999) *Journal of Leukocyte Biology,* **65**, 757-63

Zhao, Z., Murasko, D. M. and Ross, A. C. (1994) *Natural Immunity,* **13**, 29-41

12

NUTRITION AND IMMUNITY IN FARM ANIMALS

J.C.MEIJER
Nutreco Swine Research Centre, P.O.Box 240, 5830 AE Boxmeer, The Netherlands

Introduction

During the last decade the feed industry had to cope with changing rules, laws and ethical constraints on animal production. The ban on in-feed use of antibiotics for growth promoting purposes has awakened considerable interest in feed additives promising the same positive effects. As a result, many products have been promoted or claimed as the long-awaited non-antibiotic growth promoter. Among them are products that, by virtue of immune stimulation, could help animals to cope with the conditions of modern intensive animal husbandry. It is common knowledge that poorly nourished animals are more susceptible to infectious diseases. Undernutrition, as well as imbalanced nutrition, impairs the immune system, resulting in less effective protection of the host from viral and bacterial infections. In addition, the effects of infection may have a greater effect on the host when nutrition is imbalanced.

The immune system and its intricate functions play a critical role in the relationship between infection and malnutrition. This applies to all life stages but especially susceptible are very young and very old animals.

Another focus of interest and current research is the possibility of stimulating the immune system of healthy animals by nutritional means in the hope of improving health. By improving health the need for antibiotic therapy in animal production will decrease. Thus, high-health livestock will provide safe food.

Since food animals generally are short lived as a matter of economics, the effects of proper functioning of the immune system on longevity, cardiovascular disease, cancer and diseases of old age are considered less important. For companion animals, however, these aspects are extremely interesting since allergy, asthma, immune diseases, cardiovascular disease and cancer are diagnosed frequently and may have their roots in disorders of the immune system.

The functioning of the immune system, general principles

IMMUNITY

The animal body has several defence systems against potentially harmful foreign agents. Skin, mucous membranes and mucous secretions keep away most micro-organisms. The inner body uses the immune system to combat invaders by means of molecules, cells and tissues widely dispersed throughout the body. The lymphoid organs include the bone marrow, thymus, spleen, tonsils, lymph nodes and the aggregates of lymphoid cells known as Gut Associated Lymphoid Tissue (GALT) and Bronchus Associated Lymphoid Tissue. The GALT is of particular importance for pick-up of antigens from the gut contents and is a major lymphoid organ.

The bone marrow produces all white blood cells including lymphocytes, monocytes and granulocytes. The lymphocytes are developed further in the thymus or in the bone marrow, where they differentiate and mature into either T-cells or B-cells. In birds B-cells mature in the Bursa of Fabricius instead of the bone marrow. Monocytes develop into macrophages when they penetrate tissues. Granulocytes are phagocytes and can chemically destruct foreign material. Concentrations of immune cells are present in the lymphoid organs, and large numbers circulate through the body as white blood cells. Several types have different tasks and specializations.

T-cells destroy infected cells, carry antigen memory and regulate the activity of B-cells.They constitute cellular immunity. B-cells produce antibodies and constitute humoral immunity. T-helper 1 cells activate the generation of lymphocytes and a cellular response, T-helper 2 cells activate the B cells to produce antibodies, thus the humoral response. "Suppressor" T-cells can reduce the activity of other functional cells. The cells of the immune system communicate by means of messenger molecules, cytokines or interleukins. Th1 and Th2 cells produce different cytokines. The Th1/Th2 cytokine balance dictates the final type of immune response (Roitt, Brostoff and Male,1985; Shearer and Clerici,1997).

Non-specific immunity

Macrophages present a primary line of defence by engulfing and chemically destroying any foreign material invading the body. These cells reside in many organs (reticulo endothelial system, Kupfer cells in the liver) or are mobile cells that are attracted to sites of inflammation. Natural killer cells are lymphocytes that recognize and kill abnormal cells.

Specific immunity

Specific immunity is possible because the system can distuinguish the body's own components (self) from foreign invaders (non-self), and because the system can recognize and respond to an unlimited number of different molecules. A fast and enhanced reaction occurs upon re-exposure to a previously encountered foreign agent (the system has "memory").

Self is recognized by the presence of specific marker molecules on the surfaces of all cells in the body. In normal circumstances the cells of the immune system do not attack cells which carry these distinctive "self" molecules. An example is the bloodgroup molecules (glycoproteins) present on the surface of the red blood cells.

However, any encounter with foreign marker molecules activates cells of the immune system into a defensive response. During this response, immature immune cells are tuned to respond to the specific antigen via receptor structures that recognize and interact with the specific targets. A few of those specialized cells remain after the response to a foreign agent is completed and then function as memory cells. Thus, the next time the body encounters the same antigen, the immune system can respond to it quickly and effectively.

Immunocompetence: animals are born with some temporary immunity derived from maternal antibodies. Their cellular system is not yet fully differentiated but is in an active stage of development. When the immune system is able to respond to antigens with the formation of antibodies, the host is called immunocompetent. Shortly before hatching (birds) or birth most animals are immunocompetent. However, fullblown competence is normally reached when maternal immunity has subsided completely.

RESISTANCE

Resistance to infectious diseases has many facets. Some varieties of viruses, bacteria, protozoa, helminths etc. are infectious to some species of host animals but not to others. Well-nourished animals can withstand larger numbers of pathogens. Integrity of skin, mucosa, skin pH, stomach pH, bronchial cleansing all play a part in resistance. The chemical attack of micro-organisms by enzymes and reactive molecules in tears, saliva, mucus and digestive juices also plays a significant role.

RESILIENCE

Coping with disease is a timely process as is recovery from (sub)clinical disease.

Quick recovery and short duration of susceptibility are best. Feeding, as opposed to fasting, shortens the recovery period of hospital patients. Enhancing feed intake helps to shorten recovery time from infectious disease.

Factors which alter the immune response: infection and nutrition

Infection and undernutrition are the interrelated major causes of disease and mortality in the developing world (Calder and Jackson, 2000). This statement is also true for animals especially very young and very old ones kept under unhygienic conditions. Undernutrition is a major risk factor for disease and mortality because many nutrients are required to fight off infections. Effective protection against invasion of the host by micro-organisms requires an intact skin and intact linings of organs in contact with the outer world. Since cell turnover rate in these linings is high, nutrients for cell replication and cell growth must be available. The immune response to infection involves a vast increase in cell replication, in the production of acute phase proteins and immunoglobulins and in the production of messenger molecules as cytokines and eicosanoids. An appropriate supply of nutrients is needed to optimize the response. Another component of the response to infection is the chemical destruction of foreign material by reactive O-species. Protection of the host from this damage requires sufficient antioxidant mechanisms, including antioxidant enzymes that all require metal ions, such as Fe, Zn, Cu, Mn, Se as active components, antioxidant vitamins, such as vitamin C and vitamin E, and glutathione. Thus, the host needs a supply of a range of nutrients to maintain protection against infective agents and to mount a succesful response if infected. Deficiencies of some vitamins and minerals impair immunity and disease resistance (Langseth, 1999: Calder and Jackson, 2000).

INFECTION

Infections can cause malnutrition and hence deficiencies of several nutrients. The low feed intake of sick animals may be caused by the appetite depressant action of cytokines released during inflammation or an active state of the immune system (Kelley, Kent and Dantzer, 1996).

High and low energy intakes as such can adversely affect the immune response. Especially weight loss or, more general, a katabolic state is associated with decreases in several measures of T-cell, B-cell and other immune cell functions. Severe malnutrition and wasting syndromes lead to impaired immune function. Wasting syndromes are often caused by viral infections, especially immunosuppressive viruses. They are more a problem of young animals under conditions of bad hygiene and high disease pressure. Chronic immune system activation as such leads to less growth and less efficient feed utilization (Williams, Stahly, Zimmerman, 1997a,b,c; Sauber and Stahly, 1996).

ENERGY AND FAT NUTRITION

Most defence mechanisms are impaired in protein-energy malnutrition, even if the nutritional deficiency is only moderate. T-cells are especially affected, resulting in low circulating numbers.

High energy intake results in obesity, which is also associated with impaired immune function, for example reduced responsiveness of T- and B-cells.

Too much fat in the diet can influence immune functions, possibly via a relative protein deficiency.

Animal studies indicate that diets high in fat may depress the immune response and increase the risk of infection, and high fat diets often result in diminished T-cell activity. In typical Western human diets, fat usually contributes some 40% or more of the energy intake. In humans, reduction in dietary fat intake resulted in increased T-cell responsiveness and increased capacity of natural killer cells to destroy tumour cells (Langseth, 1999).

In animal diets the fat content is usually much lower and fat contributes some 15 till 25 % of dietary energy. Obesity is not common in production animals; however, in companion animals overt obesity does occur quite frequently and is recognized as a "disease".

Fat composition in itself can influence immune function because fatty acids are structural components of cell membranes. The fluidity of membranes is affected by the chain length and degree of saturation of the incorporated fatty acids. Fluidity is important for the expression of cell surface structures such as receptors, which play crucial roles in immune function. Polyunsaturated fatty acids (PUFA) can alter the functioning of immune cells considerably. There are two classes of PUFA: the n-6 series found primarily in vegetable oils and the n-3 series found in fish oils and also in certain vegetable oils such as canola, soy, and linseed.

Immune cells use these fatty acids also to produce tissue hormones known as prostaglandins. Depending on the type of PUFA in the diet, immune cells produce different quantities and kinds of prostaglandins with very different effects on the immune response. In addition to their effects on tissue hormone levels, PUFA also exert effects by other mechanisms.

In general, diets rich in n-3 PUFA tend to inhibit the immune response, whereas those rich in n-6 PUFA tend to promote immune responses that lead to inflammation. The ratio of n-6 to n-3 PUFA may be more important than the absolute amount of each of these classes of fatty acids in the diet. This ratio can be influenced greatly by the choice of vegetable oils and animal fats in the diet of man and animal.

The ideal ratio of n-6 to n-3 PUFA is actually unknown, but estimates range from 5:1 to 10:1.

VITAMINS

Vitamin A deficiency is associated with increased morbidity and mortality in children as well as in animals. Vitamin A deficiency is rare in animal nutrition because it is amply available and cheap Vitamin A plays a very important role in the body's defence against some viruses. An adequate supply is needed for the normal development of many types of epithelial cells and of blood cells, including lymphocytes. Networks of cytokines, which influence immune responses, may also be altered during vitamin A deficiency, and antibody responses to antigens may be modified.

Carotenoids are yellow, orange and red compounds in fruits, vegetables and some algae. Like vitamins C and E, carotenoids are antioxidants. Carotenoids increase specific subsets of lymphocytes, enhance the activity of natural killer cells, stimulate the production of cytokines and activate phagocytic cells (Kim et al., 2000). Of particular interest is the capacity of lycopene to enhance the potency of the immune system.

Vitamin D is also active in the cell nucleus where its receptor is linked with the vitamin A receptor. The complex regulates RNA transcription. Immunomodulating effects of 1,25-dihydroxycholecalciferol have been reported with a shift towards a mucosal IgA response (Van der Stede et al., 2000). The use of 25-hydroxycholecalciferol ("HyD"), which displaces vitamin D3 from its carrier protein in the circulation, possibly results in more vitamin D3 entering the cell nucleus and a stronger immunomodulating effect.

Vitamin E occurs in a variety of plant foods; rich sources are wheat germ oil, maize oil and soyabean oil. Deficiency of vitamin E impairs several aspects of the immune response, including B-cell and T-cell activity. Supplementation of the diet with higher than recommended levels has been shown to enhance certain aspects of immune function in humans and animals, and to increase resistance to infectious diseases. The suggested mechanism by which vitamin E exerts its action on the immune system is by lowering of the prostaglandin synthesis and by preventing the oxidation of PUFA in cell mebranes.

Vitamin C is present in many vegetables and fruits. It is a water-soluble antioxidant, found in body fluids rather than in cellular lipids and membranes. Vitamin C acts as a major antioxidant in the aequous phase and also reinforces the effects of other antioxidants, such as vitamin E, by regenerating their active forms after they have reacted with free radicals. Certain organs in the body concentrate vitamin C to levels far higher than found in the blood. One of these organs is the thymus.

Vitamin C plays an important role in the function of phagocytes. High doses of vitamin C are often applied in an attempt to prevent respiratory infections. It is unclear whether this is effectively reducing common cold and flu.

Vitamin B6 is widely present in a variety of foods including cereals and pulses. Vitamin B6 is essential for synthesis and metabolism of amino acids and protein. Animal studies indicate that deficiencies of vitamin B6 modify T-cell activity and antibody production. High doses above the recommended daily allowance (RDA) do not seem to produce additional benefits.

MINERALS

Zinc deficiency is associated with skin problems, respiratory infections and diarrhoea.

Cells of the immune system require many enzymes that need zinc to function, thus zinc deficiency has profound effects on immune function. Deficiency suppresses the function of the thymus and lymphocytes become less responsive to chemical signals to grow. Antibody formation and resistance to infections are impaired. Wound healing is impaired in zinc deficiency. Too much zinc also impairs immunity; both phagocytic cell and lymphocyte function are involved, but the mechanism of these effects is not clear.

Iron deficiency is associated with gastrointestinal and respiratory infections. The influence of Fe status on infection is a complicated somewhat by the fact that micro-organisms also require Fe for their metabolism. Low Fe status should protect against infection, and the presence of Fe binding lactoferrin in milk probably plays a role protecting the suckling animal against bacterial infections. Thus, there is competition for Fe between the invading pathogens, the host stores and the host immune system.

Symptoms of copper deficiency are anaemia, diarrhoea, weight loss and impaired neutrophil function.

Manganese deficiency results in impaired synthesis of mucopolysaccharides and bone matrix; the mucosal barrier function can also be impaired.

Selenium activates glutathione peroxidase involved in neutralizing peroxides. Se deficiency impairs the antioxidant system of the body. Supplementation with Se and vitamin E has potential benefits but responses vary between animal species (Finch and Turner, 1996).

Some of the basic immune responses and related nutrients are presented in Table 12.1.

Table 12.1. Basic metabolism of an immune response and nutrients involved

Immune system	Response	Metabolic event	Nutrients required
Adaptive			
Stem cells	proliferation	protein synthesis	amino acids, vitamins A,D, B_6
Lymphoid cells	proliferation	protein synthesis	
T-lymphocytes	cytokines	protein synthesis	
B-lymphocytes	antibodies	protein synthesis	
Innate system			
phagocytes	cell activity, enzymes, cytokines,	enzyme activity	vitamins C,E, arginine, ornithine, Se, Cu, Zn, cofactors
	proliferation	protein synthesis	
lysozyme	increase	protein synthesis	
plasma enzymes	increase	protein synthesis	
complement	increase	protein synthesis	
acute-phase proteins	increase	protein synthesis	
skin, mucosa	turnover	general	vitamin A, Zn
mucus, sebum, HCl	turnover	general	

Immunonutrition

The use of food or feed enriched with nutrients known to be involved in protein synthesis, immunostimulation or anti-oxidant systems is called immunonutrition. The intention of immunonutrition is to back up disease resistance. The net effect on the animal is measurable, as far as disease incidence and severity are concerned, and economic aspects of growth rate and feed use can be calculated. Effects on isolated aspects of the immunological response usually cannot be used to predict the net effect on animals under conventional husbandry conditions. Therefore, most animal studies with immunomodulating products rely more on measures of animal performance and disease than on immune parameters.

Glucans have been studied in fish (Raa, 1996) as well as in pigs (Dritz et al., 1995; Decuypere, Dierick and Boddez, 1998; Vancaeneghem et al., 2000) with promising results. Unpublished results of our own research confirm increased disease resistance in weaned piglets fed special b-glucans but not plain yeast cell walls. Herbal products receive much attention and some plant extracts and preparations certainly contain immunomodulatory components (Wills, Bone and Morgan, 2000). Evidence that these products may improve disease resistance under practical husbandry conditions is accumulating; however, reports are still largely anecdotical. Our own research indicates that some herbal products are effective in increasing feed intake and growth in weanling piglets but most of the tested products had no positive effect on either performance or disease.

Probiotics may affect the immune system (McCracken and Gaskins, 1999) but certainly not all probiotics are equally effective under current husbandry conditions. Some lactobacillus species appear to be effective as immunostimulants or as microflora-modulating probiotics. Since they are only active when alive, dry feed is not the ideal carrier because survival of lactobacilli during pelleting and drying of feed is low. Other feed ingredients containing selected strains of lactobacilli can possibly be used to good effect. Fermented feedstuffs offer promise to enhance health and to increase disease resistance.

A summary of immunonutrition is presented in Table 12.2.

Table 12.2. Products eliciting non-specific adjuvant effects

Adjuvant effects	Bacterial cell wall components	Feed and plant components
Inflammatory responses	Bacterial capsule	Glucans from yeast, fungi, grains
Activation of complement	Lipopolysaccharide(Gram-negative BCW)	Echinacea extracts
Activation of macrophages	Arabinogalactan	Ginseng extracts
Activation of B-cells	Peptidoglycan (murein)	Triterpene-saponins
Activation of T-cells	Heat stress proteins	Phorbolesters
Modified antigen processing	Endotoxins	Cichoric acid

Conclusions

Since the immune system serves to safeguard the body from any foreign invading micro-organism, it is always active. Functioning of the system requires cell proliferation, production of protein molecules and enzyme activity to destroy phagocytized material (infected body cells, bacteria, viruses). Nutrients required for the immune system and its responses are for protein synthesis and for oxidative radical production; moreover, the body requires anti-oxidants for protection. Nutrient allowances for growth are also expected to cover the metabolic needs of the immune system. Supplementing the diet with the anti-oxidant vitamins C and E improves resilience and is favoured whenever disease and infection are present.

In general, all cell-toxic elements will be immunosuppressive. Food should be free from lead, cadmium, mercury, mycotoxins, dioxin, pesticides and organic solvents. As several antibiotics inhibit protein synthesis, they also suppress the immune system; tetracyclines and chloramphenicol are examples, and they should not be used as feed ingredients.

Feed will contain a number of immunomodulators including b-glucans and arabino(rhamno)galactans of plant, fungal, yeast or microbial origin, saponins and phenolic derivatives. ß-glucans are taken up by M-cells in the gut and stimulate macrophages as well as the complement system. Thus, selected feed ingredients as well as herbal extracts may enhance or modulate immune responses.

The balance between cellular and humoral responses can be altered by feed components; vitamin D may shift the response towards humoral; enhanced cellular activity may result from selected plant extracts. This area offers potential for special feeds for young animals in a stage of imbalance between acquired and maternal immunity, i.e. around weaning when activation of the innate system may increase disease resistance.

The gut microflora is an ever-present source of microbial antigens and adjuvant-active components. Segmented filamentous bacteria stimulate immunity and become part of the flora after weaning to solid food (Snel, 1997). Once established, the complex intestinal ecological system is highly stable and resists establishment of newly introduced organisms. Probiotics may enhance macrophage activity of the innate immune system and the response to particular antigens through adjuvant activity after lysis of the cells. The potential for probiotics is in the period of weaning and, in cases of disturbed intestinal ecology, following oral antibiotic treatment.

Well-balanced quality nutrition is of the utmost importance for disease resistance and the underlying immune processes but cannot compensate for bad husbandry. Hygiene can significantly contribute to better animal performance by reducing the need of the animal for immune responses and the associated metabolic needs (Toussaint *et al.*, 2001).

References

Calder, P.C. and Jackson, A.A. (2000) Undernutrition, infection and immune function. *Nutrition Research Reviews*,**13**, 3-29

Decuypere, J., Dierick, N. and Boddez, S. (1998) The potentials for immunostimulatory substances (b-1,3/1,6 glucans) in pig nutrition. *Journal of Animal and Feed Sciences*, **7**, Supp.1: 259-265

Dritz, S.S., Shi, J., Kielian, T.L, Goodband, R.D., Nelssen, J.L., Tokach, M.D., Chengappa, M.M., Smith, J.E. and Blecha, F. (1995) Influence of dietary beta-glucan on growth performance, nonspecific immunity and resistance to Streptococcus suis infection in weanling pigs. *Journal of Animal Science*, **73**, 3341-3350

Finch, J.M. and Turner, R.J. (1996) Effects of selenium and vitamin E on the immune responses of domestic animals. *Research in Veterinary Science*, **60**, 97-106

Gaskins, H.R. (1996) Development and structure of mucosal defense in the pig intestine. In *Biotechnology in the Feed Industry,Proceedings of Alltech's 12ᵗʰ Annual Symposium*, pp 23-35. Edited by T. Lyons and K. Jacques. Nottingham: Nottingham University Press

Kelley, K.W., Kent, S. and Dantzer, R. (1996) Why sick animals don't grow: an immunological explanation. In *Growth of the Pig*, pp 119-132. Edited by G.R.Hollis. Wallingford: CAB INTERNATIONAL

Kim, H.W., Chew, B.P., Wong, T.S., Park, J.S., Weng, B.B., Byrne, K.M., Hayek, M.G., Reinhardt (2000) Dietary lutein stimulates immune response in the canine. *Veterinary Immunology and Immunopathology*, **74**, 315-327

Kolb, E. Vitamins and the immune system. Basel: Hoffmann-La Roche

Langseth, L. (1999) Nutrition and immunity in man. *ILSI Europe Concise Monograph Series*. Brussels: International Life Sciences Institute, ISBN 12-57881-058-2

McCracken, V.J. and Gaskins, H.R. (1999) Probiotics and the immune system. In *Probiotics: A Critical Review,* pp 85-111. Wymondham: Horizon Scientific Press, ISBN 1-898486-15-8

Raa, J. (1996) The use of immunostimulatory substances in fish and shellfish farming. *Reviews in Fisheries Science*, **4**, 229-288

Roitt, I., Brostoff, J. and Male, D. (1985) Immunology. Gower Medical Publishing, London

Sauber, T.E. and Stahly, T.S. (1996) Chronic immune system activation impairs lactational performance of sows. *Feedstuffs*, July 22

Shearer, G.M and Clerici, M. (1997) Vaccine strategies: selective elicitation of cellular or humoral immunity? *Tibtech* ,**15**, 106-109

Snel, J. (1997) Symbiosis between the mouse and segmented filamentous bacteria, a gnotobiotic study. Thesis, University of Nijmegen, ISBN 90-9010578-6

Toussaint, M.J.M., Paboeuf, F., Madec, F., Meijer, J., Gruys, E. (2001) Impact of environmental condition on growth performance and health of pigs: effect of housing and transport on blood indicators of inflammation. *Journal of Animal Science* (Submitted for publication)

Vancaeneghem, S., Cox, E., Deprez, P., Arnouts, S., Goddeeris, B.M. (2000) Beta glucanen als immunostimulantia en als adjuvantia. *Vlaams Diergeneeskundig Tijdschrift*, **69**, 412-421

Van der Stede, Y., Cox, E., Goddeeris, B.M. (2000) 1-25 Dihydroxyvitamine D3, rol in het immuunsysteem.. *Vlaams Diergeneeskundig Tijdschrift*, **69**, 229-234.

Williams, N.H., Stahly, T.S. and Zimmerman, D.R. (1997) Effect of chronic immune system activation on the rate, efficiency, and composition of growth and lysine needs of pigs fed from 6 to 27 kg. *Journal of Animal Science*, **75**, 2463-2471

Williams, N.H., Stahly, T.S. and Zimmerman, D.R. (1997) Effect of chronic immune system activation on body nitrogen retention, partial efficiency of lysine utilization, and lysine needs of pigs. *Journal of Animal Science*, **75**, 2472-2480

Williams, N.H., Stahly, T.S. and Zimmerman, D.R. (1997) Effect of level of chronic immune system activation on the growth and dietary lysine needs of pigs fed from 6 to 112 kg. *Journal of Animal Science*, **75**, 2481-2496

Wills, R.B.H., Bone, K. and Morgan, M. (2000) Herbal products: active constituents, modes of action and quality control. *Nutrition Research Reviews*, **13**, 47-77

13

IDEAL CONCENTRATE FEEDS FOR GRAZING DAIRY COWS – RESPONSES TO SUPPLEMENTATION IN INTERACTION WITH GRAZING MANAGEMENT AND GRASS QUALITY

JEAN LOUIS PEYRAUD AND LUC DELABY
Dairy Production Research Unit, UMR INRA-ENSAR, 35590 St Gilles, France

Introduction

Limitations have recently appeared for intensive production systems with the introduction of milk quota, the necessity to take into account environmental concerns and the General Agreement on Trade and Tariffs (GATT) proposals. These limitations increase the pressure on the price of milk and leads to an increased emphasis on production efficiency per litre milk. Despite some differences in the grass-growing seasons between European countries, utilisation of grass by grazing should form the basis of sustainable dairying systems in the future. Grazing is the cheapest source of nutrients for dairy cows and contributes to the competitiveness of milk production, preserving the rural landscape and giving a good image of dairy production.

Full exploitation of grazed grass will require development of grazing systems designed to maximise daily herbage intake per cow and to improve the efficiency of nutrient use through the provision of supplementary feeds. Supplements are generally provided to grazing dairy cows to increase total energy intake and animal performance above that which can be produced from pasture alone. However, the efficiency of supplementation (kg increase in milk per kg increase in supplement-feed DM intake) largely depends on the effect of supplementation on herbage intake, so it is necessary to define the conditions where use of supplementary feeds will minimise the reduction in herbage intake.

Since the early 1950s, a large number of studies have quantified the effects of supplements on performance per animal or per unit of land area. The principal factors controlling the efficiency of supplementation at grazing have been identified and described. However, most of these studies have compared two levels of a given factor, so the input-output response curves that should be used to evaluate the optimal supply according to the current price ratio were not described. It is therefore still difficult to predict quantitative responses of milk yield and milk composition to supplements and to propose practices with a predictable outcome (Leaver, 1985). The objective of this chapter is to quantify the effect of supplements on performance of dairy cows according to grazing conditions and to define the conditions necessary for an efficient response to supplements.

Milk yield at grazing without supplements

From 187 lactations, Delaby, Peyraud and Delagarde (1999) examined the performance of unsupplemented grazing cows (Figure 13.1). They showed that actual milk yield (aMY) in spring-grazing cows (April to early July) averaged 22.2 kg when no concentrate was offered, but there were large differences between cows, the ranges being 10 kg for primiparous and 17 kg for multiparous cows. Cow potential is the primary cause of these differences. The authors proposed that cow potential could be estimated using the concept of expected milk yield (eMY). The eMY parameter is calculated as the reference milk yield at turnout (early April) when cows were fed *ad libitum* with maize silage and grass, corrected for the length of the experiment, assuming a weekly persistency of 0.98. The relationship between aMY and eMY is linear (aMY = 0.25 HA + 0.65 eMY + D; n=187; Syx = 1.71; R^2 = 0.83 (Equation 1), where HA is herbage allowance at 5 cm above ground level in kg DM/cow/day and D = +1.3 for multiparous and +0.6 for primiparous cows). This relationship shows that, when yielding more than 15 kg of milk per day, 0.65 of each kg of expected milk above 15 kg/day could be produced from grass. This means that a cow producing 40 kg of milk per day at turnout (i.e. 35 kg/day eMY during 12 spring weeks) is able to produce around 28 kg/day with no supplements at spring grazing and illustrates the potential to achieve quite high performance levels at pasture. However, the difference between expected and actual milk yields reflects the shortfall between the theoretical requirements of cows and the energy inputs that can be achieved from herbage alone. This shortfall increases from 4 kg/day in cows with an expected milk yield of 25 kg/day to 9 kg/day in cows with an expected milk yield of 45 kg/day. Kolver and Muller (1997) also reported a major reduction in milk yield in cows with high potential that were fed with grass only. Due to higher energy deficits, greater responses to concentrates might be expected in grazing cows with high genetic merit.

This relationship also shows that the milk yield of an unsupplemented cow increases by 1.0 kg/day per 4 kg/day increase in herbage allowance. However, increased herbage allowance in early season also increases residual sward height, which may result in a deterioration of sward quality in mid and late season. From

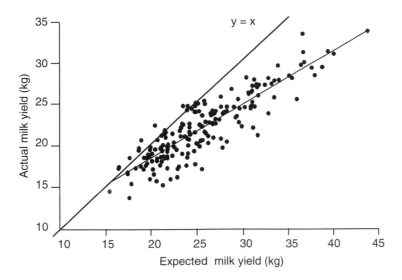

Figure 13.1 Milk yield at grazing without supplementation in relation to milk potential (after Delaby, Peyraud and Delagarde 1999).

a practical point of view, the room for manoeuvre is not very large. Hoden *et al.* (1991) and Delaby, Peyraud and Delagarde (1999) found that milk yield can be increased by no more than 1.5 kg/cow per day by increasing herbage allowance in spring without noticeable effects on sward quality later in season. Offering supplements is a more powerful tool for controlling animal performance at grass.

Intake and milk responses to incremental amounts of supplements

Following the reviews of Leaver *et al.* (1968) and Journet and Demarquilly (1979), it was generally accepted that concentrates were not used very efficiently by grazing cows; average responses were 0.4 to 0.6 kg milk per kg of dry matter (DM) from concentrates, mainly because feeding concentrates generally reduces herbage intake. Substitution rate (kg reduction in herbage intake per kg increase in concentrate DM intake) averaged 0.5 to 0.6. From two comprehensive reviews of the literature, we have recently shown that the efficiency of supplementation (Table 13.1) and substitution rate (0.4 ± 0.3; n = 57; Delagarde and Peyraud, unpublished) are quite variable, indicating that high responses can be achieved in some circumstances. Indeed, milk responses to concentrates are higher in reports published after 1990 (Table 13.1). The incremental increase in milk response averages + 0.1 kg/ kg concentrate DM every ten years. Indeed, overall efficiencies

Table 13.1 Response of milk yield, protein and fat content to concentrate supplementation (Delaby and Peyraud, unpublished)

	n	*Mean*	*Min*	*Max*
All data				
Concentrate intake (kg DM/day)	141	2.8 ± 1.2	0.90	5.90
Increase in milk yield (kg/kg concentrate DM)	141	0.66 ± 0.46	-0.56	2.39
Increase in milk protein content (g/kg per kg concentrate DM)	107	0.23 ± 0.32	-0.52	1.34
Increase in milk fat content (g/kg per kg concentrate DM)	133	-0.29 ± 0.53	-1.92	1.06
Data published after 1990				
Concentrate intake (kg DM/day)	55	2.9 ± 1.2	0.90	5.40
Increase in milk yield (kg/kg concentrate DM)	55	0.89 ± 0.43	-0.18	2.00
Increase in milk protein content (g/kg per kg concentrate DM)	54	0.20 ± 0.29	-0.51	0.81
Increase in milk fat content (g/kg per kg concentrate DM)	54	-0.21 ± 0.46	-1.25	0.80

close to or higher than 1.0 kg of milk per kg concentrate DM were recently reported when less than 4-5 kg of concentrate were provided to cows producing more than 25-30 kg milk at turnout (Table 13.2). It is probable that these higher responses are related to the genetic merit of the cow, which has increased appreciably since the earlier reviews.

Very few studies have focussed on the form of the milk and intake responses associated with increases in concentrate inputs. Milk response to concentrates tends to decrease with increasing concentrate allowance (Table 13.2), but this effect seems to be moderate, providing concentrate allowance does not exceed 6 kg/day. The marginal efficiency, which determines economic return, decreases above 3 to 4 kg concentrates in some studies, but this is not always the case. This mainly occurs when sward limitations are minimised and/or with cows of moderate genetic merit. Delaby, Peyraud and Delagarde (2001) recently reported that marginal efficiency hardly decreases between 4 and 6 kg concentrates on slightly-restricted grazing conditions. A marginal efficiency greater than 0.6 was recently reported between 5 and 10 kg of concentrates for cows producing more than 35 kg milk at turnout (Sayers Mayne and Bartram, 2000). In the range of 2 to 6 kg/day, concentrate allowance has no consistent effect on substitution rate in our data set (Figure 13.2). Several experiments also failed to observe a consistent effect

Table 13.2 Effect of concentrate allowance on milk response to concentrate supplements in grazing dairy cows

Concentrate allowance (kg DM/day)			Overall efficiency (kg milk/kg concentrates)		Marginal efficiency (kg milk/kg concentrates)	
Control	*Medium*	*High*	*0 to Medium*	*0 to High*	*Medium to High*	*Authors*
0.9	2.6	4.3	0.8	0.5	0.2	Meijs and Hoekstra (1984)
0	1.8	3.5	1.4	0.9	0.4	Wilkins *et al.* (1994)
0	1.8	3.6	0.9	0.7	0.6	O'Brien, Crosse and Dillon (1996)
0	1.8	3.6	1.3	0.9	0.5	Delaby and Peyraud (1997)
0	1.8	3.6	0.7	0.6	0.5	Dillon, Crosse, and O'Brien (1997)
0	3.4	6.7	0.9	0.8	0.7	Robaina *et al.* (1998)
0	2.7	5.4	1.1	1.0	1.0	Delaby, Peyraud and Delagarde (2001)

of increasing amounts of concentrates on substitution (Meijs and Hoekstra, 1984; Kibon and Holmes, 1984 ; Opatpatanakit, Kekkaway and Lean, 1993), probably because high-producing dairy cows rarely approach their maximum voluntary intake under grazing conditions. Indeed, substitution rate increases slightly with concentrate allowance (Meijs, 1981; Hijink *et al.*, 1982) when unrestricted fresh grasses are offered to cows indoors.

Responses in milk yield are usually accompanied by a steady increase in milk protein content (Table 13.1), indicating an improvement in energy status of the cows. In a series of experiments conducted in Rennes (Delaby, Peyraud and Delagarde, 2001), the increase in protein content averaged 0.2 g/kg per kg of concentrate DM and was linear up to 6 kg of concentrates. Increasing concentrate allowance generally produces a reduction in the concentration of milk fat, although there is a wide variation in response (Table 13.1). In the study of Delaby, Peyraud and Delagarde (2001), the reduction in milk fat content was consistent and averaged -0.6 g/kg per kg concentrate DM between 0 and 5.4 kg concentrate DM. This was primarily due to a dilution effect on milk fat, which increased less rapidly than milk yield. It might also have been a consequence of the ruminal fermentation pattern, because the acetic to propionic ratio in the rumen decreased in supplemented cows (Delagarde, Peyraud and Delaby, 1999 and unpublished). High levels of concentrate feeding can cause a severe depression in the fat content of milk. Stockdale, Callaghan and Trigg (1987) reported a 15 to 20 g/kg decrease in milk fat content when 9 kg of concentrates were fed to cows. Feeding concentrates always increases live-weight gain; in our experiments, the mean increase was 60 g per kg of concentrate DM.

The response in milk yield to incremental levels of supplements does not vary with the potential of the animals at turnout (from 25 to 40 kg milk) when allocation of concentrate is constant between cows, at least up to levels of 4-6 kg of concentrates (Delaby, Peyraud and Delagarde, 2001). This probably reflects the inability of cows to meet their nutrient requirements from herbage alone, even moderate-producing animals that are considerably below their expected milk yield (see equation 1), and the ability of high-producing cows to increase herbage intake according to their potential milk yield. Incremental increases in intake average 250 g OM/kg eMY (Peyraud *et al.* 1996). Whether these identical responses among cows also apply at higher concentrate allocations is unknown, but responses to concentrates might progressively decrease for higher levels of concentrates in low-producing animals when they reach their expected milk yield.

Interaction between supplementation and grazing conditions

Herbage allowance has been recognised for a long time as a major factor affecting the substitution between grass and concentrates. Substitution rate decreases from around 0.6 to 0.1 when herbage allowance is restricted (Meijs and Hoeskstra 1984; Stakelum 1986a,b,c; Kibon and Holmes 1987; Grainger and Mathews, 1989). Delagarde and Peyraud (unpublished) summarised the responses of 48 grazing experiments in which the net energy balance of unsupplemented cows was calculated from measured herbage intake, grass digestibility and milk yield. They demonstrated that substitution rate between grass and concentrates is poorly related to the level of concentrates, but is primarily a function of net energy balance (EB in MJ/day) of the unsupplemented cows (Figure 13.2). Similar relationships have been described for supplementation of conserved forages (Faverdin *et al.* 1991). According to the relationship obtained (SR = 0.32 + 0.010 EB, rsd = 0.19, n = 48), the lower the energy balance, the lower the substitution rate. Substitution rate is only 0.1 when energy needs are far from being met from grass alone (EB = -21 MJ) and increases to 0.6 when sward limitations are minimised (EB = 28 MJ). Energy balance is affected by stage of lactation; thus, Jennings and Holmes (1984) reported a higher milk response in early-lactation cows than in mid-lactation animals (1.2 versus 0.7 kg milk/kg concentrate DM).

In a similar way, the efficiency of supplementation appeared to be closely related to the proportion of requirements that was met from grass alone. In a comprehensive review, Delaby and Peyraud (unpublished) characterised the

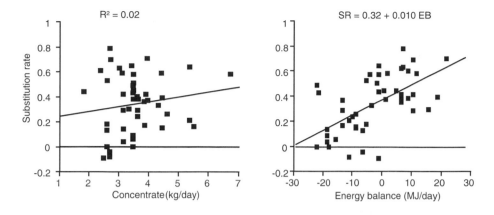

Figure 13.2 Effect of level of supplementation and net energy balance on substitution rate between fresh grass and concentrates (Delagarde and Peyraud, unpublished)

severity of grazing conditions in 95 experiments by calculating the difference between actual milk yield of unsupplemented cows and their expected yield (eMY), assuming that the greater the difference between actual and expected milk yield (e.g. eMY-aMY for unsupplemented cows), the more adverse the grazing conditions. Response in milk yield to increasing levels of concentrates was linear, but highly variable (Figure 13.3). The precision of the prediction sharply increases when grazing conditions are taken into account. The efficiency of supplementation is only 0.3 when energy needs are met from grass alone, but reaches 0.9 in more severe grazing conditions. When pasture intake is restricted, much larger responses are observed when extra energy is provided. This principle has also been demonstrated in cows fed indoors with fresh grass. Stockdale and Trigg (1989) reported a marginal response of 1.8 kg milk per kg concentrate DM when cows consumed 6.8 kg of pasture DM each day; the response dropped to 0.6 when cows consumed 11.6 kg of pasture DM.

In practice, energy balance may differ according to grass intake. This explains why numerous studies have concluded that substitution rate is positively related to herbage allowance. For the same reason, substitution rate between fresh grass and conserved forages dramatically increases when sward surface height increases. Phillips and Leaver (1985) reported an increase in the substitution rate between grass and grass silage from 0.7 to 1.3, when herbage height was increased from 7.2 to 9.6 cm. As a consequence, several papers have reported that milk response increases with increasing grazing severity. This occurs when stocking rate is increased on rotational grazing; Hoden *et al.* (1991) reported an increase in milk-production response from 0.5 to 0.8 with increasing stocking rate from 2.3 to 3.0 cows per ha. It also occurs when sward height is reduced in set-stocking

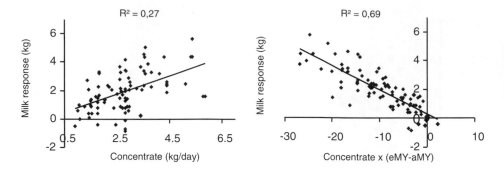

Figure 13.3 Effect of the severity of grazing conditions on the milk response to supplementation (Delaby and Peyraud, unublished)

management; milk response increased from 0.3 to 0.9 when sward surface height was reduced from 6 to 4.5 cm (Wilkins, Gibb and Huckle, 1995). The effects on milk response to concentrates, of interactions between level of concentrates and herbage allowance, are illustrated in Figure 13.4. When herbage is restricted, there is a linear response in milk up to 6 kg of concentrates, whereas with high allowances, the response reached a plateau after 4 kg of concentrates.

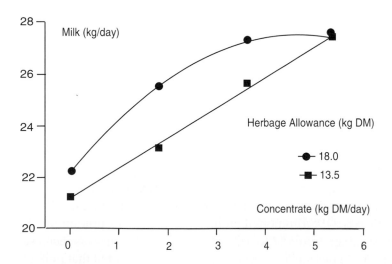

Figure 13.4 Interaction between level of concentrate and herbage allowance on milk response to supplementation (after Delaby, Peyraud and Delagarde, 2001 – Experiment 1).

Energy balance may also differ according to the quality of grass, which is why substitution rate is positively related to herbage digestibility (Grainger and Mathews, 1989). This also explains why milk responses can increases during the grazing season, when grass quality and availability are reduced. Gleeson (1981) reported a milk response increasing from 0.2 in spring to 0.6 in summer and 0.9 in autumn, when pasture provided less energy because grass quality and availability (fouled area, moisture) were reduced. Stakelum (1986 a,b,c) also reported a higher response in summer compared with spring. For the same reason, milk response to concentrates increases between vegetative and heading stages during the first growth of grass (Garel and Hoden, 1985). For a given stocking rate, grazing system does not affect performance, and hence energy balance, of dairy cows (Hoden *et al*. 1987, Le Du, 1980). Therefore, responses to concentrates are similar between rotational grazing systems, in which fresh pasture is allocated either daily (strip grazing) or over several days (paddock system) (Hoden *et al*. 1987). Responses are also similar between rotational and continuous grazing systems (Arriaga Jordan and Holmes, 1986).

Feeding concentrates generally decreases pH and acetate to propionate ratio in the rumen. However, supplements and grazing conditions also interact to affect digestive processes in the rumen. Feeding 4 kg of cereal grains marginally decreased the acetate: propionate ratio from 2.7 to 2.5 when herbage allowance was restricted, whereas the same supplement decreased the acetate: propionate ratio from 2.6 to 1.8 on a high herbage allowance (Delagarde and Peyraud, unpublished). Milk fat content followed a similar trend, with a larger decrease for a high than a low allowance. On a high herbage allowance, cows have access to higher quantities of a more leafy diet, which is fermented more rapidly.

Nature of the supplements

Herbage intake is more drastically lowered when cows were supplemented with forages than with concentrates (Mayne and Wright, 1988). Under good grazing conditions, when the supply of fresh grass was large, giving conserved forages (grass silage, hay or maize silage) as a buffer feed, resulted in high substitution rates, often over 1.0 (Leaver, 1985; Phillips and Leaver, 1985; Phillips, 1988). In these situations, very low milk responses or even a decrease in milk yield, compared with control cows, were obtained (Bryant and Donnelly, 1974; Leaver, 1985) because net energy content is lower in conserved than in fresh forages. The substitution rate between fresh grass and buffer forages decreases to 0.3 when the availability of fresh forage is restricted. Therefore, forages supplements must be provided only during periods of grass shortage, or in areas where the availability of grass is restricted. This explains why response to supplementary forages is much higher in summer than in spring (Phillips and Leaver, 1985). The large substitution rates obtained with forages appear to be mediated by a large reduction in grazing time, which can reach 40 min/kg of silage DM (Mayne, 1991), whereas the reduction in grazing time per kg of concentrate DM averages 10 to 15 min (Combellas, Baker and Hodgson 1979; Jennings and Holmes 1984; Kibon and Holmes 1987). Sarker and Holmes (1974) reported higher values (29 min/kg), but for non-lactating cows offered large amounts of concentrates.

In the current context of dairy production, it is generally assumed that energy is the first factor that limits animal performance during grazing. Readily-fermentable starch (e.g. barley or wheat) are generally considered to decrease milk fat content and acetate:propionate ratio in the rumen compared with high-fibre concentrates or less readily fermentable starch (maize). Indeed, energy source in the concentrate has little effect on milk yield and composition when moderate levels of concentrates are fed. Compared with 3.5 kg of wheat, feeding 3.5 kg of a concentrate rich in soya-bean hulls, which is a source of slowly-degraded cellulose, increased milk fat content (+ 1.3 g/kg, Table 13.3) and marginally decreased milk protein content (- 0.5 g/kg) (Delaby and Peyraud, 1994). The difference between starch and fibre is even lower when using a concentrate rich in sugar-beet pulp and citrus pulps, which are sources of rapidly-fermentable pectins (Table 13.3 and Meijs, 1986). Using 5 kg of concentrates, Sayers Mayne and Bartram (2000) reported a small decrease in milk protein content, a 3 g/kg reduction in milk fat content and a no effect on milk yield, with high-fibre compared with high-starch concentrates. These results agree with those reported by Coulon *et al.* (1989) for fresh and conserved forages, showing that the nature of the carbohydrate has a significant effect when carbohydrate source represents more than half of the total DM intake.

Table 13.3 Effect of energy source on milk yield by dairy cows grazing vegetative perennial ryegrass swards (after Delaby et al. 1994)

	Wheat	Dried sugar-beet pulp	Soya bean hulls
Milk yield (kg/day)	26.7	27.2	27.3
Milk protein content (g/kg)	30.0	29.4	29.5
Milk fat content (g/kg)	36.2	36.6	37.5

Intake of concentrates : 3.5 kg DM/day

The nature or the rate of degradation of carbohydrate does not affect substitution rate when fresh grass is fed indoors (Spörnly, 1991; Schwartz, Haffner and Kirchgessner, 1995). Under good grazing conditions, some data suggest a marginally higher herbage intake with fibre compared with rapidly-degradable starch (Kibon and Holmes, 1987; Fisher, Dowdeswell and Perrott, 1996; Sayers Mayne and Bartram (2000). Herbage intake is about 0.5 to 1 kg higher when cows are supplemented with 5 kg of sugar-beet pulp than with similar amounts of cereals. This is probably because the sources of carbohydrate are compared at a same DM intake, whereas the net energy content is lower for fibrous than for starchy concentrates (Faverdin *et al.* 1991). When grass was probably restricted, there was no substitution, whatever the origin of the carbohydrate (Kibon and Holmes, 1987; Delagarde, Peyraud and Delaby, 1999). These results suggest that when less than 5 kg of concentrates are offered, the nature of the energy source does not necessarily produce enough digestive perturbations to affect animal performance.

The effects of highly-fermentable carbohydrates increased when high doses were used. Sayers Mayne and Bartram (2000) compared high-starch and high-fibre concentrates, either at 5 or 10 kg DM per cow per day. They reported a dramatic fall in milk fat content with increasing amounts of starch (29.9 versus 36.6 g/kg), but not with increasing amounts of fibre (36.2 *vs* 39.4 g/kg). This effect was probably associated with modifications in ruminal fermentation profile. Because the effect of supplementation on rumen digestion is much higher on a high than on a low allowance (Delagarde and Peyraud, unpublished), the effects of a large dose (i.e. more than 5 kg/day) of readily-fermented starch might be affected by sward characteristics. The higher the leaf content of the sward, the higher will be the effect of highly-fermentable starch on ruminal digestion and milk fat content. However, as far as we know, no trial has been designed to combine these two factors.

In most cases, milk production from pasture is not limited by metabolisable protein (MP) supply but, in some circumstances, the crude protein content of grass can decrease and supplementation with MP may be beneficial. This might occur when nitrogen (N) fertilisation is low, or during summer grazing. Delaby *et al.* (1996) described the response curve of milk yield when increasing supply of MP was provided by progressively replacing 3 kg of wheat by protected soya-bean meal. On a highly-fertilised sward, with a crude protein content greater than 160 g/kg DM, milk yield marginally increased with MP supply, whereas on low-fertilised sward, with a crude protein content lower than 130 g/kg DM, the response was markedly greater (Figure 13.5). The provision of 600 g/day of supplementary MP produced a similar milk yield to that obtained with a high-fertilised sward without MP supplementation; the lower the protein content of grass, the higher the response to the MP supplementation.

On a low-N sward, replacement of a carbohydrate concentrate by protected soya-bean meal increased forage intake. Delagarde, Peyraud and Delaby (1999) reported an increase in herbage intake (+ 0.8 kg / kg concentrate) when cows

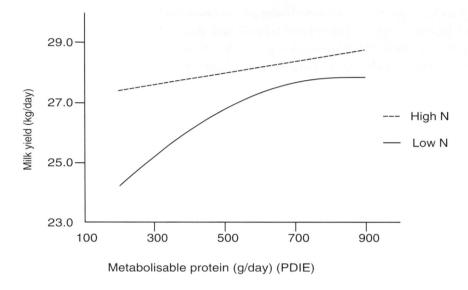

Figure 13.5 Interaction between crude protein content of grass and response to supplementation with metabolisable protein (after Delaby *et al.* 1996).

were supplemented with protected soya-bean meal (Table 13.4). This positive effect of MP supplementation on grass intake is clearly attributable to the low protein content of grass, as previously observed with poor-quality conserved forages (Journet *et al.* 1983). Thus, on low-N swards, MP supply increases milk yield because it alleviates a shortfall in MP supply, but part of the increase in milk yield can also be assigned to an increase in herbage intake (see Table 13.4). For pastures with high crude protein contents, MP supply in the concentrates generally has little effect on herbage intake (Vadiveloo and Holmes, 1979; Delagarde, Peyraud and Delaby, 1997).

Supplementation strategies

Supplements can be administered as constant single doses to all cows, regardless of milk potential, or computed as a function of cow potential: the higher the cow potential, the higher the concentrate allocation (Hoden *et al.* 1991). This latter strategy assumes that a cow is not able to adjust grass consumption according to her requirements. Delaby and Peyraud (1997) compared 3 kg of concentrates, administered either at a constant rate for all cows, or at a rate of 1 kg of concentrate for each 3 kg of milk above 20 kg at turnout, and observed similar animal performances. Thus, it is possible to use a single dose of concentrates for all cows, at least under grazing conditions favouring maximum grass intake.

Table 13.4 Effect of energy and protein source on herbage intake and digestion by dairy cows grazing on plots of perennial ryegrass with a low CP content (after Delagarde, Peyraud and Delaby, 1999).

	No concentrate	*Cereal-based concentrates*	*Soya-bean meal*
Concentrate intake (kg DM/day)	0	2.8	2.8
Grass intake (kg DM/day)	14.6	14.9	17.2
Grass digestibility	0.774	0.761	0.793
Rumen VFA (mmoles/l)	99	101	111
NH_3 (mg/l)	11	11	21
Protein flowing into the duodenum (kg/day)	2.2	2.5	3.5
Milk yield (kg/day)	19.6	22.0	24.8

Theoretically, increased frequency of feeding concentrates should result in less diurnal variations in rumen pH (Sutton *et al.* 1986) which, in turn, might increase the amount of grass a cow can consume and increase animal performance. On pasture, grain is normally fed twice-daily to cows, during or after milking. Recently, mobile computerised grain feeders were developed to manipulate the temporal pattern of supplementation for cows fed with more than 8 to 10 kg concentrate DM. Cows tended to consume less concentrates with the grain feeder, due to missing occasional meals, and this resulted in lower milk yields (Hongerholt, Muller and Buckmaster, 1997; Gibb, Huckle and Nuthall, 2000). However, even in cows that consumed the same amount of concentrates in parlour or from the grain feeder, increasing the frequency of concentrate meals did not improve animal performance.

Conclusion

Because grazed grass represents the cheapest source of nutrients for dairy cows, the objective of supplementation is to achieve high performance while maintaining full exploitation of grass potential. Recent developments in the description of the input-output response curves to supplements, together with a better understanding of the factors that influence animal responses, allow the utilisation of concentrates to be rationalised. There is scope for substantial improvement in response to supplementary feed inputs. Milk and grass intake responses are mainly determined by the energy balance of unsupplemented cows. Recent data show that milk responses to concentrate supplementation are generally high, mainly because cows rarely approach their potential intake at grazing.

Even if it is possible to manage high-producing dairy cows at grazing without supplementation, these cows can achieve satisfactory levels of performance, with high economic returns, with a supply of concentrates not exceeding 6 kg/day. The level of supplementation should be defined according to grazing conditions for a substantial improvement in milk response. However, little improvement is expected from modifying the nature of carbohydrate when low amounts of concentrate are given. It is recommended not to use concentrates rich in highly-fermentable carbohydrates when concentrate allowances are high or animals have access to high amounts of leafy swards.

References

Arriaga Jordan, C.M. and Holmes, W. (1986) The effect of concentrate supplementation on high yielding dairy cows under two systems of grazing. *Journal of Agricultural Science, Cambridge*, **107**, 453-461.

Bryant, A.M. and Donnelly, P.E. (1974) Yield and composition of milk from cows fed pasture herbage supplemented with maize and pasture silages. *New Zealand Journal of Agricultural Research*, **27**, 491-493.

Combellas, J., Baker, R.D. and Hodgson, J. (1979) Concentrate supplementation and the herbage intake and milk production of heifers grazing Cenchrus ciliaris. *Grass and Forage Science*, **34**, 303-310.

Coulon, J.B., Faverdin, P., Laurent, F. and Cotto, G. (1989) Influence de la nature de l'aliment concentré sur les performances des vaches laitières. *INRA Productions Animales*, **2**, 47-53.

Delaby, L. and Peyraud, J.L. (1994) Influence de la nature du concentré énergétique sur les performances des vaches laitières au pâturage. *Rencontres autour des Recherches sur les Ruminants,* **1**, 113-116, Paris, France (Abstract in English).

Delaby, L., Peyraud, J.L., Vérité, R. and B. Marquis, (1996) Effects of protein content in the concentrate and level of nitrogen fertilisation on the performance of dairy cows in pasture. *Annales de Zootechnie*, **45**: 327-341.

Delaby, L. and Peyraud, J.L. (1997) Influence of concentrate supplementation strategy on grazing dairy cows' performance. *XVIII International Grassland Congress*, pp 1111-1112, Saskatoon, Canada,

Delaby, L., Peyraud, J.L. and Delagarde, R. (1999) Production des vaches laitières au pâturage sans concentré. *Rencontres autour des Recherches sur les Ruminants*, **6**,123-126. Paris, France (Abstract in English).

Delaby, L., Peyraud, J.L. and Delagarde, R. (2001) Effect of the level of concentrate supplementation, herbage allowance and milk yield at turnout on the performances of dairy cows in mid-lactation at grazing. *Animal Science, in press.*

Delaby, L., Peyraud, J.L., Bouttier, A. and Peccatte, J.R. (1999) Effect of grazing conditions and supplementation on herbage intake by dairy cows. *Rencontres autour des Recherches sur les Ruminants*, **6**,139

Delagarde, R., Peyraud, J.L. and Delaby, L. (1997) The effect of nitrogen fertilisation level and protein supplementation on herbage intake, feeding behaviour and digestion in grazing dairy cows. *Animal Feed Science and Technology*, **66**, 165-180.

Delagarde, R., Peyraud, J.L. and Delaby, L. (1999) Influence of carbohydrate or protein supplementation on intake, behaviour and digestion in dairy cows strip grazing low nitrogen fertilised perennial ryegrass. *Annales de Zootechnie*, **48**, 81-96

Dillon, P., Crosse, S. and O'Brien, B. (1997) Effect of concentrate supplementation of grazing dairy cows in early lactation on milk production and milk processing quality. *Irish Journal of Agricultural and Food Research*, **36**, 145-159.

Faverdin, P., Dulphy, J.P., Coulon, J.B., Vérité, R., Garel, J.P., Rouel, J. and Marquis, B. (1991) Substitution of roughage by concentrate for dairy cows. *Livestock Production Science*, **27**, 137-156.

Fisher, G.E.J., Dowdeswell, A.M. and Perrott, G. (1996) The effects of sward characteristics and supplement type on the herbage intake and milk production of summer-calving cows. *Grass and Forage Science*, **51**, 121-130.

Garel, J.P. and Hoden, A. (1985) Pâturage en zone de montagne : niveaux de chargement et de complémentation pour des vaches laitières. *Bulletin Technique du CRZV de Theix*, INRA, **62**, 35-46.

Gibb, M.J., Huckle, C.A. and Nuthall, R. (2000) Effect of temporal pattern of supplementation on grazing behaviour and herbage intake by dairy cows. In *Grazing management – 2000*, pp 91-96. Edited by A.J. Rook and P.D. Penning, BGS Occasional Symposium no. 34.

Gleeson, P.A. (1981) Concentrate supplementation for spring calving cows. *Grass and Forage Science*, **36**, 138-149.

Grainger, C. and Mathews, G.L. (1989). Positive relation between substitution rate and pasture allowance for cows receiving concentrates. *Australian Journal of Experimental Agriculture*, **29**, 355-360.

Hijink, W.F., Le Du, Y.L.P., Meijs, J.A.C. and Meijer A.B. (1982). Supplementation of the grazing dairy cow. Report IVVO N° 141, Lelystad, The Netherlands.

Hoden, A., Fiorelli, J.L., Jeannin, B., Huguet, L., Muller, A. and Weiss, P. (1987) Le pâturage simplifié pour vaches laitières : synthèse de résultats expérimentaux. *Fourrages*, **111**, 239-257.

Hoden, A., Peyraud, J.L., Muller, A., Delaby, L. and Faverdin, P. (1991) Simplified rotational grazing management of dairy cows: effects of rate of stocking and concentrate. *Journal of Agricultural Science, Cambridge*, **116**, 417-428

Hongerholt, D.D., Muller, L.D and Buckmaster, D.R. (1997) Evaluation of a mobile computerized grain feeder for lactating cows grazing grass pasture. *Journal of Dairy Science*, **80**, 3271-3282.

Jennings, P.G. and Holmes W. (1984) Supplementary feeding of dairy cows on continuously stocked pasture. J*ournal of Agricultural Science, Cambridge*, **103**, 161-170.

Journet, M. and Demarquilly, C. (1979) Grazing. In *Feeding Strategy for the High Yielding Dairy Cow.- 1979*, pp 295-321. Edited by W.H. Broster and H. Swan. St Albans, Canada Publishing Co.

Journet. M., Faverdin, P., Rémond, B. and Vérité, R. (1983) Niveau et qualité des apports azotés en début de lactation. *Bulletin Technique du CRZV de Theix*, INRA, **51**, 7-17.

Kibon, A. and Holmes, W. (1987) The effect of height of pasture and concentrate composition on dairy cows grazed on continuously stocked pastures. *Journal of Agricultural Science, Cambridge*, **109**, 293-301

Kolver, E.S. and Muller, L.D. (1997) Performance and nutrient intake of high producing holstein cows consuming pasture or a total mixed ration. *Journal of Dairy Science* **81**, 1403-1411.

Leaver, J.D. (1968) Use of supplementary feeds for grazing dairy cows. *Dairy Science Abstracts*, **30** (7), 355-361.

Leaver, J.D. (1985) Milk production from grazed temperate grassland. *Journal of Dairy Research,* **52**, 313-344.

Le Du, J. (1980) Le pâturage continu : l'expérience anglaise. *Fourrages*, **82**: 31-43.

Mayne, C.S. (1991) Effect of supplementation on the performance of both growing and lactating cattle at pasture. In *Management Issues for Grassland Farmers in the 1990s*, pp 55-71. Edited by C.S. Mayne. British Grassland Society, Occasional Symposium n°35.

Mayne, C.S. and Wright, I.A. (1988) Herbage intake and utilization by the grazing dairy cows. In *Nutrition and Lactation in the Dairy Cow*, pp 280-293. Edited by P.C. Garnsworthy. Butterworths, London.

Meijs J.A.C. (1981). Herbage intake by grazing dairy cows. Agricultural Research Reports 909. Center Agricultural Publishing and Documentation, Wageningen, Netherlands, pp 264

Meijs J.A.C. (1986) Concentrate supplementation of grazing dairy cows. 2. Effect of concentrate composition on herbage intake and milk production. *Grass and Forage Science*, **41**, 229-35.

Meijs, J.A.C. and Hoekstra, J.A. (1984) Concentrate supplementation of grazing dairy cows. 1. Effect of concentrate intake and herbage allowance on herbage intake. *Grass and Forage Science*, **39**, 59-66.

Opatpatanakit, Y., Kekkaway, R.C. and Lean, L.J. (1993) Substitution effects of feeding rolled barley grain to grazing dairy cows. *Animal Feed Science and Technology*, **42**, 25-38.

O'Brien, B., Crosse, S. and Dillon, P. (1996) Effects of offering a concentrate or silage supplement to grazing dairy cows in late lactation on animal performances and on milk processability. *Irish Journal of Agricultural and Food Research*, **35**, 113-125.

Peyraud, J.L., Comeron, E.A., Wade, M. and Lemaire G. (1996) The effect of daily herbage allowance, herbage mass and animal factors upon herbage intake by grazing dairy cows. *Annales de Zootechnie*, **45**: 201-217.

Phillips, C.J.C. (1988) The use of conserved forage as a supplement for grazing dairy cows. *Grass and Forages Science*, **43**, 215-230.

Phillips, C.J.C. and Leaver, J.D. (1985) Supplementary feeding of forage to grazing dairy cows. 2. Offering grass silage in early and late season. *Grass and Forage Science*, **40**, 193-199.

Robaina, A.C., Grainger, C., Moate, P., Taylor, J. and Stewart, J. (1998) Responses to grain feeding by grazing dairy cows. *Australian Journal of Experimental Agriculture*, **38**, 541-549.

Rook, A.J., Huckle, C.A. and Wilkins, R.J. (1994) The effects of sward height and concentrate supplementation on the performance of spring calving dairy cows grazing perennial ryegrass-white clover swards. *Animal Production*, **58**, 167-172.

Sarker, A.B. and Holmes, W. (1974) The influence of supplementary feeding on the herbage intake and grazing behaviour of dry cows. *Journal of the British Grassland Society*, **29**, 141-143.

Sayers, H.J., Mayne, C.S. and Bartram, C.G. (2000) The effect of level and type of supplement and change in the chemical composition of herbage as the season progresses on herbage intake and animal performance of high yielding dairy cows. In *Grazing management – 2000*, pp 85-90. Edited by A.J. Rook and P.D. Penning, BGS Occasional Symposium n°34.

Schwartz, F.J., Haffner, J., Kirchgessner, M. (1995) Supplementation of zero-grazed dairy cows with molassed beet pulp, maize or a cereal-rich concentrate. *Animal feed Science and Technology*, **54**, 237-248.

Spörndly, E. (1991) Supplementation of dairy cows offered freshly cut herbage ad libitum with concentrates based on unmolassed sugar beet pulp and wheat bran. *Swedish Journal of Agricultural Research*, **21**, 131-139.

Stakelum, G. (1986a) Herbage intake of grazing dairy cows. 3. Effect of autumn supplementation with concentrates and herbage allowance on herbage intake. *Irish Journal of agricultural Research*, **25**,31-40.

Stakelum, G. (1986b) Herbage intake of grazing dairy cows. 2. Effects of herbage allowance, herbage mass and concentrates feeding on the intake of cows grazing primary spring grass. *Irish Journal of agricultural Research*, **25**, 41-51.

Stakelum G. (1986c) Herbage intake of grazing dairy cows. 3. Effects of herbage mass, herbage allowance and concentrate feeding on the herbage intake of dairy cows grazing on mid-summer pasture. *Irish Journal of agricultural Research*, **25**, 179-189.

Stockdale, C.R. and Trigg, T.E. (1989) Effect of pasture allowance and level of concentrate feeding on the productivity of dairy cows in late lactation. *Australian Journal of Experimental Agriculture*, **25**, 739-744.

Stockdale, C.R., Callaghan A. and Trigg, T.E. (1989) Feeding high energy supplements to pasture fed dairy cows. Effects of stage of lactation and level of supplements. *Australian Journal of Experimental Agriculture*, **29**, 601-611.

Sutton, J.D., Hart, I.C., Broster, W.H., Elliot, R.J. and Schuller, E/ (1986) Feeding frequency for lactating cows: effects on rumen fermentation, blood metabolites and hormones. *British Journal of Nutrition*, **56**, 181-191.

Vadiveloo, J., Holmes, W. (1979) Supplementary feeding of grazing beef cattle. *Grass and Forage Science*, **34**, 173-179.

Wilkins, R.J., Gibb, M.J., Huckle C.A. and Clements, A.J. (1994) Effect of supplementation on production by spring-calving dairy cows grazing swards of differing clover content. *Grass and Forage Science*, **49**, 465-75.

Wilkins, R.J., Gibb, M.J. and Huckle, C.A. (1995) Lactation performance of spring-calving dairy cows grazing mixed perennial ryegrass/white clover swards of differing composition and height. *Grass and Forage Science*, **50**, 199-208.

14

CONJUGATED LINOLEIC ACID (CLA) AND THE DAIRY COW

D.E. BAUMAN[1], B.A. CORL[1], L.H. BAUMGARD[1] and J.M. GRIINARI[2]
[1]*Department of Animal Science, Cornell University, Ithaca, NY 14853, USA*
[2]*Department of Animal Science, University of Helsinki, Helsinki, Finland*

Introduction

Milk is a valuable source of nutrients providing energy, high quality protein, and essential minerals and vitamins. Fat is the foremost energy component in milk and it accounts for many of the physical properties, manufacturing characteristics and organoleptic qualities of milk and milk products. Milk fat also represents the major energy cost in milk production and changes in its economic value favour increased production in some circumstances and decreased production in others. In addition, fat is the most variable among the major milk components and its synthesis is affected by many factors - especially dietary and environmental factors (Sutton, 1989; Palmquist, Beaulieu and Barbano, 1993; Doreau, Chilliard, Rulquin and Demeyer, 1999).

As the biological roles of specific fatty acids become better understood, there is interest in designing milk fat to improve its healthfulness as a food. This is of particular importance because it is increasingly recognised that foods can be a contributing factor in the prevention as well as in the development of some disease conditions. The generic term "functional foods" has been used to describe ingested foods that have beneficial effects beyond their traditional nutritive values (Milner, 1999). Conjugated linoleic acids (CLA) are functional food components of milk fat that may have positive effects on human health and disease prevention (Parodi, 1997). CLA in human diets are derived almost exclusively from food products of ruminant origin. In the following sections we will provide background on CLA, examine its biosynthesis in dairy cows, and discuss recent developments establishing biological roles for CLA.

Background

CLA represent a mixture of octadecadienoic acid ($C_{18:2}$) isomers in which the two double bonds are conjugated. The presence of fatty acids with conjugated double bonds was first reported in milk fat from cows grazing spring pasture (Booth, Kon, Dann and Moore, 1935), and subsequent work by Parodi (1977) demonstrated that these were primarily *cis*-9, *trans*-11 CLA. Theoretically, different positional isomers (n = 8; e.g. 7-9, 8-10, 9-11, 10-12 and so forth) and geometric isomers (n = 4 for each; *cis-cis, cis-trans, trans-cis,* and *trans-trans*) can exist. In Figure 14.1, the structures of two key CLA isomers, *cis*-9, *trans*-11 CLA and *trans*-10, *cis*-12 CLA, are contrasted with linoleic acid.

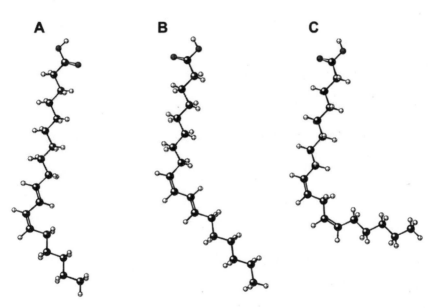

Figure 14.1 Chemical structure of conjugated linoleic acid isomers (CLA) and linoleic acid. Fatty acids are *trans*-10, *cis*-12 CLA (A), *cis*-9, *trans*-11 CLA (B) and *cis*-9, *cis*-12 octadecadienoic acid (linoleic acid) (C). From Bauman *et al.* (2000b).

Dairy products are the major dietary source of CLA. Food chemistry investigations have established that the CLA content of products is relatively stable during processing, manufacturing and storage conditions that are typical for the dairy food industry (Shantha, Ram, O'Leary, Hicks and Decker, 1995; Banni and Martin, 1998). Therefore, the CLA concentration in dairy products is essentially a function of the concentration in raw milk fat. The major CLA in milk fat is the *cis*-9, *trans*-11 isomer, and it generally constitutes 80 to 90% of total CLA. However, detectable levels of many different CLA isomers exist in milk fat (Sehat, Kramer, Mossoba, Yurawecz, Roach, Eulitz, Morehouse and Ku, 1998).

Pariza and co-workers were the first to discover a "functional food" role for CLA when they observed that conjugated dienoic isomers of linoleic acid possessed anticarcinogenic effects (see review by Pariza, 1999). Subsequent work has identified additional beneficial health effects and these are listed in Table 14.1 (see reviews by Banni and Martin, 1998; Pariza, 1999; McGuire and McGuire, 2000; Whigham, Cook and Atkinson, 2000). The range of physiological processes affected by CLA is impressive and raises interesting questions as to possible mechanisms of action. In general, effects have been identified using biomedical studies with animal models and chemically synthesised CLA supplements containing a variety of isomers. Thus, it is of interest to establish which biological effects are related to specific CLA isomers.

Table 1. Beneficial health effects of CLA reported from biomedical studies with animal models.

Biological effect
Anticarcinogenic (*in vivo* and *in vitro* studies)
Antiatherogenic
Altered nutrient partitioning and lipid metabolism
Antidiabetic (type II diabetes)
Immune modulation
Improved bone mineralisation

The anticancer effect is the most extensively investigated response to CLA. Studies conducted by Ip and colleagues, using a rat model involving chemical induction of mammary tumours, are recognised as seminal research in this area (see reviews by Ip, Scimeca and Thompson, 1994; Banni and Martin, 1998). This series of investigations examined different aspects of mammary cancer including establishing diet as an effective route to provide cancer protection and that the magnitude of the reduction in mammary tumours was directly proportional to the amount of CLA consumed. Subsequent research verified that CLA was an anticarcinogen and extended the biomedical studies to show that CLA was cytotoxic to many types of human cancer cell lines and reduced the incidence of tumours in animal models for mammary, skin, stomach, intestinal and prostate cancer (see review by Scimeca, 1999). Much of the current biomedical research is focused on establishing mechanisms of action for the anticarcinogenic effects of CLA. The uniqueness of CLA was recognised in a National Academy report on "Carcinogens and Anticarcinogens in the Human Diet" which stated that "conjugated linoleic acid (CLA) is the only fatty acid shown unequivocally to inhibit carcinogenesis in experimental animals" (National Research Council, 1996). Of particular importance, studies have established that *cis*-9, *trans*-11 CLA, the major isomer in milk fat, is anticarcinogenic (Ip, Banni, Angioni, Carta, McGinley, Thompson, Barbano and Bauman, 1999; Ip, Ip, Loftus, Shoemaker and Shea-Eaton, 2000).

The effects of CLA on nutrient partitioning and lipid metabolism are elicited by the *trans*-10, *cis*-12 CLA isomer (Park, Storkson, Albright, Liu and Pariza, 1999; Baumgard, Corl, Dwyer, Saebo and Bauman, 2000b). This is sometimes referred to as an antiobesity effect because CLA supplements cause a reduction in body fat content of growing animals. Although the *trans*-10, *cis*-12 CLA isomer is normally very low in milk fat, it does play an important role in lipid metabolism of the dairy cow, as will be discussed later. Recent studies also demonstrate that the antidiabetic effect of CLA is due to the *trans*-10, *cis*-12 isomer (Ryder, Bauman, Portocarrero, Song, Yu, Barbano, Zierath and Houseknecht, 1999). In contrast, the *cis*-9, *trans*-11 CLA isomer does not elicit the biological responses in nutrient partitioning and lipid metabolism, or the antidiabetic effects. Thus, small differences in position and geometry of the double bonds results in tremendous differences in the biological effect of CLA isomers. Other CLA-related health effects and other CLA isomers have not been investigated in a similar manner, but this is an active area of research around the world.

The Animal Dimension

RUMEN BIOHYDROGENATION

Dietary lipids consumed by ruminants include phospholipids and glycolipids derived primarily from forages, and triglycerides originating from seed oils and concentrate feeds. Fatty acids are esterified in these lipid classes, and the first transformation that occurs in the rumen is hydrolysis of the ester linkages by microbial lipases to produce free fatty acids (Keeney, 1970; Dawson, Hemington and Hazlewood, 1977). This is a prerequisite to the second transformation, biohydrogenation of the unsaturated fatty acids. Rumen biohydrogenation is extensive, which explains why ruminant fats are relatively more saturated than fats from non-ruminants, regardless of the fatty acid composition of the diet.

Bacteria are largely responsible for the biohydrogenation of unsaturated fatty acids in the rumen; protozoa are of only minor importance (Keeney, 1970; Dawson *et al.*, 1977). *Butyrivibrio fibrisolvens* is the bacterium that has been most extensively investigated (Kepler and Tove, 1967; Kepler, Tucker and Tove, 1970). However, a diverse range of rumen bacteria have been isolated that are capable of some reactions in the biohydrogenation of unsaturated fatty acids (see review by Harfoot and Hazlewood, 1988). Two major polyunsaturated fatty acids in diets of ruminants are linoleic and linolenic acids. The predominant pathways for their biohydrogenation are illustrated in Figure 14.2. The first reaction of biohydrogenation for both fatty acids is isomerisation of the *cis*-12 double bond to form a *trans*-11 double bond. In the case of linoleic acid, this produces *cis*-9, *trans*-11 CLA. Thus, a CLA isomer is an intermediate in the biohydrogenation of linoleic acid, but not linolenic acid (Figure 14.2). The next step involves hydrogenation of the *cis*-9 double bond, resulting in *trans*-11 $C_{18:1}$ and *trans*-11, *cis*-15 $C_{18:2}$ for linoleic and linolenic acids, respectively. An additional step to hydrogenate the *cis*-15 double bond of linolenic acid produces *trans*-11 $C_{18:1}$, the common intermediate in the biohydrogenation of both fatty acids (Figure 14.2). The final reaction is hydrogenation of the *trans*-11 double bond to produce stearic acid.

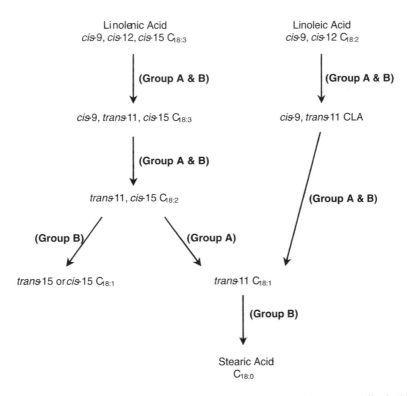

Figure 14.2 Predominant pathways for the ruminal biohydrogenation of linoleic and linolenic acids with bacterial populations (Groups A and B) for each reaction shown. Based on Kemp and Lander (1984) and adapted from Harfoot and Hazlewood (1988).

Polan, McNeill and Tove (1964) were among the first to suggest that different populations of rumen bacteria were necessary for the complete hydrogenation of polyunsaturated fatty acids. This was supported by subsequent studies with pure cultures of rumen bacteria causing Kemp and Lander (1984) to group bacteria based on the reactions and end products of biohydrogenation (Figure 14.2). Group A bacteria were able to hydrogenate linoleic and linolenic acid with *trans*-11 $C_{18:1}$

being their major end product. Group B bacteria utilised *trans*-11 C$_{18:1}$ as a substrate with stearic acid being the end product. A listing of bacteria species in Groups A and B, and details on the specific bacteria in rumen biohydrogenation, are provided in the review by Harfoot and Hazlewood (1988).

Other pathways of rumen biohydrogenation must also exist, although the predominant pathway for linoleic acid clearly involves isomerisation and reduction steps to produce *cis*-9, *trans*-11 CLA and *trans*-11 C$_{18:1}$. Milk fat contains a number of other CLA isomers (e.g. *trans*-7, *cis*-9; *trans*-10, *cis*-12; and *cis*-12, *trans*-14) and *trans*-C$_{18:1}$ isomers (e.g. *trans*-9 C$_{18:1}$; *trans*-10 C$_{18:1}$; and *trans*-12 C$_{18:1}$) suggesting that several specific isomerases and reductases exist. Changes in the diet often result in bacterial population shifts that alter the pattern of fermentation end products. Working with gnotobiotic lambs, Leat, Kemp, Lysons and Alexander (1977) were among the first to provide evidence that shifts in rumen bacteria populations must be modifying biohydrogenation pathways based on observed alterations in the *trans*-C$_{18:1}$ profile in rumen lipids and body fat. More recently, this was demonstrated with diet-induced changes in the *trans*-C$_{18:1}$ profile in milk fat of dairy cows (Griinari, Chouinard and Bauman, 1997; Griinari, Dwyer, McGuire, Bauman, Palmquist and Nurmela, 1998). In particular, feeding a high grain/low roughage diet caused a shift in biohydrogenation pathways so that *trans*-10 C$_{18:1}$ replaced *trans*-11 C$_{18:1}$ as the predominant *trans*-octadecenoic fatty acid in milk fat. Putative pathways for the formation of *trans*-10 C$_{18:1}$ are presented in Figure 14.3. The initial reaction involves isomerisation of the *cis*-9 double bond to form an intermediate with *trans*-10, *cis*-12 conjugated double bonds. In the case of linoleic acid this intermediate is *trans*-10, *cis*-12 CLA. The biological effects of this CLA isomer on fat synthesis will be discussed in a later section, but suffice to say that milk fat concentrations of both *trans*-10 C$_{18:1}$ and *trans*-10, *cis*-12 CLA are increased with high concentrate/low roughage diets or diets supplemented with plant oils or fish oils (Bauman and Griinari, 2000a, 2000b).

Cis-9, *trans*-11 CLA is the major CLA isomer in milk fat, and it has been generally assumed that this reflects its escape from complete rumen biohydrogenation. Although isomerisation and reduction reactions proceed in a stepwise fashion, different relative amounts of intermediates and products reach the small intestine for absorption. For linoleic acid, the conversion of *cis*-9, *trans*-11 CLA to *trans*-11 C$_{18:1}$ is more rapid than the hydrogenation of *trans*-11 C$_{18:1}$ so that this intermediate accumulates. Even at short incubation intervals or when incubating linoleic acid with pure strains of group B microbes, bacteria capable of hydrogenating linoleic acid to stearic acid, *trans*-11 C$_{18:1}$ was still the major product (Harfoot, Noble and Moore, 1973; Kemp, White and Lander, 1975; Hazlewood, Kemp, Lander and Dawson, 1976; Kellens, Goderis and Tobback, 1986). In addition, other studies have demonstrated that increasing concentrations of linoleic

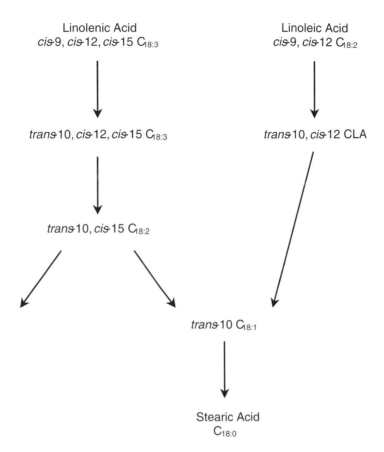

Linolenic Acid
cis-9, *cis*-12, *cis*-15 C$_{18:3}$

Linoleic Acid
cis-9, *cis*-12 C$_{18:2}$

trans-10, *cis*-12, *cis*-15 C$_{18:3}$

trans-10, *cis*-12 CLA

trans-10, *cis*-15 C$_{18:2}$

trans-10 C$_{18:1}$

Stearic Acid
C$_{18:0}$

Figure 14.3 Putative pathways of ruminal biohydrogenation of linoleic and linolenic acids involving *cis*-9, *trans*-10 isomerase. Adapted from Griinari and Bauman (1999).

acid prevented the conversion of *trans*-11 C$_{18:1}$ to C$_{18:0}$, but had little effect on the conversion of linoleic acid to *trans*-11 C$_{18:1}$ (Polan *et al.*, 1964; Kemp *et al.*, 1975; Tanaka and Shigeno, 1976). Overall, kinetic studies of linoleic and linolenic acid biohydrogenation indicate conjugated intermediates are transient and *trans*-11 C$_{18:1}$ accumulates.

ENDOGENOUS SYNTHESIS

A linear relationship between the milk fat content of *trans*-11 C$_{18:1}$ and *cis*-9, *trans*-11 CLA has been observed across a range of diets (Jiang, Bjöerck, Fonden and Emanuelson, 1996; Jahreis, Fritsche and Steinhart, 1997; Precht and Molkentin,

1997; Griinari and Bauman, 1999; Solomon, Chase, Ben-Ghedalia and Bauman, 2000). This has been generally attributed to their common source as fatty acid intermediates that have escaped complete biohydrogenation in the rumen. However, a linear relationship is also consistent with a precursor-product relationship. Based on this and the kinetics of rumen biohydrogenation indicating accumulation of *trans*-11 $C_{18:1}$, we hypothesised that endogenous synthesis of *cis*-9, *trans*-11 CLA would be an important source of milk fat CLA (Griinari *et al.*, 1997). This would involve the absorption of *trans*-11 $C_{18:1}$ and its subsequent conversion to *cis*-9, *trans*-11 CLA by the enzyme Δ^9-desaturase (Figure 14.4). For desaturation, fatty acids must first be activated to acyl-CoA by acyl-CoA synthetase. The introduction of a *cis*-double bond between carbons 9 and 10 requires a system that includes NADH-cytochrome b_5 reductase, cytochrome b_5, and the terminal Δ^9-desaturase enzyme (see review by Bauman, Baumgard, Corl and Griinari, 2000b). Desaturated fatty acids have a lower melting point than their more saturated precursors and represent important determinants of fluidity characteristics of milk fat, depot lipids and cell membranes. The CoA esters of palmitic and stearic acids are common substrates for mammary Δ^9-desaturase, contributing a large portion of the palmitoleic and oleic acids found in milk fat.

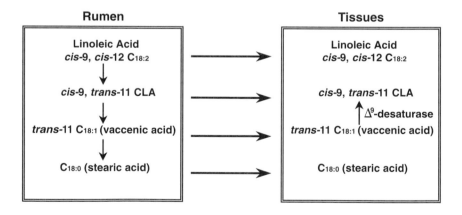

Figure 14.4 Pathways for ruminal and endogenous synthesis of conjugated linoleic acid (CLA). Adapted from Bauman *et al.* (2000b).

Our initial investigation with lactating cows established that endogenous synthesis of milk fat CLA was possible when we observed a 31% increase in milk fat CLA with abomasal infusion of 12.5 g/d of *trans*-11 $C_{18:1}$ (Griinari, Corl, Lacy, Chouinard, Nurmela and Bauman, 2000b). To examine the quantitative significance of endogenous synthesis, we used sterculic acid, a potent inhibitor of Δ^9-desaturase due to its cyclopropene ring at carbons 9 and 10 (Bickerstaffe and Annison, 1970;

Jeffcoat and Pollard, 1977). We have used different approaches and in all cases observed a dramatic reduction in the milk fat content of *cis*-9, *trans*-11 CLA when sterculic oil was abomasally infused (Corl, Baumgard, Bauman and Griinari, 2000; Griinari *et al.*, 2000b). When a correction was applied for the extent of desaturase inhibition, it was clear that endogenous synthesis accounted for over 75% of the *cis*-9, *trans*-11 CLA found in milk fat. The mammary gland is the apparent site of endogenous CLA synthesis for lactating ruminants based on the tissue activity of Δ^9-desaturase (Bickerstaffe and Annison, 1970; Kinsella, 1972). *In vivo* studies are also consistent with the mammary gland being the major site for endogenous synthesis of CLA. Bickerstaffe and Johnson (1972) demonstrated that intravenous infusion of sterculic acid to lactating goats resulted in a marked decrease in the oleic acid:stearic acid ratio in milk fat, but only minimal differences in plasma fatty acid ratio. Given that circulating sterculic acid can permeate all organs except the small intestine, the authors concluded the mammary gland must be the major site of desaturation for fatty acids found in milk fat.

Investigations of Δ^9-desaturase have typically utilised rats and rat liver preparations. These studies have established that gene transcription and enzyme activity are responsive to changes in diet, hormonal balance and physiological state (see reviews by Ntambi,1995; Tocher, Leaver and Hodgson, 1998; Ntambi, Choi and Kim, 1999). Only a limited number of investigations have examined Δ^9-desaturase in ruminant animals. Recently, Medrano and DePeters (2000) identified 3 genetic polymorphisms for the Δ^9-desaturase gene in dairy cattle and suggested a possible relationship with degree of saturation and CLA concentration of milk fat. In the case of sheep, Ward, Travers, Richards, Vernon, Salter, Buttery and Barber (1998) demonstrated that the onset of lactation resulted in a decrease in mRNA abundance for Δ^9-desaturase in adipose tissue and an increase in mammary tissue. They also demonstrated that insulin increased Δ^9-desaturase gene-expression in explants of sheep adipose tissue. Based on changes in the pattern of milk fatty acids, a similar influence of insulin on Δ^9-desaturase activity occurred in lactating dairy cows during a hyperinsulinaemic-euglycaemic clamp (L'Esperance, Bernier, Bauman, Corl and Chouinard, 2000).

DIET EFFECTS

The content of CLA in milk fat is dependent on ruminal production of both CLA and *trans*-11 $C_{18:1}$, and the tissue activity of Δ^9-desaturase. There is a substantial variation in milk content of CLA among individual cows consuming the same diet (Jiang *et al.*, 1996; Kelly, Berry, Dwyer, Griinari, Chouinard, Van Amburgh and Bauman, 1998a; Kelly, Kolver, Bauman, Van Amburgh and Muller, 1998b; Lawless, Murphy, Harrington, Devery and Stanton, 1998; Solomon *et al.*, 2000). However, seasonal and herd variation as well as results from feeding studies indicate that diet has a major influence (Riel, 1963; Kelly and Bauman, 1996; Griinari and Bauman, 1999; Bauman *et al.*, 2000b; Chilliard, Ferlay, Mansbridge and Doreau, 2000). For example, Precht and Molkentin (1997) analysed 927 milk samples from cows fed various diets in confinement and reported that CLA averaged 0.45 ± 0.14% (mean ± SD) and ranged from 0.10 to 1.05% of total milk fatty acids.

Milk fat content of CLA can be increased several fold by dietary means. Thus far, the highest reported value for a group of cows is 4.1% of total fatty acids (Bauman, Barbano, Dwyer and Griinari, 2000a). Dietary factors that affect CLA in milk fat are grouped into categories relative to the potential mechanisms and summarised in Table 14.2. The first category includes dietary factors that provide lipid substrates for rumen production of CLA or *trans*-11 $C_{18:1}$. The second group consists of dietary factors that affect rumen bacteria involved in biohydrogenation either directly or via changes in rumen environment. The third category includes dietary factors that involve a combination of lipid substrates and modification of rumen biohydrogenation. Diet supplements of CLA or *trans*-$C_{18:1}$ fatty acids provide additional means of altering milk fat content of CLA, and these represent the last category.

Dietary addition of plant oils often results in substantial increases in milk fat concentration of CLA (Table 14.2). The form in which plant oils are introduced in the diet has a distinct effect. In general, addition of free oils increases CLA content in milk fat more than plant oils incorporated in feed materials (Table 14.2). Linoleic acid content of the plant oil has sometimes been a major determinant of the response (Kelly *et al.*, 1998a), but other studies find little or no difference between plant oils rich in linoleic and linolenic acid (Dhiman, Satter, Pariza, Galli, Albright and Tolosa, 2000; Shingfield, Ahvenjärvi, Toivonen and Griinari, 2000). The effects of plant oils are probably mediated by direct effects on biohydrogenating bacteria and modulated by the composition of the basal diet. Harfoot *et al.* (1973) report that high levels of linoleic acid irreversibly inhibited the biohydrogenation of *trans*-11 octadecenoic acid *in vitro*. *In vivo*, this inhibition would result in the production and absorption of substrate for endogenous synthesis of *cis*-9, *trans*-11 CLA (Figure 14.4). Polan *et al.* (1964) demonstrated that *trans*-$C_{18:1}$ fatty acids, rather than stearic acid, were the major end product when linoleic acid was added to rumen cultures and attributed the mechanism to linoleic acid acting as a competitive inhibitor for monoenoic acid biohydrogenation. This also suggests that bacteria involved in biohydrogenation of *trans*-$C_{18:1}$, Group B bacteria (Figure 14.2) are specifically inhibited by linoleic acid.

Use of plant oils in ruminant diets is limited because they produce inhibitory effects on rumen microbial growth (Jenkins, 1993). A method to minimise this effect is to feed Ca salts of plant oil fatty acids. Ca salts of unsaturated fatty acids are largely biohydrogenated in the rumen (Fotouhi and Jenkins, 1992), but

Table 14.2 Summary of dietary factors which affect concentrations of conjugated linoleic acids (CLA) in milk fat[a]

Dietary factor	Effect on the content of CLA in milk fat	Reference[b]
Lipid substrate		
Unsaturated *vs* saturated fat	Increase by addition of unsaturated fat	10
Plant oils		
Type of plant oil	Variable increase	2, 11, 28, 34, 37
Level of plant oil	Non-linear, dose dependent increase	2, 4, 25, 28
Ca salts of plant oils	Increase	37
High oil plant seeds		
Raw seeds	No effect	37, 28
Processed seeds	Variable increase	8, 14, 21, 37
High oil corn grain and silage	Minimal effect	20, 37
Animal fat by-products	Minimal effect	37
Modifiers of rumen biohydrogenation		
Forage:concentrate ratio	Variable effect	3, 9, 10
Nonstructural carbohydrate level	Minimal effect	9, 35
Restricted feeding	Variable effect	1, 3, 8
Fish oils	Increase	13, 17, 26, 29, 32, 37
Fish meal	Minimal effect	20
Marine algae	Increase	22
Ionophores	Variable effect	9, 16, 20
Dietary buffers	Minimal effect	9

Table 14.2 Contd.

Combination		
Pasture	Higher than on conserved forages	1, 5, 6, 7, 12, 20
Plant oil + high NSC diet	Maximal effect, but transient	27
Processed seeds + fish oil	Additive effect	36
Growth stage of forage	Increased with less-mature forage	9
Dietary Supplements		
CLA Supplement	Linear, dose-dependent increase	15, 18, 19, 24, 31, 33
trans-11 C$_{18:1}$ supplement	Non-linear, dose-dependent increase	23, 30

[a]Adapted from Griinari and Bauman (1999) [b]Symbols are as follows: 1 = Timmen and Patton, 1988; 2 = Tesfa et al., 1991; 3 = Jiang et al., 1996; 4 = McGuire et al., 1996; 5 = Zegarska et al., 1996; 6 = Jahreis et al., 1997; 7 = Precht and Molkentin, 1997; 8 = Stanton et al., 1997; 9 = Chouinard et al., 1998; 10 = Griinari et al., 1998; 11 = Kelly et al., 1998a; 12 = Kelly et al., 1998b; 13 = Lacasse and Ahnadi, 1998; 14 = Lawless et al., 1998; 15 = Loor and Herbein, 1998; 16 = Sauer et al., 1998; 17 = Chilliard et al., 1999; 18 = Chouinard et al., 1999a; 19 = Chouinard et al., 1999b; 20 = Dhiman et al., 1999a; 21 = Dhiman et al., 1999b; 22 = Franklin et al., 1999; 23 = Griinari et al., 1999b; 24 = Giesy et al., 1999; 25 = Mir et al., 1999; 26 = Offer et al., 1999; 27 = Bauman et al., 2000a; 28 = Dhiman et al., 2000; 29 = Donovan et al., 2000; 30 = Griinari et al., 2000a; 31 = Gulati et al., 2000; 32 = Jones et al., 2000; 33 = Medeiros et al., 2000; 34 = Shingfield et al., 2000; 35 = Solomon et al., 2000; 36 = Whitlock et al., 2000; 37 = Chouinard et al., 2001.

apparently this occurs in a manner that minimises negative effects on bacterial fermentation and fibre digestion. The slow ruminal release of unsaturated fatty acids from Ca salts also creates favourable conditions for accumulation of *trans*-$C_{18:1}$ fatty acids and a subsequent increase in milk fat content of CLA (Chouinard, Corneau, Butler, Chilliard, Drackley and Bauman, 2001).

Dietary factors that affect milk fat CLA by modifying rumen biohydrogenation generally affect both Group A and Group B bacteria (Figure 14.2). The *trans*-11/*trans*-10 shift described earlier indicates an effect on Group A bacteria, whereas ruminal accumulation of *trans*-$C_{18:1}$ indicates an effect on Group B bacteria. Modifiers of rumen biohydrogenation will increase concentration of CLA in milk fat only in the presence of lipid substrate, and the magnitude of the increase is determined by their combined effects on Group A and Group B bacteria.

Dietary addition of fish oils and feeding high concentrate/low forage diets increase the CLA in milk fat (Table 14.2) by inhibiting biohydrogenation of *trans*-octadecenoic acids. Both diets may also change the biohydrogenation balance away from the *trans*-11 $C_{18:1}$ type towards the *trans*-10 $C_{18:1}$ pathway (Figure 14.3), although the specific mechanisms may differ. As described earlier, the *trans*-11 $C_{18:1}$ to *trans*-10 $C_{18:1}$ shift in rumen biohydrogenation induced by feeding high concentrate/low forage diets is associated with alterations in bacterial populations. Similar effects on bacterial populations have not been established for fish oil diets. A decrease in rumen pH has been implicated as the main factor causing *trans*-$C_{18:1}$ to accumulate when high concentrate/low forage diets are fed (Kalscheur, Teter, Piperova and Erdman, 1997). Consistent with this, decreasing pH in rumen cultures will result in increased accumulation of *trans*-$C_{18:1}$ (Qiu, Eastridge, Griswold and Firkins, 2000). Although no mechanism has been established for the increased accumulation of *trans*-$C_{18:1}$ fatty acids in the rumen when fish oils are fed, they may have an inhibitory effect on the hydrogenation of *trans*-octadecenoic acid similar to that of linoleic acid, as described earlier. Consistent with this, Wonsil, Herbein and Watkins (1994) demonstrated that feeding fish oil increased rumen production of *trans*-octadecenoic acids. The specific mechanism could involve an inhibition of the growth of biohydrogenating bacteria or a specific inhibition of the reductases in the biohydrogenation pathways. The effect of marine algae on concentration of CLA in milk fat is similar to the effect of fish oils (Franklin, Martin, Baer, Schingoethe and Hippen, 1999), and attributable to the increased formation of *trans*-11 $C_{18:1}$ in the rumen.

Alterations in feed intake have had variable effects on milk fat content of CLA. Restricting feed intake by about 30% resulted in milk fat concentration of CLA being increased in one study (Jiang *et al.*, 1996) and decreased in another (Stanton, Lawless, Kjellmer, Harrington, Devery, Connolly and Murphy, 1997). Timmen and Patton (1988) more severely restricted feed intake and observed a doubling in milk fat content of CLA. Alterations in feed intake would obviously affect substrate supply and change the rumen environment, and both factors would contribute to a change in the ruminal biohydrogenation process. In addition, underfeeding would increase the supply of CLA and *trans*-11 $C_{18:1}$ from mobilised body fat stores, and the magnitude of this increase would relate to the extent of the negative energy balance.

Ionophores inhibit the growth of gram-positive bacteria. Several gram-positive bacteria are involved in rumen biohydrogenation, including *Butyrivibrio fibrisolvens*. Using a continuous flow-through rumen fermenter, Fellner, Sauer and Kramer (1997) observed that ionophores inhibited linoleic acid biohydrogenation resulting in decreased stearic acid and increased monounsaturated $C_{18:1}$ concentrations in ruminal contents. However, including ionophores in dairy cattle diets has had variable effects (Table 14.2). Sauer, Fellner, Kinsman, Kramer, Jackson, Lee and Chen (1998) reported an increase, whereas Chouinard, Corneau, Kelly, Griinari and Bauman (1998) and Dhiman, Anand, Satter and Pariza (1999a) observed no effect on milk fat concentration of CLA in cows receiving monensin. Differences may relate to ruminal adaptations in which ionophore-resistant species replace ionophore-sensitive bacteria responsible for ruminal biohydrogenation. In addition, variable levels of dietary polyunsaturated fatty acids could explain the differences, although these data were not reported. In this case, bacterial populations involved in biohydrogenation may be altered, but substrate supply would be insufficient to allow the change in biohydrogenation to be expressed.

The effects of pasture on milk fat concentration of CLA have been described in a number of studies (Table 14.2). Generally, pasture feeding increases milk fat content of CLA, compared with feeding either a total mixed ration with a similar lipid content or conserved forages. However, forage lipid composition appears to only partly explain effects on milk fat content of CLA. It is apparent that the lipid content of pasture forage and other pasture components altering rumen biohydrogenation produce a synergistic effect. The highest reported level of enrichment of CLA in milk fat was produced by a diet with the combination of low forage:concentrate ratio and sufficient level of lipid substrate (Bauman *et al.* 2000a). This study also demonstrated that it is difficult to maintain a markedly elevated level of CLA enrichment for more than one week. The transient nature of the CLA response in this study was attributed to the *trans*-11/*trans*-10 shift (i.e. loss of substrate for endogenous synthesis of *cis*-9, *trans*-11 CLA) induced by feeding a basal diet with low ratio of forage:concentrate (Griinari, Nurmela, Dwyer, Barbano and Bauman, 1999a). When fish oil was fed in conjunction with extruded soyabeans (Table 14.2) the effect on milk fat concentration of CLA was greater than expected (Whitlock, Schingoethe, Hippen, Baer, Ramaswamy and Kasperson, 2000). Feeding of fish oil was associated with a decrease in milk fat concentration, so we can predict that fish oils induced a *trans*-11/*trans*-10 shift and responses in milk-fat CLA were not maximised.

Concentration of CLA in milk fat in lactating ruminants can be increased significantly by postruminal infusions or dietary supplements of rumen-protected CLA (Table 14.2). CLA have been protected by producing Ca salts (Giesy, Viswanadha, Hanson, Falen, McGuire, Skarie and Vinci, 1999; Medeiros, Oliveira, Aroeira, McGuire, Bauman and Lanna, 2000) or by formaldehyde treatment of CLA encapsulated in a casein matrix (Gulati, Kitessa, Ashes, Fleck, Byers, Byers and Scott, 2000). Apparent transfer efficiencies for CLA isomers varied from 22 to 34% for abomasal infusions in cows (Chouinard, Corneau, Barbano, Metzger and Bauman, 1999a; Chouinard, Corneau, Saebo and Bauman, 1999b) and from 36 to 41% in goats fed formaldehyde-protected CLA (Gulati *et al.*, 2000). Milk fat CLA was also increased by abomasal infusion of *trans*-11 $C_{18:1}$, the substrate for endogenous synthesis of CLA (Griinari *et al.*, 2000b). Partially hydrogenated vegetable oils contain variable amounts of *trans*-11 $C_{18:1}$, and a dietary supplement of Ca salts of partially-hydrogenated fatty acids from soyabean oil (17% *trans*-11 $C_{18:1}$) increased milk fat content of CLA more than two-fold (Griinari, Tesfa, Tuori and Holma, 1999b).

Based on the above summary, we can conclude that a successful strategy to produce and maintain a high level of CLA in milk fat involves three components: sufficient dietary supply of C_{18} polyunsaturated fatty acids to serve as substrate, maintenance of the biohydrogenation pathway which makes *trans*-11 $C_{18:1}$ as an intermediate, and inhibition of further hydrogenation of *trans*-11 $C_{18:1}$ in the biohydrogenation processes. However, in many cases described in the literature, at least one of the three components is missing and a substantial enrichment in CLA content of milk fat is not obtained. Even large amounts of plant oil supplements fail to increase milk fat CLA if the basal diet is resistant to alterations in rumen biohydrogenation (McGuire, McGuire, Guy, Sanchez, Shultz, Harrison, Bauman and Griinari, 1996). If the biohydrogenation pathway for formation of *trans*-11 $C_{18:1}$ cannot be maintained, then enhanced milk fat CLA will be transient (Bauman *et al.*, 2000a; Griinari *et al.*, 1999a). Preventing a shift in the biohydrogenation pathways from *trans*-11 $C_{18:1}$ to *trans*-10 $C_{18:1}$ by feeding buffers (Piperova, Teter, Bruckental, Sampugna, Mills, Yurawecz, Fritsche, Ku and Erdman, 2000) will not be a successful strategy to increase milk fat CLA, because this treatment also decreases rumen accumulation of *trans*-11 $C_{18:1}$ (Kalscheur *et al.* 1997). Dietary fish oil supplementation will effectively inhibit further reduction of *trans*-$C_{18:1}$ intermediates, but it also results in a shift in biohydrogenation pathways from *trans*-11 $C_{18:1}$ to *trans*-10 $C_{18:1}$ as demonstrated by decreased milk fat concentration when fish oils are fed (Lacasse and Ahnadi, 1998; Chilliard, Chardigny, Chabrot, Ollier, Sebedio and Doreau, 1999; Offer, Marsden, Dixon, Speake and Thacker, 1999; Griinari, Bauman, Chilliard, Peräjoki and Nurmela, 2000a).

Lipid metabolism

CLA EFFECTS

CLA also markedly affects lipid metabolism. Loor and Herbein (1998) were the first to show that a short-term (24 hour) infusion of CLA reduced milk fat secretion in dairy cows. Our initial studies also observed a dramatic reduction in milk fat yield; a 4-day abomasal infusion of 50 g/d of a CLA supplement to lactating cows resulted in a 52% reduction in milk fat yield (Chouinard *et al.*, 1999a). Effects appear to be specific for the fat component of milk. Milk yield and milk protein have generally been unchanged in short term studies with well-fed cows (see reviews by Bauman *et al.*, 2000b; Baumgard, Corl and Bauman, 2000a). Dietary supplements of CLA have also been shown to reduce the milk fat content of nursing women (Masters, McGuire and McGuire, 1999) and lactating sows (Harrell, Phillips, Jerome, Boyd, Dwyer and Bauman, 2000).

The CLA supplements used in the aforementioned studies contained a mixture of isomers. Based on the increase in milk fat content of *trans*-10 $C_{18:1}$ observed with diet-induced milk fat depression (Griinari *et al.*, 1997), we hypothesised that CLA isomers containing a *trans*-10 double bond were the cause of the milk fat reduction and examined this by abomasally infusing relatively pure CLA isomers (Figure 14.5). After 4 days of infusion to dairy cows, the *trans*-10, *cis*-12 CLA isomer resulted in over a 40 % reduction in milk fat percentage and yield whereas *cis*-9, *trans*-11 CLA had no effect (Baumgard *et al.*, 2000b). We recently extended this work and observed a curvilinear relationship between the increase in *trans*-10, *cis*-12 CLA dose and the reduction in milk fat yield (Baumgard *et al.*, 2000a). This isomer is a very potent inhibitor of milk fat synthesis with a dose of 3.5 g/d eliciting a 25% reduction in milk fat yield.

Dietary supplements of CLA also cause a reduction in body fat in growing animals as demonstrated for mice (Park, Albright, Liu, Storkson, Cook and Pariza, 1997; DeLany, Blohm, Truett, Scimeca and West, 1999), rats (Azain, Hausman, Sisk, Flatt and Jewell, 2000; Stangl, 2000), hamsters (Matsuba, Gavino, Tuchweber and Gavino, 2000) and pigs (Dugan, Aalhus, Schaefer and Kramer, 1997; Ostrowska, Muralitharan, Cross, Bauman and Dunshea, 1999). Although some have claimed CLA is a repartitioning agent that also increases protein accretion (Dugan *et al.*, 1997; Stangl, 2000; Whigham *et al.*, 2000), close examination revealed that the increases in carcass protein content were merely a consequence of the reduction in body fat (Baumgard *et al.*, 2000a). These investigations have used a commercial CLA supplement containing a mixture of isomers and, similar to lactating dairy cows, the reduced body fat appears to be due to the *trans*-10, *cis*-12 isomer, as shown by studies with growing mice (Park *et al.*, 1999). Overall, available studies demonstrate that CLA alters nutrient partitioning by reducing fat

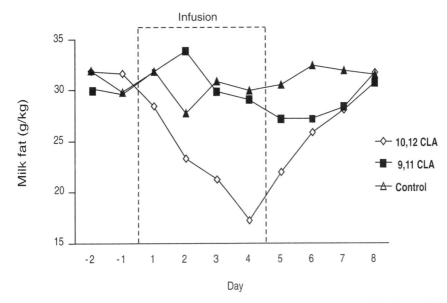

Figure 14.5 Temporal pattern of milk fat content during abomasal infusion of conjugated linoleic acid (CLA) isomers. Infusions were 10 g/d of *cis*-9, *trans*-11 CLA or *trans*-10, *cis*-12 CLA. Adapted from Baumgard *et al.* (2000b).

accretion, but there is little or no support for effects on protein accretion or for CLA functioning as a repartitioning agent.

Most investigations with growing animals have used a CLA dose of 0.5 to 2.0% of the diet and measured body fat content as the end point. Ostrowska *et al.* (1999) conducted the only study to examine CLA effects on accretion rates using growing pigs. They observed a linear decrease in fat accretion over the CLA dose range, with a 31% reduction at the highest dose (CLA isomers equalled 0.6% of diet). This amounts to a *trans*-10, *cis*-12 CLA dose of 1.7 g/kg diet for the growing pigs, and contrasts with a *trans*-10, *cis*-12 CLA dose of 0.16 g/kg dietary dry matter to get a similar reduction in milk fat yield of dairy cows. Overall, milk fat synthesis in cows is about 10 to 20-fold more sensitive to dietary CLA than observed for body-fat reduction in growing animals (Baumgard *et al.*, 2000a).

DIET-INDUCED MILK FAT DEPRESSION

An interesting aspect of CLA is its relationship to diet-induced milk fat depression (MFD). For over 50 years, animal scientists have sought to identify the physiological basis for the reduction in milk fat yield that occurs with dietary situations such as high concentrate/low roughage and supplements of plant and fish oils. Many theories were proposed and subsequently shown to be inadequate (see reviews by Doreau *et al.*, 1999; Bauman and Griinari, 2000a, 2000b). Davis and Brown (1970) were the first to suggest that rumen-derived *trans*-$C_{18:1}$ fatty acids might be responsible, based on the increase in milk fat content of *trans*-$C_{18:1}$ fatty acids that occurred during MFD. As the data base expanded, it became clear that MFD was related to an increase in the *trans*-$C_{18:1}$ content of milk fat across a wide range of diets. However, with improved analytic methods we discovered the relationship was specific for the milk fat content of *trans*-10 $C_{18:1}$ rather than total *trans*-$C_{18:1}$ fatty acids (Griinari *et al.*, 1997, 1998). *Trans*-10, *cis*-12 CLA is an intermediate in the formation of *trans*-10 $C_{18:1}$ from linoleic acid as previously described (Figure 14.3), and we have shown that milk fat concentrations of *trans*-10, *cis*-12 CLA and *trans*-10 $C_{18:1}$ are linearly related (Griinari *et al.*, 1999a). Although the effect of diet on ruminal production of *trans*-10, *cis*-12 CLA has not been quantified, we have observed a curvilinear relationship between the reduction in milk fat yield and the increase in milk fat content of *trans*-10, *cis*-12 CLA in cows fed various diets that cause MFD (Figure 14.6; Griinari *et al.*, 1999a, 2000a).

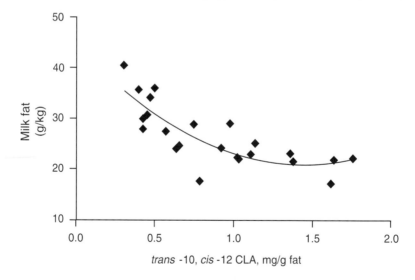

Figure 14.6 Relationship between milk fat content of *trans*-10, *cis*-12 conjugated linoleic acid (CLA) and milk fat concentration for cows fed a low-fibre diet supplemented with sunflower oil. Adapted from Griinari *et al.* (1999).

Based on the above results, we modified the *trans* theory and referred to it as the "biohydrogenation theory" of MFD (Bauman and Griinari, 2000a, 2000b). We chose this name because under certain conditions rumen biohydrogenation results in unique fatty acids that are potent inhibitors of milk fat synthesis. The two conditions needed for MFD across all diets, altered rumen microbial processes and the presence of dietary unsaturated fatty acids, are essential features of this theory. Thus, rumen microbial processes are altered with certain diets, so that a portion of biohydrogenation produces *trans*-10, *cis*-12 CLA from linoleic acid, and possibly related intermediates from linolenic acid and other polyunsaturated fatty acids. Our work has clearly shown the inhibitory effects of *trans*-10, *cis*-12 CLA, and other unique biohydrogenation intermediates might also be inhibitory, e.g. *trans*-10, *cis*-12, *cis*-15 conjugated octadecatrienoic acid, an intermediate formed in the isomerisation of the *cis*-9 double bond of α-linolenic acid (Figure 14.3). Based on other studies using various isomer enrichments of CLA, the c/t 8,10 CLA isomer also appears to inhibit milk fat synthesis (Chouinard *et al.*, 1999b).

Implications and applications

The list of beneficial effects of CLA continues to expand as additional types of animal models are examined. The anticarcinogenic effects are of special interest because they involve the predominant CLA isomer in milk fat. Only a few naturally occurring anticarcinogens have been identified and most are of plant origin and present at trace levels (National Research Council, 1996). *Cis*-9, *trans*-11 CLA is unique because it is potent at low concentrations and the only one present in food products derived from animals. To date there have been no epidemiological studies relating CLA to the risk of breast cancer or other diseases. Knekt and Järvinen (1999) recently summarised investigations relating cancer risk to the intake of dairy products and concluded results were inconsistent. They discussed the basis for the inconsistencies and emphasised the need to "carry out epidemiological studies on serum concentrations of conjugated linoleic fatty acid and breast cancer risk". The first study to examine this relationship found a significant preventive effect for CLA. The odds ratio for breast cancer in the highest quintile vs. the lowest was 0.4 (95% CI 0.2-0.9) for CLA in serum (Aro, Männistö, Salminen, Ovaskainen, Kataja and Uusitupa, 2000)

Cancer is a complex disease with many causes. The animal models typically used in CLA studies involve chemically induced carcinogenesis. We recently produced CLA-enhanced butter by dietary manipulation of dairy cows (Bauman *et al.*, 2000a) and compared it to two chemically synthesised CLA sources in a rat mammary cancer model (Ip *et al.*, 1999). Animals receiving diets fortified with CLA, including the CLA-enriched butter diet, had a lower tumour incidence and fewer mammary tumours than those receiving the control diet (Table 14.3). The *cis*-9, *trans*-11 CLA isomer represented 91% of the total CLA in the CLA-enriched butter, further demonstrating this isomer is anticarcinogenic. These results are particularly exciting because they are among the first to demonstrate that a natural anticarcinogen as a component of a natural food effectively reduces tumours in an animal cancer model.

Table 14.3. Incidence of mammary tumours in rats fed on diets containing differing concentrations and sources of CLA and treated with a chemical carcinogen[a].

Treatment group	Dietary CLA g/100 g	Tumor incidence	Tumor numbers
Control	0.1	28/30 (93%)	92
CLA-enriched butter	0.8	15/30 (50%)	43
CLA Supplement 1[b]	0.8	16/30 (53%)	46
CLA Supplement 2[c]	0.8	17/30 (57%)	48

[a]Rats were fed treatment diets for four weeks and then given a single dose of a carcinogen (methylnitrosourea). After carcinogen administration, rats were fed on a diet containing maize oil without CLA and were killed 24 weeks later. Adapted from Ip *et al.* (1999).
[b]9,11 CLA.
[c]Mixture of CLA isomers (8-10, 9-11, 10-12 and 11-13).

There have been an increasing number of lay articles about milk CLA. In addition to the well-recognised nutritional value of milk, an awareness of CLA can only improve the public perception of milk and other dairy products. Several US dairy cooperatives are exploring the possibility of a niche market for CLA-enriched dairy products. The fact that diet can markedly alter CLA, and the large variation among cows, demonstrates the clear potential to substantially enhance milk fat content of CLA. To date, we know of no niche marketing of CLA-enriched dairy products. However, in the long term the impact of CLA on public perception of dairy products may be of greatest importance to the dairy industry, and this should not be underestimated.

The effect of CLA on lipid metabolism also has many implications, but in this case they apply to the dairy cow herself. The discovery that *trans*-10, *cis*-12 CLA is an extremely potent inhibitor of milk fat synthesis and that it is also produced in the rumen under certain dietary conditions is significant. Diet-induced MFD was first described over a century ago and has been an active area of interest to producers and researchers for the last 50 years (Bauman and Griinari, 2000a, 2000b). Additional work will be needed to establish if the biohydrogenation theory represents a unifying concept that applies across diets to explain the basis for MFD. Several different fatty acid isomers could be involved and our results with *trans*-10, *cis*-12 CLA provide an example of the dramatic effect such isomers can have in regulating milk fat synthesis.

Dietary supplements of *trans*-10, *cis*-12 CLA to decrease milk fat synthesis could also be of commercial interest. In some instances, a reduction in milk fat yield would be of economic value to producers. Examples include countries having production quotas based on milk fat yield, and situations where production costs associated with the fat component of milk are greater than its value. Dietary CLA supplements could also be used to improve energy balance at specific times during the lactation cycle. For example, at the onset of lactation intake is inadequate to meet nutrient requirements, and this transition period is when cows are most susceptible to metabolic disorders and other health problems. Therefore, dietary supplements of *trans*-10, *cis*-12 CLA to reduce milk fat yield in the transition cow might be beneficial to animal well-being. In early lactation, energy status can also affect milk yield and reproduction. Reducing milk fat yield during this period could improve energy status, resulting in an increased milk yield, as well as a more rapid return to oestrus and improved fertility. A recent study by Giesy *et al.* (2000) observed a 6% post-peak increase in milk yield when cows in early lactation received a rumen-protected CLA supplement, but the study was too limited to evaluate reproduction variables. Synthesis of milk and milk protein can also be limited by energy availability under other dietary situations. Working with grazing cows, Medeiros *et al.* (2000) demonstrated that a dietary supplement of rumen-protected CLA reduced milk fat yield while increasing the milk protein content by 10% and milk yield by 13%. Key features in all of these examples are that CLA effects are specific for milk fat synthesis and reversible by omitting the diet supplement. We would emphasise that to date there have not been adequate long-term studies of dietary CLA supplementation of dairy cows. However, investigations are currently in progress in many countries to more fully evaluate each of the potential applications.

References

Aro, A., Männistö, S., Salminen, I., Ovaskainen, M.-L., Kataja, V. and Uusitupa, M. (2001) Inverse association between dietary and serum conjugated linoleic acid and risk of breast cancer in postmenopausal women. *Nutrition and Cancer*, (in press).

Azain, M.J., Hausman, D.B., Sisk, M.B., Flatt, W.P. and Jewell, D.E. (2000) Dietary conjugated linoleic acid reduces rat adipose tissue cell size rather than cell number. *Journal of Nutrition*, **130**, 1548-1554.

Banni, S. and Martin, J.C. (1998) Conjugated linoleic acid and metabolites. In *Trans Fatty Acids in Human Nutrition*, pp. 261-302. Edited by J.J. Sebedio and W.W. Christie. Oily Press, Dundee, Scotland.

Bauman, D.E., Barbano, D.M., Dwyer, D.A. and Griinari, J.M. (2000a) Technical Note: Production of butter with enhanced conjugated linoleic acid for use in biomedical studies with animal models. *Journal of Dairy Science*, **83**, 2422-2425.

Bauman, D.E., Baumgard, L.H., Corl, B.A. and Griinari, J.M. (2000b) Biosynthesis of conjugated linoleic acid in ruminants. *Proceedings of the American Society of Animal Science,* (1999). Available at: http://www.asas.org/jas/symposia/proceedings/0937.pdf

Bauman, D.E. and Griinari, J.M. (2000a) Historical perspective and recent developments in identifying the cause of diet-induced milk fat depression. *Proceedings Cornell Nutrition Conference for Feed Manufacturers,* pp. 191-202. Cornell University, Ithaca, NY, USA.

Bauman, D.E. and Griinari, J.M. (2000b) Regulation and nutritional manipulation of milk fat: low-fat milk syndrome. In *Biology of the Mammary Gland,* pp. 209-216. Edited by J.A. Mol and R.A. Clegg. Kluwer Academic/Plenum Publishers, New York, NY, USA.

Baumgard, L.H., Corl, B.A. and Bauman, D.E. (2000a) Effect of CLA isomers on fat synthesis during growth and lactation. *Proceedings Cornell Nutrition Conference for Feed Manufacturers,* pp. 180-190. Cornell University, Ithaca, NY, USA.

Baumgard, L.H., Corl, B.A., Dwyer, D.A., Saebo, A. and Bauman, D.E. (2000b) Identification of the conjugated linoleic acid isomer that inhibits milk fat synthesis. *American Journal of Physiology,* **278**, R179-R184.

Bickerstaffe, R. and Annison, E.F. (1970) The desaturase activity of goat and sow mammary tissue. *Comparative Biochemistry and Physiology,* **35**, 653-665.

Bickerstaffe, R. and Johnson, A.R. (1972) The effect of intravenous infusions of sterculic acid on milk fat synthesis. *British Journal of Nutrition,* **27**, 561-570.

Booth, R.G., Kon, S.K., Dann, W.J., and Moore, T. (1935) A study of seasonal variation in butter fat. A seasonal spectroscopic variation in the fatty acid fraction. *Biochemical Journal,* **29**, 133-137.

Chilliard, Y., Chardigny, J.M., Chabrot, J., Ollier, A., Sebedio, J.L. and Doreau, M. (1999) Effects of ruminal or postruminal fish oil supply on conjugated linoleic acid (CLA) content of cow milk fat. *Proceedings of the Nutrition Society,* **58**, 70A.

Chilliard, Y., Ferlay, A., Mansbridge, R.M. and Doreau, M. (2000) Ruminant milk fat plasticity: nutritional control of saturated, polyunsaturated, trans and conjugated fatty acids. *Annales de Zootechnie,* **49**, 181-205.

Chouinard, P.Y., Corneau, L., Barbano, D.M., Metzger, L.E. and Bauman, D.E. (1999a) Conjugated linoleic acids alter milk fatty acid composition and inhibit milk fat secretion in dairy cows. *Journal of Nutrition,* **129**, 1579-1584.

Chouinard, P.Y., Corneau, L., Butler, W.R., Chilliard, Y., Drackley, J.K. and Bauman, D.E. (2001) Effect of dietary lipid source on conjugated linoleic acid concentrations in milk fat. *Journal of Dairy Science,* (in press).

Chouinard, P.Y., Corneau, L., Kelly, M.L., Griinari, J.M. and Bauman, D.E. (1998) Effect of dietary manipulation on milk conjugated linoleic acid concentrations. *Journal of Dairy Science*, **81** (Suppl. 1) 233.

Chouinard, P.Y., Corneau, L., Saebo, A. and Bauman, D.E. (1999b) Milk yield and composition during abomasal infusion of conjugated linoleic acids in dairy cows. *Journal of Dairy Science*, **82**, 2737-2745.

Corl, B.A., Baumgard, L.H., Bauman, D.E. and Griinari, J.M. (2000) Role of Δ^9-desaturase in the synthesis of the anticarcinogenic isomer of CLA and other milk fatty acids. *Proceedings Cornell Nutrition Conference for Feed Manufacturers,* pp. 203-212. Cornell University, Ithaca, NY, USA.

Davis, C. L. and Brown, R.E. (1970) Low-fat milk syndrome. In *Physiology of Digestion and Metabolism in the Ruminant,* pp. 545-565. Edited by A.T. Phillipson. Oriel Press, Newcastle Upon Tyne, UK.

Dawson, R. C., Hemington, N. and Hazlewood, G.P. (1977) On the role of higher plant and microbial lipases in the ruminal hydrolysis of grass lipids. *British Journal of Nutrition,* **38**, 225-232.

DeLany, J.P., Blohm, F., Truett, A.A., Scimeca, J.A. and West, D.B. (1999) Conjugated linoleic acid rapidly reduces body fat content in mice without affecting energy intake. *American Journal of Physiology,* **276**, R1172-R1179.

Dhiman, T.R., Anand, G.R., Satter, L.D. and Pariza, M.W. (1999a) Conjugated linoleic acid content of milk from cows fed different diets. *Journal of Dairy Science*, **82**, 2146-2156.

Dhiman, T.R., Helmink, E.D., McMahon, D.J., Fife, R.L. and Pariza, M.W. (1999b) Conjugated linoleic acid content of milk and cheese from cows fed extruded oilseeds. *Journal of Dairy Science*, **82**, 412-419.

Dhiman, T.R., Satter, L.D., Pariza, M.W., Galli, M.P., Albright, K. and Tolosa, M.X. (2000) Conjugated linoleic acid (CLA) content of milk from cows offered diets rich in linoleic and linolenic acid. *Journal of Dairy Science*, **83**, 1016-1027.

Donovan, D.C., Schingoethe, D.J., Baer, R.J., Ryali, J., Hippen, A.R. and Franklin, S.T. (2000) Influence of dietary fish oil on conjugated linoleic acid and other fatty acids in milk fat from lactating cows. *Journal of Dairy Science*, **83**, 2620-2628.

Doreau, M., Chilliard, Y., Rulquin, H. and Demeyer, D.I. (1999) Manipulation of milk fat in dairy cows. In *Recent Advances in Animal Nutrition-1995,* pp. 81-109. Edited by P.C. Garnsworthy and D.J.A. Cole. Nottingham University Press, Nottingham, UK.

Dugan, M.E.R., Aalhus, J.L., Schaefer, A.L. and Kramer, J.K.G. (1997) The effect of conjugated linoleic acid on fat to lean repartitioning and feed conversion in pigs. *Canadian Journal of Animal Science,* **77**, 723-725.

Fellner, V., Sauer, F.D. and Kramer, J.K.G. (1997) Effect of nigericin, monensin, and tetronasin on biohydrogenation in continuous flow-through ruminal fermenters. *Journal of Dairy Science*, **80**, 921-928.

Fotouhi, N. and Jenkins, T.C. (1992) Ruminal biohydrogenation of linoleoyl methionine and calcium linoleate in sheep. *Journal of Animal Science*, **70**, 3607-3614.

Franklin, S.T., Martin, K.R., Baer, R.J., Schingoethe, D.J. and Hippen, A.R. (1999) Dietary marine algae (*Schizochytrium sp.*) increases concentrations of conjugated linoleic, docosahexaenoic and transvaccenic acids in milk of dairy cows. *Journal of Nutrition*, **129**, 2048-2052

Giesy, J.G., Viswanadha, S., Hanson, T.W., Falen, L.R., McGuire, M.A., Skarie, C.H. and Vinci, A. (1999) Effects of calcium salts of conjugated linoleic acid (CLA) on estimated energy balance in Holstein cows early in lactation. *Journal of Dairy Science*, **82** (Suppl. 1), 74.

Griinari, J. M. and Bauman, D.E. (1999) Biosynthesis of conjugated linoleic acid and its incorporation into meat and milk in ruminants. In *Advances in Conjugated Linoleic Acid Research*, pp. 180-200. Edited by M.P. Yurawecz, M.M. Mossoba, J.K.G. Kramer, M.W. Pariza and G.J. Nelson. AOCS Press, Champaign, IL, USA.

Griinari, J.M., Bauman, D.E., Chilliard, Y., Peräjoki, P. and Nurmela, K. (2000a) Dietary influences on conjugated linoleic acids (CLA) in bovine milk fat. *Meeting of the European Section of the American Oil Chemists' Society*, pp. 87.

Griinari, J. M., Chouinard, P.Y. and Bauman, D.E. (1997) *Trans* fatty acid hypothesis of milk fat depression revised. *Proceedings Cornell Nutrition Conference for Feed Manufacturers*, pp. 208-216, Cornell University, Ithaca, NY, USA.

Griinari, J.M., Corl, B.A., Lacy, S.H., Chouinard, P.Y., Nurmela, K.V.V. and Bauman, D.E. (2000b) Conjugated linoleic acid is synthesized endogenously in lactating dairy cows by Δ^9-desaturase. *Journal of Nutrition*, **130**, 2285-2291.

Griinari, J.M., Dwyer, D.A., McGuire, M.A., Bauman, D.E., Palmquist, D.L. and Nurmela, K.V.V. (1998) *Trans*-octadecenoic acids and milk fat depression in lactating dairy cows. *Journal of Dairy Science*, **81**, 1251-1261.

Griinari, J.M., Nurmela, K., Dwyer, D.A., Barbano, D.M. and Bauman, D.E. (1999a) Variation of milk fat concentration of conjugated linoleic acid and milk fat percentage is associated with a change in ruminal biohydrogenation. *Journal of Animal Science*, **77** (Suppl. 1), 117-118.

Griinari, J.M., Tesfa, A.T., Tuori, M. and Holma, M. (1999b) Effect of feeding graded levels of partially hydrogenated soybean oil fatty acids to lactating dairy cows on concentrations of conjugated linoleic acid (CLA) in milk fat. *Journal of Dairy Science,* **82** (Suppl. 1), 84.

Gulati, S.K., Kitessa, S.M., Ashes, J.R., Fleck, E., Byers, E.B., Byers, Y.G. and Scott, T.W. (2000) Protection of conjugated linoleic acids from ruminal hydrogenation and their incorporation into milk fat. *Animal Feed Science Technology,* **86**, 139-148.

Harfoot, C.G. and Hazlewood, G.P. (1988) Lipid metabolism in the rumen. In *The Rumen Microbial Ecosystem,* pp. 285-322. Edited by P.N. Hobson. Elsevier Applied Science Publishers, London, UK.

Harfoot, C.G., Noble, R.C. and Moore, J.H. (1973) Factors influencing the extent of biohydrogenation of linoleic acid by rumen micro-organisms *in vitro*. *Journal of the Science of Food and Agriculture,* **24**, 961-970.

Harrell, R.J., Phillips, O., Jerome, D.L., Boyd, R.D., Dwyer, D.A. and Bauman, D.E. (2000) Effects of conjugated linoleic acid on milk composition and baby pig growth in lactating sows. *Journal of Animal Science,* **77** (Suppl. 1) , 137-138.

Hazlewood, G.P., Kemp, P., Lander, D. and Dawson, R.M.C. (1976) C_{18} unsaturated fatty acid hydrogenation patterns of some rumen bacteria and their ability to hydrolyse exogenous phopholipid. *British Journal of Nutrition,* **35**, 293-297.

Ip, C., Banni, S., Angioni, E., Carta, G., McGinley, J., Thompson, H.J., Barbano, D. and Bauman, D. (1999) Conjugated linoleic acid-enriched butter fat alters mammary gland morphogenesis and reduces cancer risk in rats. *Journal of Nutrition,* **129**, 2135-2142.

Ip, C., Ip, M.M., Loftus, T., Shoemaker, S. and Shea-Eaton, W. (2000) Induction of apoptosis by conjugated linoleic acid in cultured mammary tumor cells and premalignant lesions of the rat mammary gland. *Cancer Epidemiology, Biomarkers and Prevention,* **9**, 689-696.

Ip, C., Scimeca, J.A. and Thompson, H.J. (1994) Conjugated linoleic acid: a powerful anticarcinogen from animal fat sources. *Cancer,* **74**, 1050-1054.

Jahreis, G., Fritsche, J. and Steinhart, H. (1997) Conjugated linoleic acid in milk fat: high variation depending on production system. *Nutrition Research,* **17**, 1479-1484.

Jeffcoat, R. and Pollard, M.R. (1977) Studies on the inhibition of the desaturases by cyclopropenoid fatty acids. *Lipids,* **12**, 480-485.

Jenkins, T.C. (1993) Lipid metabolism in the rumen. *Journal of Dairy Science,* **76**, 3851-3863.

Jiang, J., Bjöerck, L., Fonden, R. and Emanuelson, M. (1996) Occurrence of conjugated *cis*-9, *trans*-11 octadecadienoic acid in bovine milk: effects of feed and dietary regimen. *Journal of Dairy Science,* **79**, 438-445.

Jones, D.F., Weiss, W.P. and Palmquist, D.L. (2000) Short communication: Influence of dietary tallow and fish oil on milk composition, *Journal of Dairy Science*, **83**, 2024-2026.

Kalscheur, K.F., Teter, B.B., Piperova, L.S. and Erdman, R.A. (1997) Effect of dietary forage concentration and buffer addition on duodenal flow of *trans*-$C_{18:1}$ fatty acids and milk fat production in dairy cows. *Journal of Dairy Science,* **80**, 2104-2114.

Keeney, M. (1970) Lipid metabolism in the rumen. In *Physiology of Digestion and Metabolism in the Ruminant,* pp. 489-503. Edited by A.T. Phillipson. Oriel Press, Newcastle upon Tyne, UK.

Kellens, M.J., Goderis, H.L. and Tobback, P.P. (1986) Biohydrogenation of unsaturated fatty acids by a mixed culture of rumen microorganisms. *Biotechnology Bioengineering,* **28**, 1268-1276.

Kelly, M.L. and Bauman, D.E. (1996) Conjugated linoleic acid: a potent anticarcinogen found in milk fat. *Proceedings Cornell Nutrition Conference for Feed Manufacturers,* pp. 124-133, Cornell University, Ithaca, NY, USA.

Kelly, M.L., Berry, J.R., Dwyer, D.A., Griinari, J.M., Chouinard, P.Y., Van Amburgh, M.E. and Bauman, D.E. (1998a) Dietary fatty acid sources affect conjugated linoleic acid concentrations in milk from lactating dairy cows. *Journal of Nutrition,* **128**, 881-885.

Kelly, M.L., Kolver, E.S., Bauman, D.E., Van Amburgh, M.E. and Muller, L.D. (1998b) Effect of intake of pasture on concentrations of conjugated linoleic acid in milk of lactating dairy cows. *Journal of Dairy Science,* **81**, 1630-1636.

Kemp, P. and Lander, D.J. (1984) Hydrogenation *in vitro* of α-linolenic acid to stearic acid by mixed cultures of pure strains of rumen bacteria. *Journal of General Microbiology,* **130**, 527-533.

Kemp, P., White, R.W. and Lander, D.J. (1975) The hydrogenation of unsaturated fatty acids by five bacterial isolates from the sheep rumen, including a new species. *Journal of General Microbiology,* **90**, 100-114.

Kepler, C.R. and Tove, S.B. (1967) Biohydrogenation of unsaturated fatty acids: III. Purification and properties of a linoleate Δ^{12}-*cis*, Δ^{10}-*trans*-isomerase from *Butyrivibrio fibrisolvens. Journal of Biological Chemistry,* **242**, 5686-5692.

Kepler, C.R., Tucker, W.P. and Tove, S.B. (1970) Biohydrogenation of unsaturated fatty acids. IV. Substrate specificity and inhibition of linoleate Δ^{12}-*cis*, Δ^{11}-*trans* isomerase from *Butyrivibrio fibrisolvens. Journal of Biological Chemistry,* **245**, 3612-3620.

Kinsella, J.E. (1972) Stearyl CoA as a precursor of oleic acid and glycerolipids in mammary microsomes from lactating bovine: possible regulatory step in milk triglyceride synthesis. *Lipids,* **7**, 349-355.

Knekt, P. and Järvinen, R. (1999) Intake of dairy products and breast cancer risk. In *Advances in Conjugated Linoleic Acid Research*, pp. 444-470. Edited by M.P. Yurawecz, M.M. Mossoba, J.K.G. Kramer, M.W. Pariza and G.J. Nelson. AOCS Press, Champaign, IL, USA.

Lacasse, P. and Ahnadi, C.E. (1998) Feeding protected and unprotected fish oil to dairy cows: I Effect on animal performance. *Journal of Dairy Science,* **81** (Suppl. 1), 231.

Lawless, F., Murphy, J.J., Harrington, D., Devery, R. and Stanton, C. (1998) Elevation of conjugated *cis*-9, *trans*-11-octadecadienoic acid in bovine milk because of dietary supplementation. *Journal of Dairy Science, 81*, 3259-3267.

Leat, W.M.F., Kemp, P., Lysons, R.J. and Alexander, T.J.L. (1977) Fatty acid composition of depot fats from gnotobiotic lambs. *Journal of Agricultural Science, Cambridge,* **88**, 175-179.

L'Esperance, A., Bernier, J.F., Bauman, D.E., Corl, B.A. and Chouinard, P.Y. (2000) Effect of plasma insulin concentrations on milk fatty acid profile in dairy cows. *Journal of Dairy Science*, **83** (Suppl. 1), 168.

Loor, J.J. and Herbein, J.H. (1998) Exogenous conjugated linoleic acid isomers reduce bovine milk fat concentration and yield by inhibiting de novo fatty acid synthesis. *Journal of Nutrition,* **128**, 2411-2419.

Masters, N., McGuire, M.A. and McGuire, M.K. (1999) Conjugated linoleic acid supplementation and milk fat content in humans. *Federation of the American Society of Experimental Biology Journal*, **13**, A697.

Matsuba, K., Gavino, G., Tuchweber, B. and Gavino, V. (2000) Pure dietary cis-9, trans-11-octadecadienoic and an isomeric mixture of conjugated linoleic acids have different effects in hamsters. *Federation of the American Society of Experimental Biology Journal*, **14**, A726.

McGuire, M.A. and McGuire, M.K. (2000) Conjugated linoleic acid (CLA): A ruminant fatty acid with beneficial effects on human health. *Proceeedings, American Society of Animal Science (1999)* Available at: http://www.asas.org/jas/symposia/proceedings/0938.pdf.

McGuire, M.A., McGuire, M.K., Guy, M.A., Sanchez, W.K., Shultz, T.D., Harrison, Y., Bauman, D.E. and Griinari, J.M. (1996) Short-term effect of dietary lipid concentration on content of conjugated linoleic acid (CLA) in milk from dairy cattle. *Journal of Animal Science*, **74** (Suppl. 1), 266.

Medeiros, S.R., Oliveira, D.E., Aroeira, L.J.M., McGuire, M.A., Bauman, D.E. and Lanna, D.P.D. (2000) The effect of long-term supplementation of conjugated linoleic acid (CLA) to dairy cows grazing tropical pasture. *Journal of Dairy Science*, **83** (Suppl. 1) , 169.

Medrano, J.F. and DePeters, E.J. (2000) Genetic modification of the composition of milk fat. *Proceedings International Conference on the Biology of the Mammary Gland.* pp. 28. Tours, France.

Mir, Z., Goonewardene, L.A., Okine, E., Jaegar, S. and Scheer, H.D. (1999) Effect of feeding canola oil on constituents, conjugated linoleic acid (CLA) and long chain fatty acids in goats milk. *Small Ruminant Research,* **33**, 137-143.

Milner, J. (1999) Functional foods and health promotion. *Journal of Nutrition,* **129**, 1395S-1397S.

National Research Council. (1996) *Carcinogens and Anticarcinogens in the Human Diet.* National Academy Press, Washington, DC, USA.

Ntambi, J.M. (1995) The regulation of stearoyl-CoA desaturase (SCD). *Progress in Lipid Research,* **34**, 139-150.

Ntambi, J.M., Choi, Y. and Kim, Y.C. (1999) Regulation of stearoyl-CoA desaturase by conjugated linoleic acid. In *Advances in Conjugated Linoleic Acid Research,* pp. 340-347. Edited by M.P. Yurawecz, M.M. Mossoba, J.K.G. Kramer, M.W. Pariza and G.J. Nelson. AOCS Press, Champaign, IL, USA.

Offer, N.W., Marsden, M., Dixon, J., Speake, B.K. and Thacker, F.F. (1999) Effect of dietary fat supplements on levels of n-3 polyunsaturated fatty acids, *trans* fatty acids and conjugated linoleic acid in bovine milk. *Journal of Animal Science,* **69**, 613-625.

Ostrowska, E., Muralitharan, M., Cross, R.F., Bauman, D.E. and Dunshea, F.R. (1999) Dietary conjugated linoleic acids increase lean tissue and decrease fat deposition in growing pigs. *Journal of Nutrition,* **129**, 2037-2042.

Palmquist, D.L., Beaulieu, A.D. and Barbano, D.M. (1993) Feed and animal factors influencing milk fat composition. *Journal of Dairy Science,* **76**, 1753-1771.

Pariza, M.W. (1999) The biological activities of conjugated linoleic acid. In *Advances in Conjugated Linoleic Acid Research,* pp. 12-20. Edited by M.P. Yurawecz, M.M. Mossoba, J.K.G. Kramer, M.W. Pariza and G.J. Nelson. AOCS Press, Champaign, IL, USA.

Park, Y., Albright, K.J., Liu, W., Storkson, J.M., Cook, M.E. and Pariza, M.W. (1997) Effect of conjugated linoleic acid on body composition in mice. *Lipids,* **32**, 853-858.

Park, Y., Storkson, J.M., Albright, K.J., Liu, W. and Pariza, M.W. (1999) Evidence that the *trans*-10, *cis*-12 isomer of conjugated linoleic acid induces body composition changes in mice. *Lipids,* **34**, 235-241.

Parodi, P.W. (1977) Conjugated octadecadienoic acids of milk fat. *Journal of Dairy Science,* **60**, 1550-1553.

Parodi, P.W. (1997) Cows' milk fat components as potential anticarcinogenic agents. *Journal of Nutrition,* **127**, 1055-1060.

Piperova, L.S., Teter, B.B., Bruckental, I., Sampugna, J., Mills, S.E., Yurawecz, M.P., Fritsche, J., Ku, K. and Erdman, R.A. (2000) Mammary lipogenic enzyme activity, *trans* fatty acids and conjugated linoleic acids are altered in lactating dairy cows fed a milk fat-depressing diet. *Journal of Nutrition,* **130**, 2658-2574.

Polan, C.E., McNeill, J.J. and Tove, S.B. (1964) Biohydrogenation of unsaturated fatty acids by rumen bacteria. *Journal of Bacteriology,* **88**, 1056-1064.

Precht, D. and Molkentin, J. (1997) Effect of feeding on conjugated *cis*-Δ9, *trans*-Δ11 octadecadienoic acid and other isomers of linoleic acid in bovine milk fats. *Nahrung,* **41**, 330-335.

Qiu, X., Eastridge, M.L., Griswold, K.E. and Firkins, J.L. (2000) Effects of solid passage rate, pH, and level of linoleic acid on the production of *cis*-9, *trans*-11 octadecadienoic acid (CLA) in continuous culture. *Journal of Dairy Science,* **83** (Suppl. 1), 285.

Riel, R.R. (1963) Physico-chemical characteristics of Canadian milk fat unsaturated fatty acids. *Journal of Dairy Science*, **46**, 102-106.

Ryder, J., Bauman, D.E., Portocarrero, C., Song, X., Yu, M., Barbano, D.M., Zierath, J. and Houseknecht, K.L. (1999) Anti-diabetic effects of dietary conjugated linoleic acid (CLA): Isomer-specific effects on glucose tolerance and skeletal muscle glucose transport. *Journal of Animal Science*, **77** (Suppl. 1), 157.

Sauer, F.D., Fellner, V., Kinsman, R., Kramer, J.K.G., Jackson, H.A., Lee, A.J. and Chen, S. (1998) Methane output and lactation response in Holstein cattle with monensin or unsaturated fat added to the diet. *Journal of Animal Science,* **76**, 906-914.

Scimeca, J.A. (1999) Cancer inhibition in animals. In *Advances in Conjugated Linoleic Acid Research,* pp. 340-347. Edited by M.P. Yurawecz, M.M. Mossoba, J.K.G. Kramer, M.W. Pariza and G.J. Nelson. AOCS Press, Champaign, IL, USA.

Sehat, N., Kramer, J.K.G., Mossoba, M.M., Yurawecz, M.P., Roach, J.A.G., Eulitz, K., Morehouse, K.M. and Ku, Y. (1998) Identification of conjugated linoleic acid isomers in cheese by gas chromatography, silver ion high performance liquid chromatography and mass spectral reconstructed ion profiles. *Lipids*, **33**, 963-971.

Shantha, N.C, Ram, L.N., O'Leary, J., Hicks, C.L. and Decker, E.A. (1995) Conjugated linoleic acid concentrations in dairy products as affected by processing and storage. *Journal of Food Science,* **60**, 695-697.

Shingfield, K.J., Ahvenjärvi, S., Toivonen, V. and Griinari, J.M. (2000) Conjugated linoleic acid biosynthesis in the lactating dairy cow. *Meeting of the European Section of the American Oil Chemists' Society,* pp. 35.

Solomon, R., Chase, L.E., Ben-Ghedalia, D. and Bauman, D.E. (2000) The effect of nonstructural carbohydrate and addition of full fat extruded soybeans on the concentration of conjugated linoleic acid in the milk fat of dairy cows. *Journal of Dairy Science,* **83**, 1322-1329.

Stangl, G.I. (2000) Conjugated linoleic acids exhibit a strong fat-to-lean partitioning effect, reduce serum VLDL lipids and redistribute tissue lipids in food-restricted rats. *Journal of Nutrition,* **130**, 1140-1146.

Stanton, C., Lawless, F., Kjellmer, G., Harrington, D., Devery, R., Connolly, J.F. and Murphy, J. (1997) Dietary influences on bovine milk *cis*-9, *trans*-11 conjugated linoleic acid content. *Journal of Food Science,* **62**, 1083-1086.

Sutton, J.D. (1989) Altering milk composition by feeding. *Journal of Dairy Science,* **72**, 2801-2814.

Tanaka, K. and Shigeno, K. (1976) The biohydrogenation of linoleic acid by rumen micro-organisms. *The Japanese Journal of Zootechnology and Science,* **47**, 50-53.

Tesfa, A.T., Tuori, M. and Syrjälä-Qvist, L. (1991) High rapeseed oil feeding to lactating dairy cows and its effect on milk yield and composition in ruminants, *Finnish Journal of Dairy Science,* **49**, 65-81.

Timmen, H. and Patton, S. (1988) Milk fat globules: fatty acid composition, size and in vivo regulation of fatty liquidity. *Lipids,* **23**, 685-689.

Tocher, D.R., Leaver, M.J., and Hodgson, P.A. (1998) Recent advances in the biochemistry and molecular biology of fatty acyl desaturases. *Progress in Lipid Research,* **37**, 73-117.

Ward, R.J., Travers, M.T., Richards, S.E., Vernon, R.G., Salter, A.M., Buttery, P.J. and Barber, M.C. (1998) Stearoyl-CoA desaturase mRNA is transcribed from a single gene in the ovine genome. *Biochimica et Biophysica Acta,* **1391**, 145-156.

Whigham, L.D., Cook, M.E. and Atkinson, R.L. (2000) Conjugated linoleic acid: implications for human health. *Pharmacological Research,* **42**, 503-510.

Whitlock, L.A., Schingoethe, D.J., Hippen, A.R., Baer, R.J., Ramaswamy, N. and Kasperson, K.M. (2000) Milk production and composition from cows fed fish oil, extruded soybeans or their combination. *Journal of Dairy Science,* **83** (Suppl. 1), 134.

Wonsil, B.J., Herbein, J.H., Watkins, B.A., (1994) Dietary and ruminally derived *trans*-18:1 fatty acids alter bovine milk lipids. *Journal of Nutrition,* **124**, 556-565.

Zegarska, Z., Paszczyk, B. and Borejszo, Z. (1996) *Trans* fatty acids in milk fat. *Polish Journal of Food and Nutrition Sciences,* **5**, 89-96.

15

DEVELOPMENTS IN RUMEN FERMENTATION - THE SCIENTIST'S VIEW

C. JAMIE NEWBOLD, COLIN S. STEWART AND R. JOHN WALLACE

The Rowett Research Institute, Bucksburn, Aberdeen AB21 9SB, UK

Introduction

The importance of rumen fermentation in governing the response of productive ruminants to dietary changes is well accepted. Over the last five decades considerable efforts have been made to understand and ultimately manipulate the rumen microbial population to improve animal productivity (Nagaraja, Newbold, Van Nevel and Demeyer, 1997). As a result the rumen is one of the better-studied microbial ecosystems and it might be argued that little benefit would arise from continued intensive study of it. However, we contend, and intend to demonstrate, that ongoing advances will continue to identify new ways in which the rumen microbes might be manipulated to alter ruminant production.

The framework within which the work of the rumen microbiologist can be applied to practical agriculture is changing. There is growing resistance to the use of antibiotic growth promoters within the food chain and the short-term possibility of using genetically modified organisms, either by modifying the plants which the animal eats or by introducing modified "superbugs" into the rumen, seems unlikely in the face of current consumer concern. Indeed the technologies most likely to be adopted are those that are perceived to be based on natural or green products. It is however important to remember that just because a product might be based on naturally occurring microbial or plant extracts that is no guarantee of either its efficacy or its safety.

The targets for manipulation are also changing and no longer can the productivity of the animal be considered in isolation. There is a growing awareness of the health, safety and environmental issues associated with animal agriculture. It will be beholden to the rumen microbiologist to address these concerns while simultaneously trying to maintain and improve the productivity and profitability of ruminant agriculture. Three examples will be used to demonstrate how rumen microbiology might help in the production of healthy, safe and environmentally acceptable products from ruminants to the ultimate benefit of the farmer and consumer. These are:

- The role of rumen microbes in the production of conjugated linoleic acid,
- The role of the rumen as a barrier preventing the spread of pathogens at a farm level, and
- The manipulation of methane production by rumen microbes.

Conjugated linoleic acids – good news at last for the livestock industry

The possible health benefits of conjugated linoleic acids (CLA) has been covered elsewhere in this volume (Bauman, 2001). It is now important to capitalise on these findings, by understanding how CLA are formed and how their concentration in ruminant products can be enhanced.

OCCURRENCE AND STRUCTURE OF CLA

Many foods, particularly vegetable oils, contain abundant 18:2 acids (Kritchevsky, 2000). Total 18:2 fatty acids are abundant in sunflower oil and salmon, for example, but the CLA content is less than 0.4% of the 18:2 acids present. In ruminant meats and dairy products, on the other hand, 18:2 acids are relatively plentiful and up to 25% are present as CLA. The total CLA content of ruminant products is virtually always higher than in non-ruminant and vegetable products (Table 15.1). It is worth noting that the data are usually presented, as in Table 15.1, as a proportion of fat, rather than total product; thus, one might obtain 1 g of CLA by eating 200 g of cheese, but to obtain the equivalent quantity from turkey, which was the only non-ruminant product with high CLA content, one would have to eat perhaps 4 kg of meat.

Thus, the main dietary source of CLA is ruminant products, and in order for humans to obtain a significant intake of CLA in natural foods, these products are the best ones available.

INFLUENCE OF DIET ON CLA CONTENT OF MEAT AND DAIRY PRODUCTS

It is clear from the limited number of studies carried out to date that altering the diet of animals can have a major influence on the CLA content of meat and dairy

Table 15.1 Proportions of conjugated linoleic acid (CLA) in fat associated with different foodstuffs

Ruminant products			Other products		
	CLA (mg/g fat)	cis-9,trans-11 (g/100g CLA)		CLA (mg/g fat)	cis-9,trans-11 (g/100g CLA)
Raw meat/fish					
Beef	2.9-4.3	79-85	Fish	0.3-0.6	n.d.
Veal	2.7	84	Chicken	0.9	84
Lamb	5.6	92	Pork	0.6	82
			Turkey	2.5	76
Oils/fats/dairy foods					
Milk	5.4-7.0	82-92	Egg yolk	0.6	82
Cheese	2.9-7.1	88-95	Olive oil	0.2	40
Butter	4.7	88	Corn oil	0.2	37
Cream	4.6	90	Sunflower oil	0.4	38
Beef tallow	2.6	84			
Infant foods					
Veal	6.8	78	Ham	1.3	68
Beef	6.8	74	Chicken	0.7	73
Lamb	8.8	81	Turkey	1.8	81

products, and supplementing the diet with fats in the form of oilseeds or vegetable oils also affects the CLA content of ruminant products. Among unsupplemented diets, fresh pasture causes the CLA content of products to be highest. Dhiman, Anand, Satter and Pariza (1999a) found that cows grazing pasture had 500% more CLA in milk fat (22 mg CLA/g) than cows fed other typical dairy diets; the fresh pasture was important, because feeding grass hay gave only 34% of the CLA in milk fat that was found in grazing animals. Linoleic and linolenic acids form the majority of fatty acids in fresh forages (Harfoot, 1981), while preservation of forages, particularly ensilage, causes the loss of 18:2 and 18:3 fatty acids (Noble, 1981). In the silo, micro-organisms live in an anaerobic environment comparable to the rumen, which means that disposing of reducing power (hydrogen or hydrogen equivalents) is a priority activity and, as in the rumen, unsaturated fatty acids are reduced. Thus, fresh forages contain more linoleic acid than other feedstuffs. Feeding fresh forage rather than conserved materials or concentrates provides more linoleic acid as a precursor of CLA and is a simple and effective means of enhancing the CLA content of ruminant products.

The feeding of fresh forage is not always possible, however, so judicious supplementation of the diet to increase the flow of CLA to the animal would be beneficial under many circumstances. Supplementation of the diet with fats increased the CLA content of milk (Stanton, Lawless, Kjellmer, Harrington, Devery, Connolly and Murphy. 1997; Kelly, Berry, Dwyer, Griinari, Chouinard, Van Amburgh and Bauman, 1998). Dhiman, Helmink, McMahon, Fife and Pariza (1999b) increased the CLA content of milk and cheese by 109% (to only 7 mg/g of fatty acids, however) when full-fat extruded soya beans were fed with a conserved forage/concentrate diet, and 77% when cottonseeds were fed. Roasted whole soya beans, presumably with the lipid component protected to some degree from lipolysis and subsequent saturation, increased the CLA content of milk in Holstein cows from 8.1 to 10.3 mg/g fatty acids in comparison with tallow (Sol Morales, Palmquist and Weiss, 2000a). Soyabean oil at 40 g/kg of the diet gave an increase from 8 to 21 mg/g in the CLA content of milk fat (Dhiman, Satter, Pariza, Galli, Albright, and Tolosa, 2000). Canola oil achieved an even greater increase in goats, increasing milk CLA from 10 to 32 mg/g (Mir, Goonewardene, Okine, Jaegar and Scheer, 1999). Thus the last two supplementations increased the CLA content of milk above the concentration found in forage diets. Protecting dietary CLA from hydrogenation in the rumen by encapsulation might increase the CLA content even more: incorporating CLA in a formaldehyde-casein protein matrix increased CLA flux to the small intestine of sheep substantially, to about 50 mg/g of lipid (Gulati, Kitessa, Ashes, Fleck, Byers, Byers, and Scott, 2000). Thus, CLA–protected supplements appear to hold much promise for increasing CLA in the final products, so long as the economics of the supplementation balances the added value of the products.

CLA FROM RUMINAL FERMENTATION – BIOCHEMISTRY AND MICROBIOLOGY

The best, most sustainable way of increasing the CLA content of ruminant products is to induce ruminal micro-organisms to produce more CLA in their lipids. It is therefore important to understand which micro-organisms are involved and how they form and then metabolise CLA.

In complete contrast to the recent widespread interest in the effects of CLA on health and their concentration in animal products, the metabolism of fatty acids in the rumen has received little attention for the last 20 years or so. Some outstanding research was produced in the 1960s and early 1970s, describing the metabolism and fate of lipids and fatty acids by ruminal micro-organisms. The bacteria responsible for lipolysis and hydrogenation of unsaturated fatty acids were identified, and the fundamental principles of fatty acid metabolism were established. Since then, the number of papers on the lipid metabolism of rumen micro-organisms has dwindled almost to zero. Indeed, a recent review (Harfoot and Hazlewood,

1997) concluded that "we know little more now about hydrogenators and their numbers, ecology and metabolic diversity than we knew in 1988". The same could also have been said for the previous decade. It is time to reinvestigate fatty acid metabolism by ruminal micro-organisms, paying new attention to CLA.

The principal interest of researchers in the 1960s and 1970s was in 'biohydrogenation' - understanding the mechanism by which unsaturated fatty acids, such as linolenic (*cis*-9, *cis*-12, *cis*-15, 18:3), linoleic (*cis*-9, *cis*-12, 18:2), and oleic (*cis*-9, 18:1) acids were hydrogenated by ruminal micro-organisms. CLA were occasionally observed to be formed in some bacterial cultures, but only in passing, because the health implications of CLA were unknown at the time. The first study linking fatty acid hydrogenation to a single species was that of Polan, McNeill and Tove (1964), in which the common ruminal bacterium, *Butyrivibrio fibrisolvens*, was identified as being able to to hydrogenate linoleic acid more rapidly than other species. One of the authors of that paper, S.B. Tove, subsequently pursued detailed research into the mechanisms of biohydrogenation in *B. fibrisolvens*, producing a series of excellent papers in which the enzymes responsible for the conversion of linoleic acid to CLA were characterised and purified, and the intermediates and cofactors involved in biohydrogenation identified. The papers can be traced back from Hughes and Tove (1982).

The first mention of CLA was as intermediates in the biohydrogenation of linoleic acid by *B. fibrisolvens* (Polan et al. 1964; Kepler, Hirons, McNeill, and Tove, 1966). The key word here is 'intermediates'. Normally, when *B. fibrisolvens* grows anaerobically, the CLA formed are immediately hydrogenated to 18:1. It is only when the hydrogenation is inhibited by air that CLA are seen in any quantity (Polan et al. 1964). In other words, although they are vital health-promoting nutrients, the CLA are not major end-products or cell components in the ruminal bacteria that produce them. This mechanism of biohydrogenation occurs to dispose of hydrogen, not to form fatty acids for their or our benefit!

The other main research effort was carried out at Babraham, Cambridge, where a much wider range of biohydrogenating bacteria was isolated. *B. fibrisolvens* was among these isolates (Hazlewood and Dawson, 1979), but many others were shown to hydrogenate linoleic acid (Harfoot and Hazlewood, 1997). They were divided into two Groups, A and B, by Kemp and Lander (1984), based on the ability of the isolates to take the reduction beyond the 18:1 step to stearic acid, 18:0. *B. fibrisolvens* fell into Group A. Type B bacteria appeared to be atypical of the most common ruminal species, with two of the three characterised isolates being '*Fusocillus*' (White, Kemp and Dawson, 1970; Kemp, White and Lauderet, 1975; Hazlewood et al. 1976), a genus that is not found in modern microbiological texts. It seems that none of the Group B bacteria survive (G.P. Hazlewood, personal communication), so they are not available for identification by modern molecular phylogenetic techniques. The area should be revisited, in order to determine exactly which bacterial species carry out the biohydrogenation of fatty acids.

CLA were found as minor or transient products in several studies in addition to those with *B. fibrisolvens*. *Cis*-9, *trans*-11, 18:2, occurred with *Borrelia* (Sachan and Davis, 1969), *Ruminococcus albus* (Kemp et al. 1975), two *Eubacterium* spp. (Kemp et al. 1975) and the *Fusocillus* isolates (Kemp et al. 1975). Other species also metabolise linoleic acid, however. Preliminary data from the Rowett Research Institute suggest that these species, like *B. fibrisolvens*, form butyrate as their major fermentation product. CLA formation may therefore be associated with the enzymic capability to produce butyrate. Furthermore, it has been observed that the rate of linoleic acid reduction in the rumen can change 10-fold according to the season (Polan et al. 1964); presumably changes in the diet affected the population size of the bacteria responsible. However, to date there has never been any systematic analysis of the links between microbiology in the rumen and CLA appearance in milk.

POSSIBLE MANIPULATION OF RUMINAL FERMENTATION AND MICRO-ORGANISMS THAT COULD INCREASE CLA PRODUCTION

How might rumen fermentation be manipulated to provide an increased flow of CLA to the animal? Once again, the key issues are that the rumen is a highly reducing environment and that CLA are only intermediates in the reduction process. The main opportunities might therefore be to interrupt the metabolic sequence, to disrupt the supply of reductants, or to alter the microbial population in favour of organisms that achieve the same end result (Table 15.2).

Table 15.2 Possible manipulations of ruminal fermentation that might lead to increased conjugated linoleic acid (CLA) production

Manipulation	Target
Alternative electron acceptor	Removing reductants for biohydrogenation of CLA
Providing free rather than esterified linoleic acid	Enhanced production rate of CLA, saturating biohydrogenation, and encouraging CLA incorporation into bacterial lipids
Increased protozoal population	Increasing microorganisms with naturally high CLA content
Antimicrobial feed additive	Inhibition of type B bacteria
Altered trace element status	Decreased electron transport carriers for biohydrogenation

Since the removal of CLA as an intermediate depends on its hydrogenation, it may be possible to disrupt that by providing alternative electron acceptors. Thereby, *B. fibrisolvens* and possibly other species may divert more CLA into bacterial lipids. An analogous strategy for decreasing methane formation by providing fumarate as an electron acceptor, which decreases the flow of hydrogen to methane and therefore inhibits methanogenesis, has proved successful (López, Valdes, Newbold and Wallace, 1999).

The CLA content of ruminal bacteria is low. Presumably because of its activity in reducing linoleic acid, *B. fibrisolvens* contains only 0.1% 18:2 (total, CLA content not specified) in its fatty acids (Harfoot, 1981). Other ruminal bacteria are little higher, however, but still contain 0.7% 18:2 or less (Harfoot, 1981). The 18:2 content of mixed ruminal bacteria was estimated to be a little higher, up to 5.6% (Williams and Dinusson, 1973), perhaps due to contamination with dietary material. In contrast to ruminal bacteria, ruminal ciliate protozoa contain up to 16-17% 18:2 (Emmanuel, 1974; Harfoot, 1981). Once again, the CLA content of that 18:2 was not specified, and some might be due to ingested lipids in plant material, but the observation certainly merits further investigation, with CLA composition a new focus. Enhancing the population of ciliate protozoa may therefore prove also to enhance CLA production.

Manipulation of ruminal fermentation by feed additives could be effective. Monensin is a growth-promoting feed additive, which has recently been found to increase the accumulation of CLA from soyabean oil *in vitro* (Son, Lee and Yoon, 2000). It also caused a large increase in the accumulation of 18:1, indicating that further metabolism of 18:1 had been inhibited (Son et al. 2000). One might speculate that this could occur by inhibiting the growth or activity of type B bacteria. Whether this observation translates to *in vivo* applicability is not certain: one experiment indicated a small, non-significant increase in milk CLA resulting from dietary monensin (Dhiman et al. 1999a).

Indirect manipulation of fermentation may also be possible. Depleting the copper status increased the CLA content of milk, for reasons that are not clear but appear to involve decreased biohydrogenation in the rumen (Sol Morales, Palmquist and Weiss, 2000b). The observation by Noble, Harfoot and Harfoot (1974) that *cis*-9, *trans*-11, 18:2 accumulated *in vitro* when free linoleic acid was added, but not when esterified linoleic acid was added in the form of linolein, suggests that there may be a mechanism here that could be exploited to enhance the CLA content of ruminal micro-organisms, although the mechanism itself remains unclear.

Thus, there are several possible means for increasing CLA production in the rumen, and more will undoubtedly become obvious as the precise mechanisms of CLA metabolism are elucidated.

The role of the rumen as a barrier preventing the spread of pathogens on farms

The seminal property of ruminants is the ability to use cellulosic material that is poorly digested by most monogastric animals. The ecological position of ruminants is secure (Van Soest, 1994) but the practical problems, which accompany their production especially in intensive agricultural systems, threaten the commercial viability of their continued exploitation by man. The ruminant gut provides habitats for a very wide range of different micro-organisms, including bacteria, protozoa and fungi. Most are harmless commensals, members of complex microbial communities involved in the breakdown, assimilation and fermentation of plant material in the rumen (Hungate, 1966: Van Soest, 1994). Unfortunately, these micro-organisms are sometimes accompanied by pathogens able to cause infectious diseases in man and capable of survival in the ruminant gut, often with no obvious signs of disease.

Among the bacterial pathogens, *Escherichia coli* serotype O157:H7 has caused the most severe consequences on transmission to humans, causing the death of over 20 people in the worst recorded outbreak so far, the Central Scotland outbreak of 1996-1997 (Ahmed and Donaghy, 1998). *E. coli* O157 is carried asymptomatically by cattle, sheep and other farm animals and may infect humans directly, or through transmission by other animals such as dogs. The commonest route of transmission is probably through the human food chain, following contamination of meat at the slaughterhouse (Gannon, 1999). Transmission may also occur via milk (Upton and Coia, 1994) or contamination of the environment, including watercourses (Wallace, 1999).

Other bacterial pathogens carried by ruminants include *Listeria* spp. (Jones, Howard and Wallace, 1999), Arcobacter (Wesley, Wells, Harmon, Green, Schroeder-Tucker, Glover and Siddique, 2000), *Campylobacter* (Stanley, Wallace, Currie, Diggle, and Jones, 1998) *Yersinia enterocolitica* and *Y. pseudotuberculosis* (Slee and Skilbeck, 1992), *Salmonella* spp. (Vaessen, Veling, Frankena, Graat and Klunder, 1998), *Mycobacterium paratuberculosis* (Collins, 1997) and *Helicobacter pylori*. The latter seems to pose a particular risk to people in contact with infected animals. In Sardinia, *Helicobacter* has been detected by PCR in samples from sheep's stomachs and isolated from sheep's milk. Sardinian shepherds in contact with these animals show very high rates of infection (Dore, 1999). In addition, sheep and cattle may carry opportunistic pathogens including *Pseudomonas aeruginosa* (Duncan, Doherty, Govan, Neogrady, Galfi and Stewart 1999) and a variety of strains of enterococci, streptococci and staphylococci (Jayne-Williams, 1979).

CARRIAGE OF PATHOGENS BY RUMINANTS.

The carriage of pathogens may be affected by factors such as the season, age, housing and diet of the animals concerned. In Australia for example, Slee and Skilbeck (1992) found that the excretion of *Y. pseudotuberculosis* was limited to the winter and spring, whilst excretion of *Y. enterocolitica* occurred year-round, with peaks in summer and autumn. Seasonality in shedding has also been noted with other pathogens, including *Campylobacter* (Stanley *et al.* 1998) and *Listeria* (Husu, 1990).

Some of the reported seasonality in shedding may reflect the effect of changes in the diet, such as the use of summer pasture. Dietary change itself has been shown to affect the shedding of *E. coli* O157 (Kudva, Hatfield and Hovde, 1995; Kudva, Hunt, Williams, Nance and Hovde, 1997). It has been argued that the use of diets containing large amounts of starch may result in an increase in shedding of *E. coli* O157. It was suggested that the more acidic conditions found in the rumen of animals fed such diets favour *E. coli* O157 (Diez-Gonzalez, Callaway, Kizoulis, and Russell, 1998). Others have argued that the use of hay-based diets may prolong shedding (Hovde, Austin, Cloud, Williams and Hunt, 1999). Feeding different grain diets to steers resulted in differences in the numbers of *E. coli* O157 shed, and in the persistence of shedding (Buchko, Holley, Olson, Gannon and Veira, 2000). Age-related effects on shedding have also been reported (Garber, Wells, Hancock, Doyle, Tuttle, Shere and Zhao, 1995).

The withdrawal of the diet immediately before slaughter may create conditions in the rumen in which pathogens such as *Salmonella* and *E. coli* may proliferate (Brownlie and Grau, 1967; Cray, Casey, Bosworth and Rasmussen, 1998). The results of fasting on proliferation of *E. coli* O157 in experimentally infected animals may be affected by the time of feed withdrawal in relation to inoculation with the pathogen (reviewed by Leitch, Duncan, Stanley and Stewart, 2001).

Relatively little is known about the ecology of pathogenic bacteria in the ruminant gastrointestinal tract. Brown, Harmon, Zhao and Doyle (1997) and Harmon, Doyle, Brown, Zhao, Tkalcic, Mueller, Parks, and Jacobsen (1999) experimentally infected calves with *E. coli* O157, which was subsequently found mainly in the digesta phase of the gut contents, with relatively few bacteria attached to the gut wall. Carriage and survival of pathogens through the gut remains an area of concern, not only in terms of direct transfer to the human food chain, but also in terms of maintaining a reservoir of infection within the farm. The possibility of cycling of pathogen from contaminated feeds, such as poorly-fermented silage via the animals gut onto fresh pasture and hence the next years silage, has been investigated extensively with *Listeria* (Fenlon, 1988, Fenlon, Wilson and Weddell, 1989) and suggested for other pathogens (Duffy, Garvey and Sheridan, 2000). However, little is known about factors affecting the survival of the pathogen within the animal's digestive tract. Thus we have investigated the flow of *Listeria* through the digestive tract of sheep fed a variety of diets (Shepherd, 2000). A typical result is shown in Table 15.3. As might be expected the, acidic conditions in the stomach (Bell, 1979; Kern, Slyter, Leffel, Weaver and Oltjen, 1974) led to a significant kill of *Listeria* between the rumen and duodenum. However, there was a numerically greater kill of *Listeria* due to passage through the rumen (Table 15.3). In subsequent experiments the survival of *Listeria* in rumen fluid was investigated, the apparent lysis of both *L. monocytogenes* and *L. innocua* was markedly reduced when they were incubated in protozoa-free rumen fluid, compared with strained rumen fluid (Figure 15.1). When the flows of *Listeria* from the rumen of defaunated (free from rumen protozoa) and refaunated (with the protozoa reintroduced) sheep were compared, the passage of listeria was 3 x higher in the defaunated compared with the faunated sheep (Shepherd, Newbold, Hillman and Fenlon, 2000). The role of protozoa in the engulfment and subsequent lysis of commensal rumen bacteria, fungi and archaea is well established (Wallace and McPherson, 1987; Newbold and Hillman, 1990; Newbold, Morvan, Fonty and Jouany 1996), however, to our knowledge this is the first demonstration that they may have an important role in limiting the survival of pathogens in the rumen. Other factors within the rumen may also limit pathogen survival; *E. coli* is known to be sensitive to the short chain fatty acids (SCFA), which are produced by the rumen fermentation, particularly at low pH, and this has been suggested as a factor that may limit proliferation (Wolin, 1969; Wallace, Falconer and Bhargava, 1989; Rasmussen, Cray, Casey and Whipp, 1993). However, at least some strains of *E. coli* O157 appear to be relatively tolerant of SCFA (Diez-Gonzalez and Russell, 1997; Diez-Gonzalez *et al*. 1998). There are also interactions between pathogens and non-pathogenic bacteria with the rumen (Duncan, Booth, Flint. and Stewart, 2000; Leitch et al. 2001). A probiotic mixture derived from cattle was able to reduce carriage of *E. coli* O157 in experimentally-infected animals (Zhao, Doyle, Harmon, Brown, Mueller and Parks, 1998), while *Pseudomonas aeruginosa* isolated from the rumen of sheep inhibited the growth of *E. coli* O157 (Duncan *et al.* 1999) *in vitro*. It is apparent that the rumen, and the rumen protozoa in particular, form an important barrier to pathogen flow through the animal.

Role of protozoa in ruminal N metabolism and the survival of pathogens

Initially the apparent role of protozoa in preventing pathogens leaving the rumen is rather disappointing, as in the current environment there are other reasons why one might wish to remove protozoa from the rumen. As noted above, *in vitro*

Table 15.3 Flow of *Listeria innocua* through the gut of sheep fed on dried grass

	Entering in diet	*Leaving rumen*	*Entering duodenum*	*Leaving ileum*	*Shed in faeces*
Flow of *L. innocua* (cfu/g)	2.6×10^9 (0)	5.0×10^8 (0.98×10^8)	7.5×10^3 (2.72×10^3)	1.4×10^7 (0.79×10^7)	9.5×10^7 (6.84×10^7)

cfu Colony forming unit
Figures in parenthesis are standard errors
Data from Shepherd (2000)

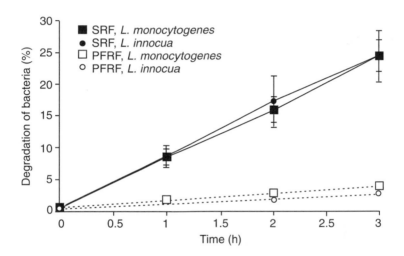

Figure 15.1 Degradation of *Listeria monocytogenes* and *Listeria innocua* incubated *in vitro* in strained rumen fluid (SRF) or protozoa-free rumen fluid (PFRF).

studies suggest that engulfment and digestion of bacteria by protozoa leads to microbial turnover in the rumen and it has been suggested that this may result in the microbial protein outflow from the rumen being less that 50% of the total microbial protein synthesised (Nolan and Stachiw, 1979). Recently, we have measured the recycling of N within the rumen (Newbold, Teferedegne, Kim, Zuur and Lobley, 2000). The data suggest that this recycling drops dramatically in the absence of protozoa (Table 15.4). Consistent with the decline in intraruminal recycling, the flow of microbial protein from the rumen increased by 50% (Table 15.4). Given the current concerns and limitations in the use of animal protein and protein from GMO crops, a significant stimulation in microbial protein supply from the rumen would be of considerable benefit to the UK livestock feed industry.

Table 15.4 Microbial protein synthesis and the kinetics of ruminal ammonia recycling in faunated and refaunated sheep fed dried grass

	Defaunated	*Refaunated*
Microbial protein flow from rumen (gN/d)	13.3	8.9
NH$_3$ pool size (g N)	0.84	2.0
Total flux rate (g N/d)	15.1	29.2
Irreversible loss rate (g N/d)	11.0	16.8
Recycling rate (g N/d)	4.1	12.4
Ruminal NH$_3$ from plasma urea (g N/d)	3.3	5.8
Intraruminal recycling (g N/d)	0.8	6.6

However, it is also apparent that such a stimulation in microbial protein flow would need to be balanced against a possible increase in the survival of pathogens within the gut. With this in mind, we have studied the breakdown of *Listeria* and other pathogens by representatives of the different protozoa species present in the rumen. It is clear that, as with non-pathogenic bacteria, different protozoa preferentially engulf and lyse different bacteria (Figure 15.2, Newbold and Jouany, 1997). It may be possible to manipulate the composition of the protozoal population in the rumen such that microbial supply is stimulated while pathogens are destroyed. We are developing methods for cloning and characterising the enzymes in different protozoa that are responsible for the breakdown of bacteria (Eschenlauer, McEwan, Calza, Wallace, Onodera and Newbold, 1998; Newbold, Eschenlauer, McEwan, Onodera and Wallace, 2000) and how they interact with different bacteria. This is, however, essentially a long-term goal and more immediate benefits may arise through the development of on-farm methods to control protozoa in the rumen coupled to appropriate management.

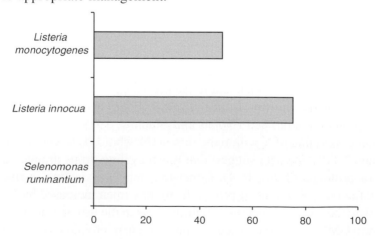

Figure 15.2 Breakdown of *Listeria monocytogenes, L. innocua* and *Selenomonas ruminantium* by *Polyplastron multivesiculatum* relative to breakdown by *Entodinium caudatum* (where breakdown by *E. caudatum* is set at 100)

CONTROL OF RUMEN PROTOZOA

A variety of techniques to remove protozoa from the rumen have been tested experimentally, but none is used routinely because of toxicity problems, either to the rest of the rumen microbial population or to the host animal (Williams and Coleman, 1992). Recently, there has been an increased interest in plant secondary metabolites for use as possible defaunating agents. Secondary metabolites in plants function both as nutrient stores and also as a mechanism for defending their structure and reproductive elements from predation by animals and insects (Haborne, 1989). A decrease in protozoal numbers in the rumen of sheep consuming the pericarp of *Sapindus saponaria* was reported by Daiz, Avendan and Escobar (1994). A methanol extract prepared from *Sapindus rarak* fruit depressed the protozoal population in the rumen of sheep by 57% (Thalib, Widiawati, Hamid, Suherman and Sabrani, 1995). Foliage from *Sesbania sesban,* a multipurpose leguminous tree from sub-saharan Africa, inhibited protozoal activity *in vitro* and transiently depressed the number of protozoa in the rumen of sheep in the UK (Newbold, El Hassan, Wang, Ortega and Wallace, 1997). In all cases the antiprotozoal action has been attributed to saponins in the plant material. Saponins are glycosides which apparently interact with the sterols present in eukaryotic membranes but not in prokaryotic cells (Cheeke, 1998). However saponins are degraded rapidly in rumen fluid *in vitro* (Makkar and Becker, 1997), although the resultant sapogenins are apparently more resistant to further degradation (Wang, McAllister, Rode, Newbold, Cheeke and Cheng, 1998). Sapogenins do not have the anti-protozoal property of the parent saponin (Teferedgne, Osuji, Odenyo, Wallace and Newbold, 1998). This observation may explain why many tropical forages are apparently toxic to protozoa in an initial *in vitro* screen, but only a few (presumably those with saponins less likely to be degraded to sapogenin) have prolonged anti-protozoal activity *in vivo* (Teferedegne, 2000). Amongst those with a prolonged activity *in vivo* (Teferedegne, 2000) *Enterolobium cyclocarpum* has been shown to increase the rate of body-weight gain and wool growth in lambs (Leng, Bird, Klieve, Choo, Ball, Asefa, Brumby, Mudgal, Chaudhry, Haryono and Hendratno, 1992; Navas-Camacho, Laredo, Cuesta, Anzola and Leon, 1993) presumably as a consequence of an increased supply of microbial protein when protozoa are suppressed. It may be that, at least until more subtle manipulation of the protozoal population is possible, it will be appropriate to include saponin-containing material in the diet during the early stages of the growth cycle, when the animal is most receptive to additional protein but to remove it prior to slaughter to allow a protozoal population to re-establish and control pathogens before the animal enters the food chain.

OTHER EFFECTS OF PLANT EXTRACTS

It is not only protozoa within the rumen that can be manipulated using plant extracts. Essential oils also inhibit pathogens (Park, Taormina. and Beauchat, 2000) and may alter rumen fermentation since reductions in ammonia production have been reported in sheep supplemented with essential oils (McIntosh, Newbold, Losa, Williams and Wallace, 2000). Preliminary observations suggest this may be due to a selective inhibitory effect of essential oils on the activity of the group of ruminal clostridia known to have very high rates of ammonia production (Newbold, McIntosh, Wallace, Williams and Sutton , 1999). Duncan *et al.* (2000) have shown that esculin, a coumarin glycoside isolated from horse chestnuts, is hydrolysed by a wide range of gut bacteria to the aglycone esculetin, which inhibits the growth of *E. coli* 0157 both in pure cultures and in incubations of mixed rumen fluid. Given the large numbers of plants with secondary compounds that may manipulate gut function, there is a need to survey such plants as they may provide the next generation of rumen manipulators.

The manipulation of methane production by rumen microbes

Methane produced during anaerobic fermentation in the rumen represents an energy loss to the host animal as well as contributing to emissions of greenhouse gases into the environment (Moss, 1993). The presence of methane in the atmosphere has been known since the 1940s (Migeotte, 1948) and numerous measurements since have demonstrated that atmospheric methane concentrations are increasing (Rodhe, 1990), although the rate of increase appears to have slowed of late (Steele, Dlugokencky, Lang, Tans, Martin and Masarie, 1992). Methane has 21 x the relative radiative effectiveness of CO_2 per molecule, in terms of reflecting energy back to earth. Thus, despite being present in the atmosphere at far lower concentrations than CO_2 (1.72 v 355 ppmv), it estimated that methane is responsible for approximately 20% of the greenhouse effect (IPCC, 1990, 1992). On a global scale agriculture, and in particular enteric fermentation in ruminants, produces between 21 and 25% of the total anthropogenic emissions of methane (Duxbury, Harper and Mosier, 1993; Isermann, 1992; Watson, Meira Filho, Sanhueza and Janetos, 1992). However, in more rural communities, such as Scotland, agriculture contributes almost 50% of annual methane emissions (compared to 28% in England and Wales) with over 90% of this due to rumen fermentation (Salway, Dore, Watterson, Murrells, 1999). Given the concerns over global warming, it is apparent that animal agriculture can play a role in control and reduction of greenhouse gas emissions.

Direct inhibition of methanogenesis by halogenated methane analogues and related compounds has been widely demonstrated *in vitro* (Van Nevel and

Demeyer, 1995) and some have been tested *in vivo* (Nagaraja *et al.,* 1997). However, the efficacy of many such additives declines with continued or repeated use (Moss, Jouany and Newbold, 2001); given the growing concern about using chemicals in animals destined for human consumption, there is a need to develop alternative strategies to control ruminal methanogenesis.

PROPIONATE ENHANCERS

Methanogenesis is a major means of disposal of hydrogen during rumen fermentation. One possible way to decrease methane formation in the rumen is to promote alternative metabolic pathways to dispose of the reducing power, competing with methanogenesis for the hydrogen uptake. Fumaric acid is a four-carbon dicarboxylic acid that is an intermediate in the propionate pathway, in which it is reduced to succinate and thence propionate by fumarate reductase. Reducing equivalents are needed in this reaction and therefore fumarate may provide an alternative electron sink for hydrogen. Indeed, Lopez *et al.* (1999a) observed that when fumarate was added to rumen simulating fermentors, propionate production increased with a stoichiometric decrease in methane production. Ouda, Newbold, Lopez, Nelson, Moss, Wallace and Omed (1999) found that acrylate, an alternative precursor of propionate, also depressed methane production in rumen simulating fermentors, but to a lesser extent than an equimolar addition of fumarate. Asanuma, Iwamoto and Hino (1999) also found that fumarate depressed methane production *in vitro* and suggested that fumarate could be an economical feed additive in Japan. Malate, which is converted to propionate via fumarate, also stimulated propionate production and inhibited methanogenesis *in vitro* (Martin and Streeter, 1995). However, Carro, Lopez, Valdes and Ovejero (1996) found that malate actually increased methane production in a rumen-simulating fermentor, although this was largely explained by stimulation of fibre digestion and methane produced per unit of dry matter fermented actually fell. Malate failed to stimulate rumen propionate concentrations in the rumen of cattle and did not affect estimated methane production (Montano, Chai, Zinn-Ware and Zinn, 1999; Martin, Streeter, Nisbet, Hill and Willamns, 1999) although malate did stimulate live-weight gain in steers (Martin *et al.*, 1999). There is now a need to identify cost-effective sources of propionate precursors for use in animal feeds.

STIMULATION OF ACETOGENS

An alternative strategy to decrease ruminal methanogenesis would be to re-channel substrates for methane production into alternative products. Acetogenic bacteria, in the hindgut of mammals and termites, produce acetic acid by the reduction of carbon dioxide with hydrogen and reductive acetogensis acts as an important hydrogen sink in hindgut fermentation (Demeyer and de Graeve, 1991; Lajoie, Bank, Miller and Wolin, 1988). Reductive acetogenesis occurs in the intestine of non-ruminants, sometimes along with methanogenesis and sometimes replacing methanogenesis (Breznak and Kane, 1990; Durand and Bernalier, 1995). Bacteria carrying out reductive acetogenesis have been isolated from the rumen (Genther, Davis and Bryant, 1981; Greening and Leedle, 1989; Morvan, Doré, Rieu-Lesme, Foucat, Fonty and Gouet, 1994), but they are few in number, and attempts to increase acetogenesis have not been successful, largely because under rumen conditions the reductive acetogens have been unable to compete with the methanogenic archaea (Demeyer, Fiedler and De Graeve, 1996; Immig, Demeyer, Fiedler, Van Nevel and Mbanzamihigo, 1996; Nollet, Demeyer and Verstraete, 1997; Nollet, Mbanzamihigo, Demeyer and Verstraete, 1998). Lopez, McIntosh, Wallace and Newbold (1999b) found that acetogens depress methane production when added to rumen fluid *in vitro* and suggested that even if a stable population of acetogens could not be established in the rumen it might be possible to achieve the same metabolic activity using the acetogens as a daily fed feed additive. Alternatively, it is possible that acetogens could be introduced into the rumen if methanogens were suppressed through chemical, biological or immunological means (Van Nevel and Demeyer, 1995, Kleive and Hegarty, 1999, Baker 1999).

METHANE OXIDISERS

Global methane accumulation is the difference between methane production and methane oxidation. Methane-oxidising bacteria have been isolated from a wide range of environments (Hanson, 1992), including the rumen (Stock and McCleskey, 1964). Studies with $^{13}CH_4$ tracers suggest that oxidation of methane to CO_2 is of little quantitative importance in the rumen (Valdes, Newbold, Hillman, and Wallace, 1996; Kajikawa and Newbold, 2000) but may be more important in the gut of pigs (Hillman, van Wyk, Milne, Stewart and Fuller, 1995). The rumen is widely considered to be anaerobic; nevertheless, rumen gas, even in non-fistulated animals, contains between 0.5 and 1.0 % oxygen (McArthur and Multimore, 1962). Oxygen is toxic to anaerobic rumen bacteria; it inhibits bacterial growth and the adhesion of cellulolytic bacteria to fibre (Loesche, 1969; Roger, Fonty, Komisarczuk-Bony and Gouet, 1990). We have previously suggested that one possible mode of action of *Saccharomyces cerevisiae* (yeast culture) when added to ruminant diets might be to remove free oxygen and thus maximise bacterial growth in the rumen (Newbold, Wallace and McIntosh, 1996). In an attempt to extend this model, we have isolated a methane-oxidising bacterium, *Brevibacillus parabrevis*, from the gut of young pigs. This bacterium decreased methane accumulation when added to rumen fluid *in vitro* (Valdes, Newbold, Hillman and Wallace, 1997). When added to an artificial rumen fermentor (Rusitec) *B. parabrevis* caused a significant decrease in methane production (Nelson, Valdes, Hillman, McEwan, Wallace and Newbold, 2000), associated with an increase in the count of total culturable bacteria, presumably due to removal of oxygen from the fermenter. We are currently evaluating this organism *in vivo*.

IMMUNISATION

Possibly the most exciting work on rumen manipulation results from work on the immunological control of ruminal methanogensis (Baker, 1999). Immunisation of sheep with an antigen prepared from a range of ruminal methanogens resulted in a reduced rate of methanogensis in rumen fluid from these sheep and an enhancement in animal productivity (Baker, 1995). Indeed the approach does not appear to be limited to the control of methanogens, as Shu, Gill, Hennessy, Leng, Bird and Rowe (1999) successfully immunised cattle against their own *Streptococcus bovis* and *Lactobacillus*, reducing the numbers of both organisms in grain-challenged animals and reducing concentrations of lactic acid in the rumen. Gnanasampanthan (1993) demonstrated the feasibility of controlling rumen protozoa through immunisation. The biggest limitation to wide-scale exploitation of such technologies is our incomplete knowledge of diversity of micro-organisms within the rumen. In order to produce the most effective antibody, the antigen in the vaccine must encompass representatives of all the bacteria species you wish to control.

PROFILING THE MICROBIAL POPULATION IN THE RUMEN

Traditionally, the diversity of gut ecosystems such as the rumen has been determined by enumerating and isolating micro-organism on specific growth media. Individual isolates would then be grown in pure culture and biochemically characterised to allow their identification (Stewart and Bryant, 1988). As a result of the limitations of this approach, in terms of labour and resources, few studies progressed past the enumeration of rumen microbes into functional groups (i.e. cellulolytic or lactic acid utilising bacteria etc). As such, it was often difficult to integrate the detailed biochemical knowledge acquired with individual pure species grown in the laboratory with the role and numbers of these organisms in the rumen. The growing use of modern molecular ecology techniques, based on sequence comparisons of nucleic acids, is greatly enhancing our ability to study microbial diversity with the rumen. 16S rRNA sequences from bacteria and 18S rRNA sequences from protozoa and fungi are being used to study the phylogenetic relationships between rumen micro-organisms, as reviewed by Mackie, Aminov, White and McSweeney (2000). Group-specific RNA target oligonucleotides are being used to quantify specific microbial genera and species within the rumen (Stahl, Flesher, Mansfield and Montgomerey, 1988; Mackie *et al*., 2000). It has been pointed out, however, that 16S rRNA may be so well conserved that it fails sometimes to distinguish between closely related populations, and that the heterogeneity of rRNA operons of a single organism may also present problems (Amman and Ludwig, 2000). It is not clear whether these potential limitations are significant for rumen micro-organisms, but analysis of 23s rDNA may prove more reliable. Genetic finger-printing techniques based on the analysis of PCR-amplified rDNA restriction fragments, offer the ability to measure the diversity of the rumen microbial population either as a whole or in terms of specific taxa (Wood, Scott, Avgustin, Newbold and Flint, 1998; Mackie *et al*., 2000; Rieu-Lesme, Godon, and Fonty, 2000). *In situ* measurements of gene expression by probing mRNA may enable combined diversity/ function studies in future (Amman and Ludwig, 2000). Such techniques will allow us to answer questions about how the type/number of microbes in the rumen relate to diet and metabolic function, providing both the fundamental information needed to develop new methods of manipulation and also the tools to investigate and characterise the impact of new additives more fully than has been possible in the past.

Conclusions

New ways in which the rumen microbes might be manipulated to alter ruminant production will continue to be identified. However, the targets for manipulation are changing and no longer can the productivity of the animal be considered in isolation: there is a need to increase the communication between the laboratory-based scientists and feed and feed supplement manufactures to make sure efforts are focussed in the areas of most concern. Molecular techniques will be central in the development of new strategies. The development of techniques to profile the rumen microbial population should allow new additives to be developed and evaluated in terms of the existing knowledge base of the metabolism of individual microbes. Such progress should allow new technologies to be rapidly integrated into existing systems by identifying when and where they will be of most use.

   In terms of identifying new additives, this review has made no attempt to be inclusive, instead focussing on areas in which we are directly involved. However, it is apparent that plant extracts and immunization deserve further attention as potential areas of progress.

Acknowledgement

The Rowett Research Institute receives financial support from Scottish Executive Rural Affairs Department.

References

Ahmed, S., and Donaghy, M . (1998) An outbreak of *Escherichia coli* O1657:H7 in Central Scotland. In: *Escherichia coli O157:H7 and Other Shiga-Toxin Producing E. coli Strains*, pp.59-65. Edited by J.B. Kaper and A.D. O'Brien. American Society for Microbiology, Washington.

Amman, R. and Ludwig, W. (2000) Ribosomal RNA-targeted nucleic acid probes for studies in microbial ecology. *FEMS Microbiology Reviews* **24**, 555-565.

Asanuma, N., Iwamoto, M., and Hino, T. (1999) Effect of the addition of fumarate on methane production by ruminal microorganisms *in vitro*. *Journal of Dairy Science,* **82**, 780-787.

Baker, S.K. (1995) *Method for improving utilization of nutrients by ruminant or ruminant like animals.* International Patent No WO9511041

Baker, S.K. (1999) Biology of rumen methanogens and stimulation of animal immunity. *Australia Journal of Agricultural Research,* **50**, 1293-1298.

Bell, F.R. (1979) The mechanisms controlling abomasal emptying and secretion. In *The Fifth International Symposium on Ruminant Physiology* pp 88-96. Edited by Y. Ruckebusch and P. Thivend. MTP Press, Lancaster, UK.

Breznak, J.A., and Kane, M.D. (1990) Microbial H_2/CO_2 acetogenesis in animal guts: nature and nutritional significance. *FEMS Microbiology Review,* **87**, 309-314

Brown, C.A., Harmon B.G., Zhao, T., and Doyle, M.P. (1997) Experimental *Escherichia coli* O157:H7 carriage in calves. *Applied and Environmental Microbiology,* **63**, 27-32.

Brownlie, L.E. and Grau, F.H. (1967) Effect of food intake on growth and survival of salmonellas and *Escherichia coli* in the bovine rumen. *Journal of General Microbiology,* **46**, 125-134.

Buchko, S.J., Holley, R.A., Olson, W.O., Gannon, V.P.J. and Veira, D.M. (2000) The effect of different grain diets on fecal shedding of *Escherichia coli* O157:H7 by steers. *Journal of Food Protection,* **63**, 1467-1474.

Carro, M.D., Lopez, S., Valdes, C. and Ovejero, F.J. (1999) Effect of D,L-Malate on mixed ruminal microorganism fermentation using the rumen simulation technique (RUSITEC). *Animal Feed Science and Technology,* **79**, 279-288.

Cheeke P.R. (1998) Natural toxicants in feeds, forages & poisonous plants. Interstate Publishers, Inc. Danville, Illnois.

Collins, M.T. (1997) *Mycobacterium paratuberculosis*: a potential food-borne pathogen. *Journal of Dairy Science,* **80**, 3445-3448.

Cray, W.C. Jr., Casey, T.A., Bosworth, B.T. and Rasmussen, M.A. (1998) Effect of dietary stress on fecal shedding of *Escherichia coli* O157:H7 in calves. *Applied and Environmental Microbiology,* **64**, 1975-1979.

Demeyer, D.I. and De Graeve, K. (1991) Differences in stoichiometry between rumen and hindgut fermentation. *Journal Animal Physiology and Animal Nutrition,* **22**, 50-61.

Demeyer, D.I., Fiedler, D. and De Graeve, K.G. (1996) Attempted induction of reductive acetogenesis into the rumen fermentation *in vitro. Reproductive Nutrition and Development,* **36**, 233-240.

Dhiman,T.R., Anand, G.R., Satter,L.D., and Pariza, M.W. (1999a) Conjugated linoleic acid content of milk from cows fed different diets. *Journal of Dairy Science,* **82**, 2146-2156.

Dhiman,T.R., Helmink, E.D., McMahon, D.J., Fife, R.L. and Pariza, M.W. (1999b). Conjugated linoleic acid content of milk and cheese from cows fed extruded oilseeds. *Journal of Dairy Science,* **82**, 412-419.

Dhiman,T.R., Satter,L.D., Pariza,M.W., Galli,M.P., Albright, K. and Tolosa, M.X.. (2000). Conjugated linoleic acid (CLA) content of milk from cows offered diets rich in linoleic and linolenic acid. *Journal of Dairy Science,* **83**,1016-1027.

Diaz, A., Avendan O., M., and Escobar, A. (1994) Evaluation of *Spinadus saponaria* as a defaunating agent and its effects on different ruminal digestion parameters. *Livestock Research for Rural Development,* **5**, 1-10.

Diez-Gonzalez, F. and Russell, J.B. (1997) The ability of *Escherichia coli* O157:H7 to decrease its intracellular pH and resist the toxicity of acetic acid. *Microbiology,* **143**, 1175-1180.

Diez-Gonzalez, F.T.R., Callaway, M., Kizoulis, G. and Russell, J.B. (1998) Grain feeding and the dissemination of acid-resistant *Escherichia coli* from cattle. *Science,* **281,** 1666-1668.

Dore, M.P., Sepulveda, A.R., Osato, M.S., Realdi, G. and Graham, D.Y. (1999) Helicobacter in sheep milk. *Lancet,* **354**, 132.

Duffy, G., Garvey, P. and Sheridan, J.J. (2000) *E coli* 0157:H7; A European perspective. *Irish Journal of Agriculture and Food Research,* **39**, 173-182.

Duncan, S.H., Flint, H.J. and Stewart, C.S. (1998) Inhibitory activity of gut bacteria against *Escherichia coli* O157 mediated by dietary plant metabolites. *FEMS Microbiology Letters,* **164**, 283-288.

Duncan, S.H., Doherty, C.J., Govan, J.R.W., Neogrady, S., Galfi, P. and Stewart, C.S. (1999) Rumen isolates of *Pseudomonas aeruginosa* inhibitory to *Escherichia coli* O157. *FEMS Microbiology Letters,* **180**, 305-310.

Duncan, S.H., Booth, I.R., Flint, H.J. and Stewart, C.S. (2000) The potential for control of *Escherichia coli* O157 in farm animals. *Journal of Applied Microbiology Symposium Supplement,* **88**, 1S-11S.

Durand, M. and Bernalier, A. (1995) Reductive acetogenesis in animal and human gut. In *Physiological and Clinical Aspects of Short-chain Fatty Acids*, pp. 57-72 Edited by J.H. Cummings, J.L. Rombeau and T Sakata. Cambridge University Press, Cambridge

Duxbury J.M., Harper, L.A. and Mosier, A.R. (1993) Contributions of agroecosystems to global climate change. In *Agroecosystem Effects on Radiatively Important Trace Gases & Global Climate Change*, pp. 1-18 Edited by L.A. Harper, A.R. Mosier, J.M. Duxbury and D.E. Rolston. ASA Special Publication No. 55., American Society of Agronomy, Madison.

Emmanuel,B. (1974) On the origin of rumen protozoan fatty acids. *Biochimica Biophysica Acta,* **337**, 404-413.

Eschenlauer, S.C.P., McEwan, N.R., Calza, R.E., Wallace, R.J., Onodera, R. and Newbold, C.J. (1998) Phyologentic position and codon usage of two centrin genes from the rumen ciliate protozoan, *Entodinium caudatum. FEMS Microbiology Letters*, **166**, 147-154.

Fenlon, D.R. (1988) Listeriosis. In *Silage and health*, pp 7-18 Edited by B.A. Stark and J.M. Wilkinson. Chalcome Publications, Marlow Bottom, UK.

Fenlon, D.R., Wilson, J. and Weddell, J.R. (1989) The relationship between spoilage and *Listeria monocytogenes* contamination in bagged and wrapped big bale silage. *Grass and Forage Science,* **44**, 97-100

Gannon, V.P.J. (1999) Control of *Escherichia coli* O157 at slaughter. In : *Escherichia coli* O157 in Farm Animals. pp 169-193 Edited by C.S. Stewart and H.J. Flint. CAB International, Wallingford

Garber, L.P, Wells, S.J., Hancock, D.D., Doyle, M.P, Tuttle, J, Shere, J.A. and Zhao, T.(1995) Risk factors for fecal shedding of *Escherichia coli* O157:H7 in dairy calves. *Journal of the American Veterinary Association,* **207**, 46-49.

Genther, B.R.S., Davis, C.L. and Bryant, M.P. (1981) Features of rumen and sewage sludge strains of *Eubacterium limosum,* a methanol and H_2/CO_2-utilizing species. *Applied Environmental Microbiology,* **42,** 12-19.

Gnanasampanthan, G (1993) Immune responses of sheep to rumen ciliates and the survival and activity of antibodies in the rumen fluid. *PhD thesis University of Adelaide, Adelaide*.

Greening, R.C. and Leedle, J.A.Z. (1989) Enrichment and isolation *of Acetitomaculum ruminis*, gen. nov., sp. nov.: acetogenic bacteria from the bovine rumen. *Archives of Microbiology* **151,** 399-406.

Gulati,S.K., Kitessa, S.M., Ashes, J.R., Fleck, E., Byers,E.B., Byers, Y.B. and Scott, T.W. (2000) Protection of conjugated linoleic acids from ruminal hydrogenation and their incorporation into milk fat. *Animal Feed Science and Technology.* **86**, 139-148.

Haborne, J. B. (1989) Biosynthesis and function of antinutritional factors in plants. *Aspects of Applied Biology,***19**, 21-28.

Hanson, S. (1992) Distribution in nature of reduced one carbon compounds and microbes that utilize them. In *Methane and Methanol Utilizers*, pp. 1-22 Edited by J.C. Murrell and H. Dalton. Plenum Press, New York.

Harfoot,C.G. (1981). Lipid metabolism in the rumen. pp 21-55 *In Lipid Metabolism in Ruminant Animals,* Edited by W.W.Christie. Pergamon, Oxford.

Harfoot,C.G. and Hazlewood, G.P. (1997). Lipid metabolism in the rumen. In *The Rumen Microbial Ecosystem.* pp 382-426. Edited by P.N.Hobson and C.S.Stewart. Chapman & Hall, London.

Harmon, B.G, Doyle, M.P, Brown, C.A., Zhao. T, Tkalcic, S, Mueller, E, Parks, A.H., and Jacobsen, K. (1999) Faecal shedding and rumen proliferation of *Escherichia coli* O157:H7 in calves: an experimental model. In*: Escherichia coli O157 in Farm Animals.* pp 71-90 Edited by C.S. Stewart and H.J. Flint. CAB International, Wallingford.

Hazlewood, G.P. and Dawson. R.M.C. (1979) Characteristics of a lipolytic and fatty acid - requiring Butyrivibrio sp. isolated from the ovine rumen. *Journal of General Microbiology,* **112**, 15-27.

Hazlewood, G.P., Kemp, P., Lauder, D. and Dawson, R.M.C. (1976). C18 unsaturated fatty acid hydrogenation patterns of some rumen bacteria and their ability to hydrolyse exogenous phospholipid. *British Journal of Nutrition,* **35**, 293-297.

Hovde, C.J., Austin, P.R., Cloud, K.A., Williams, C.J. and Hunt, C.W. (1999) Effect of cattle diet on *Escherichia coli* O157:H7 acid resistance. *Applied and Environmental Microbiology,* **65**, 3223-3235.

Hillman, K., van Wyk, H., Milne, E., Stewart, C.S. and Fuller, M.F. (1995) Oxidation of methane by digesta from the porcine ileum and colon. Society for General Microbiology (Golden Jubilee) meeting, University of Bath, April 1995. Abstracts booklet, p.24

Hughes, P.E. and Tove, S.B. (1982) Occurrence of alpha-tocopherolquinone and alpha-tocopherolquinol in microorganisms. *Journal of Bacteriology,* **151**, 1397-1402.

Hungate, R.E. (1966) The Rumen and its Microbes. Academic Press, New York.

Husu, J. R. (1990) Epidemiological studies on the occurrence of *Listeria monocytogenes* in the faeces of dairy cattle. *Zentraalblatt Veterinarmedecine,* **37**, 276-282.

Immig, I., Demeyer, D., Fiedler, D., Van Nevel, C. and Mbanzamihigo, I. (1996) Attempts to induce reductive acetogenesis into a sheep rumen. *Archives of Animal Nutrition,* **49,** 363-370.

IPPC (1990) *Climate Change: The IPPC Impact Assessment.* Australian Government Publishing Service. Canberra, Australia.

IPPC (1992) *Climate Change, 1992. The Supplementary Report to the Scientific Assessment* Edited by J.T. Houghton, B.A Callander and S.K Varney. Cambridge University Press, Cambridge.

Isserman K (1992) Territorial, continental and global aspects of C, N, P and S emissions from agricultural ecosystems. In *NATO Advanced Research Workshop (ARW) on Interactions of C, N, P and S Biochemical cycles.* Springer-Verlag. Heidelberg

Jayne-Williams, D.J. (1979) The bacterial flora of healthy and bloating calves. *Journal of Applied Bacteriology,* **47,** 271-284.

Jones, K., Howard, S and Wallace, J.S. (1999) Intermittent shedding of thermophilic campylobacters by sheep at pasture. *Journal of Applied Microbiology,* **86,** 531-536.

Kajikawa, H. and Newbold, C.J. (2000) Methane oxidation in the rumen *Journal of Animal Science,* **78,** Suppl.1, 291

Kelly, M.L., Berry, J.B., Dwyer,D.A., Griinari, J.M., Chouinard, P.Y, Van Amburgh, M.E. and Bauman, D.E. (1998) Dietary fatty acid sources affect conjugated linoleic acid concentrations in milk from lactating dairy cows. *Journal of Nutrition,* **128,** 881-885.

Kemp, P. and Lander, D.J. (1984) Hydrogenation in vitro of α-linolenic acid to stearic acid by mixed cultures of pure strains of rumen bacteria. *Journal of General Microbiology,* **130,** 527-533.

Kemp,P., White,R.W., and Lauder, D.J. (1975) The hydrogenation of unsaturated fatty acids by five bacterial isolates from the sheep rumen, including a new species. *Journal of General Microbiology,* **90,** 100-114.

Kepler,C.R., Hirons, K.P., McNeill, J.J. and Tove, S.B. (1966) Intermediates and products of the biohydrogenation of linoleic acid by *Butyrivibrio fibrisolvens. Journal of Biological Chemistry,* **241,**1350-1354.

Kern, D.L., Slyter, L.L., Leffel, E.C., Weaver, J.M. and Oltjen, R.R. (1974) Ponies vs steers: microbial and chemical characteristics of intestinal ingesta. *Journal of Animal Science,* **38,** 559-564.

Klieve, A.V. and Hegarty, R.S. (1999) Opportunities for biological control of ruminal methanogensis. *Australia Journal of Agricultural Research,* **50,** 1315-1319.

Kritchevsky,D. (2000) Antimutagenic and some other effects of conjugated linoleic acid. *British Journal of Nutrition,* **83,** 459-465.

Kudva, I.T., Hunt, C.W., Williams, C.J., Nance, U.M. and Hovde, C.J. (1997) Evaluation of dietary influences on *Escherichia coli* O157:H7 shedding by sheep. *Applied and Environmental Microbiology,* **63**, 3878-3886.

Kudva, I.T., Hatfield , P.G. and Hovde, C.J. (1995) Effect of diet on the shedding of *Escherichia coli* O157:H7 in a sheep model. *Applied and Environmental Microbiology,* **61**, 1363-1370.

Lajoie, S.F., Bank, S., Miller, T.L. and Wolin, M.J. (1988) Acetate production from hydrogen & ^{13}C carbon dioxide by the microflora of human faeces. *Applied Environmental Microbiology,* **54**, 2723-2727.

Leitch, E.C., Duncan, S., Stanley, K.N and Stewart, C.S. (2001). Dietary effect on the micropbiological safety of food. *Proceedings of the British Nutrition Society (in the press)*

Leng, R. A., Bird, S. H., Klieve, A., Choo, B. S., Ball, F. M., Asefa, G., Brumby, P., Mudgal, V. D., Chaudhry, U. B., Haryono, S. U. and Hendratno, N. (1992) The potential for tree forage supplements to manipulate rumen protozoa to enhance protein to energy ratios in ruminants fed on poor quality forages. pp. 177-191. In: *Legume Trees and Other Fodder Trees as Protein Sources for Livestock*. Edited by A. Speedy and P. L. Pugliese. FAO, Rome

Loesche, W.J. (1969) Oxygen sensitivity of various anaerobic bacteria. *Applied Microbiology,* **18**, 723-727.

Lopez, S., Valdes, C., Newbold, C.J. and Wallace, R.J. (1999a). Influence of sodium fumarate addition on rumen fermentation in vitro. *British Journal of Nutrition,* **81**, 59-64.

Lopez, S., McIntosh, F.M., Wallace, R.J. and Newbold, C.J. (1999b) Effect of adding acetogenic bacteria on methane production by mixed rumen microorganisms. *Animal Feed Science and Technology,* **78,** 1-9

McIntosh, F.M., Newbold, C.J., Losa, R., Williams, P. and Wallace, R.J. (2000) Effects of essential oils on rumen fermentation. *Reproduction Nutrition Development,* **40**, 221-222

McArthur, J.M. and Multimore, J.E. (1962) Rumen gas analysis by gas solid chromatography. *Canadian Journal of Animal Science,* **41**, 187- 192.

Mackie, R.I., Aminov, R.I., White, B.A. and McSweeney, C.S. (2000) Molecular ecology and diversity in gut microbial ecosystems. *In: Ruminant Physiology Digestion Growth and Reproduction* pp 61-77 Edited by P.B. Cronje. CABI Publishing, Oxford, UK

Makkar, H.P.S. and Becker, K. (1997) Degradation of quillaja saponins by mixed cultures of rumen microbes. *Letters in Applied Microbiology,* **25**, 243-245.

Martin, S.A. and Streeter, M.N. (1995) Effect of malate on in vitro mixed ruminal microorganism fermentation. *Journal of Animal Science,* **73**, 2141-2145.

Martin, S.A., Streeter, M.N., Nisbet, D.J., Hill, G.M. and Willamns, S.E. (1999) Effects of DL-malate on ruminal metabolism and performance of cattle fed a high-concentrate diet. *Journal of Animal Science,* **77,** 1008-1015.

Migeotte M.J. (1948) Spectroscopic evidence of methane in the earth's atmosphere. *Physical Reviews,* **73,** 519-520.

Mir,Z., Goonewardene, L.A., Okine, E., Jaegar, S. and Scheer, H.D. (1999) Effect of feeding canola oil on constituents, conjugated linoleic acid (CLA) and long chain fatty acids in goats milk. *Small Ruminant Research,* **33,** 137-143.

Montano, M.F., Chai, W., Zinn-Ware, T.E. and Zinn, R.A. (1999) Influence of malic acid supplementation on ruminal pH, lactic acid utilization, and digestive function in steers fed high-concentrate finishing diets. *Journal of Animal Science,* **77,** 780-784.

Morvan, B., Doré, J., Rieu-Lesme, F., Foucat, L., Fonty, G. and Gouet, P. (1994) Establishment of hydrogen-utilizing bacteria in the rumen of the newborn lamb. *FEMS Microbiology Letters,* **117,** 249-256.

Moss A.R. (1993) *Methane-Global Warming and Production By Animals.* Chalcombe Publications, UK.

Moss, A.R., Jouany, J.P. and Newbold, C.J. (2001) Methane production by ruminants: its contribution to global warming *Reproduction Nutrition Development (in press)*

Nagaraja, T.G., Newbold, C.J., Van Nevel, C.J. and Demeyer, D.I (1997) Manipulation of ruminal fermentation.. pp 523-632. *In : The Rumen Microbial Ecosystem.* Edited by P.N. Hobson and C.S. Stewart. Chapman & Hall, London.

Navas-Camacho A, Laredo, M.A., Cuesta, A, Anzola, H. and Leon, J.C. (1993) Effect of supplementation with a tree legume forage on rumen function. *Livestock Research for Rural Development,* **5,** 58-71.

Nelson, N., Valdes, C., Hillman, K., McEwan, N.R., Wallace, R.J. and Newbold, C.J. (2000) Effect of a methane oxidizing bacterium isolated from the gut of piglets on methane production in Rusitec. *Reproduction Nutrition Development,* **40,** 212.

Newbold, C.J. and Hillman, K. (1990) The effect of ciliate protozoa on the turnover of bacterial and fungal protein in the rumen of sheep. *Letters in Applied Microbiology,* **11,** 100-102.

Newbold, C.J. and Jouany, J.P. (1997) The contribution of individual genera in a mixed protozoal population to the breakdown of bacteria in the rumen. *Reproduction Nutrition Development,* **88,** 46.

Newbold, C.J., Wallace, R.J. and McIntosh, F.M (1996) Mode of action of the yeast, *Saccharomyc escerevisiae* as a feed additive for ruminants. *British Journal of Nutrition,* **76,** 249-261

Newbold, C.J., Morvan, B., Fonty, G. and Jouany, J.P (1996) The role of ciliate protozoa in the lysis of methanogenic archaea in rumen fluid. *Letters in Applied Microbiology,* **23**, 421-425.

Newbold, C.J., El Hassan, S.M., Wang J, Ortega, M.E. and Wallace, R.J. (1997) Influence of foliage from African multipurpose trees on activity of rumen protozoa and bacteria. *British Journal of Nutrition,* **78**, 237-249.

Newbold, C.J., McIntosh, F.M., Wallace, R.J., Williams, P. and Sutton, J.D. (1999) Effects of essential oils on ammonia production by rumen fluid in vitro. *Book of Abstracts of the VIIIth International Symposium on Protein Metabolism and Nutrition.* p 63

Newbold, C.J., Terferedegne, B., Kim, H.S., Zuur, G. and Lobely, G.E. (2000) Effects of protozoa on nitrogen metabolism in the rumen of sheep. *Reproduction Nutrition Development,* **40**, 199

Newbold, C.J., Eschenlauer, S.C.P., McEwan, N.R., Onodera, R. and Wallace, R.J. (2000) Properties of bacterolytic activities from the rumen ciliate protozoan, *Entodinium caudatum Proceedings of Japanese Society for Rumen Metabolism and Physiology* **11**, 86

Noble, R.C. (1981) Digestion, absorption and transport of lipids in ruminant animals. *In Lipid metabolism in ruminant animals.* pp 57-93 Edited by W.W.Christie. Pergamon Press Ltd, Oxford.

Noble, R.C., Harfoot, C.G. and Harfoot, C. (1974) Observations on the pattern of biohydrogenation of esterified and unesterified linoleic acid in the rumen. *British Journal of Nutrition,* **31**, 99-108.

Nolan, J. V. and Stachiw, S. (1979) Fermentation and nitrogen dynamics in Merino sheep given a low-quality-rouhage diet. *British Journal of Nutrition,* **42**, 63-79.

Nollet, L., Demeyer, D. and Verstraete, W. (1997) Effects of 2-bromoethanesulfonic acid and *Peptostreptococcus productus* ATCC 35244 addition on stimulation of reductive acetogenesis in the ruminal ecosystem by selective inhibition of methanogenesis. *Applied Environmental Microbiology,* **63,** 194-200

Nollet, L., Mbanzamihigo, L., Demeyer, D. and Verstraete, W. (1998) Effect of the addition of *Peptostreptococcus productus* ATCC 35244 on reductive acetogenesis in the ruminal ecosystem after inhibition of methanogenesis by cell-free supernatant of *Lactobacillus plantarum* 80. *Animal Feed Science and Technology,* **71**, 49-66.

Ouda, J.O., Newbold ,C.J., Lopez ,S, Nelson, N, Moss, A.R., Wallace, R.J. and Omed, H (1999) The effect of acrylate and fumarate on fermentation and methane production in the rumen simulating fermentor (Rusitec). *Proceedings of the British Society of Animal Science* **1999** p. 36

Park, C.M., Taormina, P.J. and Beauchat, LR (2000) Efficacy of allyl isothiocyanate in killing enterohemorrhagic *Escherichia coli* O157:H7 on alfalfa seeds. *International Journal of Food Microbiology,* **56**, 13-20.

Polan, C.E., McNeill, J.J. and Tove, S.B. (1964) Biohydrogenation of unsaturated fatty acids by rumen bacteria. *Journal of Bacteriology,* **88**, 1056-1064.

Rasmussen,M.A., Cray, W.C Jr., Casey, T.A. and Whipp S.C. (1993) Rumen contents as a reservoir of enterohemorrhagic *Escherichia coli. FEMS Microbiology Letters,* **114**, 79-84.

Rieu-Lesme, F., Godon, J.J. and Fonty, G. (2000) Remarkable archaeal diversity detected in the rumen of a cow. *Reproduction Nutrition Development,* **40**, 179.

Rodhe H. (1990) A comparison of the contribution of various gases to the greenhouse effect. *Science*, **248**, 1217-1219.

Roger,V., Fonty, G.,Komisarczuk-Bony, S. and Gouet, P. (1990) Effects of physiochemical factors on the adhesion to cellulose Avicel of the ruminal bacteria *Ruminicoccus flavefaciens* and *Fibrobacter succinogenes* subsp. succinogenes. *Applied and Environmental Microbiology,* **56**, 3081-3087.

Sachan,D.S.and Davis,C.L. (1969) Hydrogenation of linoleic acid by a rumen spirochete. *Journal of Bacteriology.* **98**, 300-301

Salway, A.G., Dore, C., Watterson, J. and Murrells, T. (1999) Greenhouse gas inventories for England, Scotland, Wales and Northern Ireland: 1990 and 1995. National Enviromental Technology Centre, UK.

Shepherd, J.L. (2000) The fate of Listeria in the ruminant gut. *PhD thesis Aberdeen University.*

Shepherd, J.l., Newbold, C.J., Hillman, K.and Fenlon, D.R. (2000) The effect of protozoa on the survival and flow of listeria from the rumen of sheep. *Reproduction Nutrition Development,* **40, 216.**

Shu,Q., Gill, H.S., Hennessy, D.W., Leng, R.A., Bird, S.H. and Rowe, J.B. (1999) Immunisation against lactic acidosis in cattle. *Research into Veterinary Science,* **67**, 65-71.

Slee, K.J. and Skilbeck, N.W. (1992) Epiemiology of *Yersinia pseudotuberculosis* and *Y. enterocolitica* infections in sheep in Australia. *Journal of Clinical Microbiology,* **30**, 712-715.

Sol Morales M., Palmquist, D.L. and Weiss, W.P. (2000a) Milk fat composition of Holstein and Jersey cows with control or depleted copper status and fed whole soybeans or tallow. *Journal of Dairy Science,* **83**, 2112-2119.

Sol Morales M., Palmquist, D.L. and Weiss, W.P. (2000b). Effects of fat source and copper on unsaturation of blood and milk triacylglycerol fatty acids in Holstein and Jersey cows. *Journal of Dairy Science,* **83,** 2105-2111.

Son,Y.S., Lee, J.W.C.B.R. and Yoon, J.A. (2000) Effect of soybean oil and monensin on *in vitro* lipid metabolism in the rumen. *Proceedings of Japanese Society for Rumen Metabolism and Physiology,* **11**, 44.

Stahl, D.A., Flesher, B.A., Mansfiled, H and Montgomery, L.A/ (1988) Use of phylogenteically based hybridization probes for studies of rumen microbial ecology. *Applied and Environmental Microbiology,* **54**, 1079-1084.

Stanley, K.N., Wallace, J.S, Currie, J.E., Diggle, P.J. and Jones, K. (1998) The seasonal variation of thermophilic campylobacters in beef cattle, dairy cattle and calves. *Journal of Applied Microbiology,* **85**, 472-480.

Stanton,C., Lawless, F., Kjellmer, G., Harrington, D., Devery, R., Connolly, J.F. and Murphy, J. (1997) Dietary influences on bovine milk *cis*-9, *trans*-11-conjugated linoleic acid content. *Journal of Food Science,* **62**, 1083-1086.

Steele L.P., Dlugokencky, E.J., Lang, P.M., Tans, P.P., Martin, R.C. and Masarie, K.A. (1992) Slowing down of the global accumulation of atmospheric methane during the 1980s. *Nature,* **358**, 313-316.

Stewart, C.S. and Bryant, M.P (1988) The rumen bacteria. In *The Rumen Microbial Ecosystem,* pp. 21-75 Edited by P.N. Hobson. Elsevier Applied Science. New York.

Stock, P.K. and McCleskey, C.S. (1964) Morphology & physiology of *Methamonas methanooxidans. Journal of Bacteriology,* **88**, 1071- 1077.

Teferedgne, B., Osuji, P.O., Odenyo, A., Wallace, R.J. and Newbold, C.J. (1998) Influence of saponins and sapogenins on the bacteriolytic activity of ciliate protozoa from the sheep rumen. *Proceedings of the British Society of Animal Science* **1998,** 88

Teferedgne, B (2000) The use of foliage from multipurpose trees to manipulate rumen fermentation. *PhD thesis Aberdeen University.*

Thalib, A., Widiawati, Y., Hamid, H., Suherman, D. and Sabrani, M. (1995) The effect of saponins from S*pinadus rarak* fruit on rumen microbes and host animal growth. *Annales de Zootechnie,* **44**, 161.

Vaessen, M.A., Veling, J, Frankena, K, Graat, E.A. and Klunder, J. (1998) Risk factors for *Salmonella dublin* infection on farms. *Veterinary Quarterly,* **20**, 97-99.

Valdes, C., Newbold, C.J., Hillman, K. and Wallace, R.J. (1996) Evidence for methane oxidation in rumen fluid *in vitro. Annales of Zootechnology,* **45** (Suppl), 351.

Valdes, C., Newbold, C.J., Hillman, K. and Wallace, R.J. (1997) Los microorganismos metanotrofos come agentes modifactdores de la fermentacion ruminal. *ITEA,* **18**, 157-159.

Van Nevel, C.J. and Demeyer, D.I. (1995) Feed additives and other interventions for decreasing methane emissions. In *Biotechnology in Animal Feeds & Animal Feeding,* pp. 329-349 Edited by R.J. Wallace and A.Chesson. VCH. Weinheim

Van Soest, P.J. (1994) *Nutritional Ecology of the Ruminant.* Cornell University Press, Ithaca.

Upton, P. and Coia, J.E. (1994) Outbreak of *Escherichia coli* O157 infection associated with pasteurized milk supply. *Lancet,* **344**, 1015

Wallace J.S. (1999) The ecological cycle of *Escherichia coli* O157:H7. In: *Escherichia coli O157 in farm animals* pp. 195-223. Edited by C.S. Stewart and H.J. Flint. CABI, Wallingford.

Wallace, R.J. and McPherson, C.A., (1987) Factors affecting the rate of breakdown of bacterial protein in rumen fluid. *British Journal of Science,* **58**, 313-323.

Wallace, R.J., Falconer, M.L. and Bhargava, P.K. (1989) Toxicity of volatile fatty acids at rumen pH prevents enrichment of *Escherichia coli* by sorbitol in rumen contents. *Current Microbiology,* **19**, 277-281.

Wang, Y., McAllister, T.A., Rode, L., Newbold, C.J., Cheeke, P.R. and Cheng, K.J. (1998) Effects of Yucca extract on fermentation and degradation of steroidal saponins in the Rusitec. *Animal Feed Science and Technology,* **74**, 143-153.

Watson, R.T., Meira Filho, L.G, Sanhueza, E. and Janetos T. (1992) Sources and Sinks. In *Climate Change 1992*, pp. 25-46 Edited by J.T. Houghton, B.A. Callander and S.K. Varney. Cambridge University Press. Cambridge.

Wesley, I.V., Wells, J.S. , Harmon, K.M., Green, A, Schroeder-Tucker, L, Glover, M and Siddique, I. (2000) Faecal shedding of Campylobacter and Arcobacter spp in dairy cattle. *Applied and Environmental Microbiology,* **66**, 1994-2000

White,R.W., Kemp, P. and Dawson, R.M.C. (1970) Isolation of a rumen bacterium that hydrogenates oleic acid as well as linoleic and linolenic acid. *Biochemical Journal* **116**, 767-768.

Williams, P.P. and Dinusson, W.E. (1973) Amino acid and fatty acid composition of bovine ruminal bacteria and protozoa. *Journal Animal Science,* **36**, 151-155

Williams, A. G. and Coleman, G. S. (1992) *The Rumen Protozoa*. Springer-Verlag, London.

Wood, J., Scott, K.P., Avugustin, G., Newbold, C.J. and Flint H.J. (1988) Estimation of the relative abundance od different *Bacteroides* and *Prevotella* ribotypes in gut samples by restriction enzyme profiling of PCR-amplified 16S rRNA gene sequences. *Applied and Environmental Microbiology,* **64**, 3683-3689.

Wolin, M.J. (1969). Volatile fatty acids and the inhibition of *Escherichia coli* growth by rumen fluid. *Applied Microbiology,* **17**, 83-87

Zhao, T., Doyle, M.P., Harmon, B.G., Brown, C.A., Mueller, P.O.E. and Parks, A.H. (1998) Reduction of carriage of enterohaemorrhagic *Escherichia coli* O157:H7 in cattle by inoculation with probiotic bacteria. *Journal of Clinical Microbiology,* **36**, 641-647.

16

DEVELOPMENTS IN RUMEN FERMENTATION – COMMERCIAL APPLICATIONS

Limin Kung

Department of Animal & Food Sciences, University of Delaware, Newark, Delaware 19717-1303, USA

Introduction

The use of feed additives containing live microorganisms and (or) their metabolites to alter rumen fermentation and improve animal performance has increased in response to demands for using more "natural" growth-promoting substances. The objectives of this chapter are to introduce the concept of this practice, to describe potential mechanisms of actions, and to discuss accepted and potential applications of these additives in ruminant nutrition.

Gut microbiology

Animals are sterile in the womb. After birth, their digestive tracts are naturally colonised by a variety of microorganisms from the environment (Savage, 1987). Under healthy conditions, "beneficial" microorganisms colonise the rumen and lower gut in a symbiotic relationship with the host. Gut microbes supply nutrients, aid digestion, and compete with potential pathogenic microbes. If young animals are removed and raised under sterile conditions, microorganisms are prevented from colonising their digestive tracts. These animals often have increased nutritional needs (e.g., requiring more vitamin K in the diet) and abnormal immune responses. Also, a normal, functional rumen will not develop under sterile conditions. These findings show that microbial colonisation of the digestive tract is essential for the normal development and well being of ruminants.

Direct-fed microbials

The original concept of administering microorganisms to animals involved the feeding of large amounts of "beneficial" microbes to livestock when they were "stressed" or ill. Microbial products used in this manner were originally called "probiotics", or products "for life." However, the term "probiotic" implied a curative nature. In the U.S., claims by a product to decrease mortality, to improve health, or to increase production (e.g. increased milk production or dry matter intake) cannot be made of any product unless its safety and efficacy have been documented and approved by government regulatory agencies. Thus, to overcome this requirement, the feed industry, in conjunction with regulatory agencies, has accepted the more generic term of "direct-fed microbials" (DFM) to describe microbial-based feed additives. In addition, a list of accepted microorganisms for use in animal feeds was developed.

General modes of action for DFM

Some of the major hypotheses on how DFM may benefit animals are listed in Table 16.1 and can be found in an excellent discussion by Fuller (1989). One of the most common explanations for improved animal health when ruminants are fed a DFM suggests that beneficial microbes compete with potential pathogens and prevent their establishment. Direct-fed microbials may also produce antimicrobial end products, such as acids, that limit the growth of pathogens. Additionally, metabolism of toxic compounds and production of stimulatory substances have resulted from feeding DFM to ruminants.

Table 16.1 Some proposed mechanisms of DFM when fed to animals

Proposed mechanisms

- Production of antibacterial compounds (acids, bacteriocins, antibiotics)
- Competition with undesirable organisms for colonisation space and/or nutrients (competitive exclusion)
- Production of nutrients (e.g. amino acids, vitamins) or other growth factors that stimulate other microorganisms in the digestive tract
- Production and/or stimulation of enzymes
- Metabolism and/or detoxification of undesirable compounds
- Stimulation of immune response in host animal
- Production of nutrients (e.g. amino acids, vitamins) or other growth factors that stimulate the host animal

Proof of concept of DFM for ruminants

The rumen microbial ecosystem is an extremely diverse and competitive environment. Because of this, many researchers were initially skeptical of being able to administer a DFM that would have lasting effects in the rumen. However, there have been several documented cases of improved animal performance when the rumen has been inoculated with select microbes. For example, the detoxification of the 3-hydroxy-4 (1H)-pyridone (DHP) is probably one of the most cited successes of manipulating ruminal fermentation with bacteria. The tropical forage *Leucaena leucocephala* contains mimosine, a non-protein amino acid. When consumed by ruminants in Australia and some parts of India, a metabolite of mimosine, DHP, causes goitrogenic effects. Jones and Megarrity (1986) showed that a rumen microbe, from cattle in Hawaii, was able to detoxify DHP. The specific organism responsible for detoxification, *Synergistes jonesii* (Allison *et al.*, 1990), was inoculated and established itself in the rumen of Australian cattle, thus conferring protection from DHP toxicity. Another example of a successful (but unapproved) application for a DFM is in the case of monofluroacetate poisoning of ruminants in Australia. Monofluroacetate is found in some Australian plants and can be toxic to ruminants at doses of about 0.3 mg/kg of body weight. Gregg *et al.* (1998) reported that they successfully inserted the gene encoding for fluoroacetate dehalogenase into several strains of *Butyrivibrio fibrisolvens*. When sheep were inoculated with the altered microbes, they showed reduced toxicological symptoms. The detoxification of DHP and monoflouroacetate are good evidence that ruminal fermentation can be modified with lasting effects by a DFM.

Bacterial DFM for ruminants

The general concept of inoculating ruminants with beneficial microorganisms is not a new practice. For example, many producers and veterinarians have been inoculating sick ruminants (especially those that have been off feed) with rumen fluid from healthy animals in the hope of stimulating normal rumen function and improving dry matter intake. However, there are no controlled research studies that document the efficacy of this practice and there are no commercial products based on this concept.

In contrast, there are many bacterial-based DFM that are sold for use in ruminant diets with more specific applications. These products often contain lactobacilli with *Lactobacillus acidophilus* being one of the most common microorganisms used. Other commonly used bacteria include various species of *Bifidobacterium, Enterococcus,* and *Bacillus*. Most bacterial-based DFM are probably beneficial because they have effects in the lower gut and not in the rumen. For example, *Lactobacillus acidophilus* produces lactic acid, which may lower the pH in small intestines to levels that inhibit the growth of pathogenic microbes. Early research with DFM in ruminants involved applications for young calves fed milk, calves being weaned, or cattle being shipped (Jenny *et al.*, 1991; Hutchenson *et al.*, 1980). These animals were thought to be stressed and have immature microbial ecosystems in their guts (Vandevoorde *et al.*, 1991). Cattle that are transported often have limited feed and water for prolonged periods of time during transit and may have abnormal environments in their guts that could lead to establishment of pathogenic microbes. It has been suggested that large doses of beneficial microorganisms re-colonise a stressed intestinal environment and return gut function to normal more quickly in scouring calves. However, the data supporting such claims have been inconclusive. Calves fed *L. acidophilus* had reduced incidence of diarrhoea (Beecham *et al.*, 1977) and reduced counts of intestinal coliform bacteria (Bruce *et al.*, 1979). However, feeding bacterial DFM to calves had no beneficial effects in other studies (Abu-Taroush *et al.*, 1996; Cruywagen *et al.*, 1996).

Only a few studies have documented positive effects of feeding bacterial DFM to lactating dairy cows. High producing cows in early lactation would be the best candidates for such products because these cows are in negative energy balance and are given diets containing highly fermentable carbohydrates that sometimes leads to acidosis. Jaquette *et al.* (1988) and Ware *et al.* (1988) reported increased milk production from cows fed *L. acidophilus* (1×10^9 colony-forming units per head per day). Jeong *et al.* (1998) fed Lactobacillus sp. and Streptococcus sp. to lactating cows and reported a 0.8 kg/d improvement in milk production over control cows. Supplementation of lactobacilli may be useful in the close-up dry period of lactation when intake is depressed and animals are stressed. Savoini *et al.* (2000) reported that cows fed lactobacilli in the transition period tended to produce more milk and had lower blood non-esterified fatty acids, but higher blood glucose than did untreated cows.

Experimentally, there have been several bacteria that have potential as DFM for ruminants but have not been commercialised for a number of different reasons. For example, *Megasphaera elsdenii* is the major lactate-utilising organism in the rumen of adapted cattle fed high-grain diets. However, when cattle are abruptly shifted from a high-forage to high-concentrate diet, the numbers of *Megasphaera elsdenii* are often insufficient to prevent lactic acidosis. We have shown that during a challenge with highly fermentable carbohydrates, addition of *Megasphaera elsdenii* B159 prevented an accumulation of lactic acid (Figure 16.1. Kung and Hession, 1995). Addition of a different strain of *Megasphaera elsdenii* (407A) has prevented lactic acidosis in steers (Robinson *et al.*, 1992). Success with such an organism could allow feedlot producers to decrease the time it takes to adapt cattle to a high-concentrate diet. Development of this organism for high-producing dairy cows could also be useful in reducing sub-acute lactic acidosis, but optimum timing of administration and effective dose must be determined.

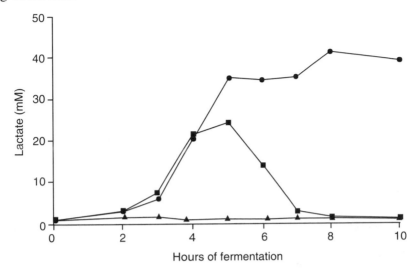

Figure 16.1 Effect of adding *Megaspahera elsdenii* (ME), a rumen bacterium that utilises lactic acid, on the prevention of accumulation of lactic acid in a mixed ruminal culture containing highly fermentable carbohydrates.   = untreated cultures, • = ME added to achieve 8.7 × 10⁵ cfu/ml,   = ME added to achieve 8.7 × 10⁶ cfu/ml. (Kung and Hession, 1995.)

Propionibacteria may also be beneficial when fed to ruminants. These bacteria are naturally found in high numbers in the rumen of animals fed forage and medium concentrate diets. They have the ability to convert lactic acid and glucose to acetic and propionic acid. *Propionibacteria* may be beneficial if inoculated into the rumen (Kung *et al.*, 1991) because higher concentrations of ruminal propionate would improve the energy status of the animal. Swinney-Flyod *et al.* (1999) reported that feedlot cattle fed a diet containing *Propionibacteria* (strain P-63, 1 × 10⁹ cfu/head/day) and *L. acidophilus* (strain 5345, 1 × 10⁸ cfu/head/day) had better feed efficiencies during adaptation to a high concentrate diet and during a 120-d feeding period. Similarly, Huck *et al.* (1999) reported that cattle fed *L. acidophilus* (5 × 10⁸ cfu/head/day) strain BG2F04, and *P. freudenrechii* (1 × 10⁹ cfu/head/day) had better feed efficiencies than those fed a control diet. Kim *et al.* (2000) reported that *Propionibacterium acidipropionici* DH42 decreased the molar percentage of acetic acid but increased the molar percentage of propionic acid when fed to steers at a minimum level of 1 × 10⁷ cfu/head/day. *Propionibacterium freudenreichii* has also been used in a commercial product that also contains several strains of lactobacilli and has improved weight gain in some studies with calves (Cerna *et al.*, 1991). Although *Propionibacteria* can metabolise lactic acid, they are probably too slow growing and acid-intolerant to prevent an acute lactic acidosis challenge. Because *Propionibacteria* can metabolise nitrates, a commercially available product based on a strain that naturally occurs in the rumen has been claimed to reduce the chance of nitrate toxicity, but definitive data are lacking.

Fungal DFM

Fungal DFM have been popular additions to ruminant diets for many years. In general, three types of additives are available. Firstly, some products contain and guarantee "live" yeast. Most of these products contain various strains of *Saccharomyces cerevisiae*. Secondly, other additives contain *Saccharomyces cerevisiae* and culture extracts, but make no guarantee for live organisms. Thirdly, there are fungal additives based on *Aspergillus oryzae* (AO) fermentation end products that also make no claim for supplying live microbes.

Fungal-based DFM are beneficial because of interactions in the rumen (Figure 16.2). Several reasons for improvements in ruminal fermentation from feeding fungal DFM have been suggested. Firstly, DFM have caused beneficial changes in activity and numbers of rumen microbes. For example, the numbers of total ruminal anaerobes (Dawson *et al.*, 1990; Newbold *et al.*, 1991) and cellulolytic bacteria (Harrison *et al.*, 1988) have been increased with fungal extracts. Beharka *et al.* (1991) reported that young calves fed an AO fermentation extract were weaned one week earlier than untreated calves and that supplementation increased the numbers of rumen bacteria and VFA concentrations. Aspergillus fermentation extracts (Chang *et al.*, 1999) and yeast cultures (Chaucheryas *et al.*, 1995b) have also directly stimulated rumen fungi, which may improve fibre digestion. Feeding *Saccharomyces cerevisiae* increased the number of rumen protozoa and increased digestion of neutral-detergent fibre (NDF) in steers fed straw-based diets (Plata *et al.*, 1994). Yeasts have also been shown to stimulate acetogenic bacteria in the presence of methanogens (Chaucheryas *et al.*, 1995b), which might result in a more efficient ruminal fermentation. Secondly, fungal DFM may prevent the accumulation of excess lactic acid in the rumen when cattle are fed diets containing highly fermentable carbohydrates. Specifically, extracts of *Aspergillus oryzae* stimulated the uptake of lactic acid by the rumen lactate-utilisers *Selenomonas ruminantium* (Nisbet and Martin, 1990) and *Megasphaera elsdenii* (Waldrip and Martin, 1993), possibly by providing a source of malic acid. Increased metabolism of lactic acid should theoretically raise ruminal pH and this may be one reason why these DFM increased numbers of rumen cellulolytic bacteria and improved fibre digestion (Arambel *et al.*, 1987). Chaucheyras *et al.* (1995c) reported that *Saccharomyces cerevisiae* was able to prevent the accumulation

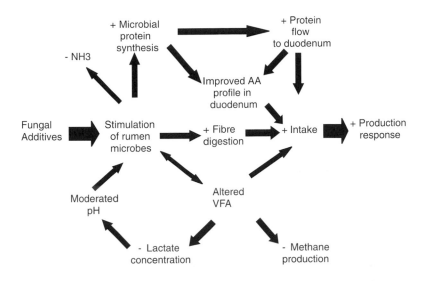

Figure 16.2 Potential effects of fungal additives on rumen fermentation.

of lactic acid production by competing with *Streptococcus bovis* for glucose and by stimulating the uptake of lactic acid by *Megasphaera elsdenii*, perhaps by supplying amino acids and vitamins. In contrast, added yeasts were unable to prevent acute episodes of lactic acidosis when fermentations were challenged with a diet rich in fermentable carbohydrates (Aslan *et al.*, 1995; Dawson and Hopkins, 1991). Yeasts may improve ruminal fermentation because they are able to scavenge excess oxygen (Newbold *et al.*, 1996) creating a more favourable environment for rumen anaerobic bacteria. *Aspergillus* extracts may improve fibre digestion because they contain esterase enzymes (Varel *et al.*, 1993). An interesting finding reported by one group of researchers showed that feeding AO to cows in hot environments decreased rectal temperatures in some (Huber *et al.*, 1994), but not all, studies (Denigan *et al.*, 1992). The reasons for these findings were unclear. Importantly, not all strains of *Saccharomyces cerevisiae* and *Aspergillus* extracts have stimulatory effects on rumen fermentation. For example, Newbold *et al.* (1995b) reported that the stimulation of rumen bacteria was different with specific strains of *Saccharomyces cerevisiae,* but the reasons for these differences were unknown.

Under farm conditions, producers are mainly interested in how a feed additive affects animal production (live-weight gain and milk yield) and feed efficiency. There have been numerous studies reporting positive effects, but also lack of effects, of *Saccharomyces cerevisiae* and *Aspergillus oryzae* on intake and milk production of lactating cows. For example, feeding *Saccharomyces cerevisiae* has increased dry matter intake in some (Williams *et al.*, 1991; Wohlt *et al.*, 1991) but not in other studies (Arambel and Kent, 1990; Kung *et al.*, 1997). Milk production has been increased in some studies (Kung *et al.*, 1997; Piva *et al.*, 1993) but not in others (Erdman and Sharma, 1989; Swartz *et al.*, 1994). Yeast cultures have also been fed to cows prepartum, improving dry matter intake in some studies (Dann *et al.*, 2000; Wohlt *et al.*, 1991), but not in others (Robinson, 1997; Soder and Holden, 1999). Similarly, feeding *Aspergillus oryzae* has improved milk production in some studies (Gomez-Alarcon *et al.*, 1991; Kellems *et al.*, 1990), but not other studies (Bertrand and Grimes, 1997; Sievert and Shaver, 1993). No single reason explains the inconsistent effects of fungal cultures on animal performance, probably because they have multiple modes of action.

Practical considerations for DFM

In general, most researchers would agree that DFM based on bacteria must be "live." Thus, they must survive processing, storage and the gut environment. In contrast, the need to provide high numbers of "live" yeast cells (*Saccharomyces cerevisiae*) has been the subject of many debates. As previously mentioned, some products guarantee live yeast cells (e.g., 1×10^9 cfu per g) and are fed at low inclusion rates (only 10-20 grams per day), but other products suggest that live organisms are not required for beneficial effects. The metabolites present in the culture extracts have been suggested to be the "active" ingredients. Newbold *et al.* (1991) reported that autoclaving, but not irradiation, decreased the ability of an *Aspergillus oryzae* extract to stimulate growth and activity of rumen bacteria. Dawson *et al.* (1990) reported that the stimulatory effect of yeast on numbers of cellulolytic bacteria in the rumen was negated when yeasts were autoclaved. Although there have been suggestions that yeasts were able to grow in continuous rumen cultures (Dawson *et al.*, 1990), *Saccharomyces cerevisiae* did not multiply in sterile ruminal fluid, despite being metabolically active (Kung *et al.*, 1996). Durand-Chaucheyras *et al.* (1998) confirmed that added *Saccharomyces cerevisiae* did not colonise the rumen of lambs and Kung *et al.* (1997) reported that yeasts were essentially washed out of continuous-flow rumen fermenters. The debate on the need for live yeasts will continue, unless more definitive studies addressing this issue are conducted.

Direct-fed microbial products are available in a variety of forms, including powders, pastes, boluses, and capsules. In some applications, DFM may be mixed with feed or administered in the drinking water. However, use of DFM in the latter manner must be managed closely, since interactions with chlorine, water temperature, minerals, flow rate, and antibiotics can affect the viability of many organisms. Non-hydroscopic whey is often used as a carrier for bacterial DFM and is a good medium to initiate growth. Bacterial DFM pastes are formulated with vegetable oil and inert gelling ingredients. Some fungal products are formulated with grain by-products as carriers. Some DFM are designed for one-time dosing, while other products are designed for feeding on a daily basis. However, there is little information comparing the efficacy of administering a DFM in a single massive dose, compared with continuous daily dosing. Lee and Botts (1988) reported that pulse dosing alone, or pulse dosing with daily feeding of *Streptococcus faecium* M74 resulted in improved performance of incoming feedlot cattle. The need for a bacterial DFM to actually attach and colonise gut surfaces in order to have a beneficial effect is also questionable. However, in certain applications, the argument could be made that a DFM organism need only produce its active component (without colonisation) to be beneficial. Dose levels of bacterial DFM have varied. Studies can be found where *L. acidophilus* have been fed at levels ranging from 10^6 to 10^{10} cfu per animal per day. Hutchenson *et al.* (1980) suggested that feeding more than 10^7 cfu per head per day may cause lower nutrient absorption due to overpopulation of the gut. Orr *et al.* (1988) reported that feeding a continuous high dose of *L. acidophilus* to feeder calves (10^{10} cfu per head/day) had no effect on gain and actually reduced feed efficiency when compared with feeding a lower dose (10^6).

Tolerance of DFM microorganisms to heat is important since many feeds are pelleted. In general, most yeasts, *Lactobacillus*, *Bifidobacterium,* and *Streptococcus* are destroyed by heat during pelleting. In contrast, bacilli form stable endospores when conditions for growth are unfavourable and are very resistant to heat, pH, moisture and disinfectants. Thus, bacilli are currently used in many applications that require pelleting. Over-blending can sometimes compensate for microbial loss during pelleting, but this is not an acceptable routine practice. Future improvements in strain development may allow use of heat-sensitive organisms in pelleted feeds. Bacterial products may or may not be compatible with use of traditional antibiotics and thus care should be taken when formulations contain both types of additives. For example, some species of bacilli are sensitive to virginiamycin, and lactobacilli are sensitive to chlortetracycline and penicillin. Information on DFM and antibiotic compatibility should be available from the manufacturer.

Viability of DFM products has improved over the past several years, but each product has its own storage recommendations that should be followed, as microorganisms can be highly sensitive to heat and moisture. Future research on new DFM products will need to address viability if oxygen-sensitive microorganisms are to be developed for commercial purposes.

The future of direct-fed microbials

Our understanding of how and when DFM improve animal production is in its infancy. In the immediate future, approaches that identify naturally-occurring microbes capable of filling specific niches within the rumen (for example, detoxification of compounds such as alkaloids, oxalates, tannins, or mycotoxins) may be fruitful. Inhibition of lactic acidosis, by selection for bacteria that utilise lactic acid, would also be useful. In the long term, rumen and traditional DFM organisms may be genetically modified through recombinant DNA technology. For example, organisms may be engineered to secrete essential amino acids or growth factors. Genetic modification of bacteria to improve fibre digestion in the rumen has also been studied. However, the future of genetically-modified organisms is in question, due to the reluctance of government agencies to approve them, and resistance from consumers and producers.

References

Abu-Taroush, H.M., Al-Saiady, M.Y. and El-Din, A.H.K. (1996) Evaluation of diet containing Lactobacilli on performance, fecal coliform, and Lactobacilli of young dairy calves. *Animal Feed Science Technology*, **57**, 39-49.

Allison, M.J., Hammond, A.C. and Jones, R.J. (1990) Detection of rumen bacteria that degrade toxic dihydroxypridine compounds produced from mimosine. *Applied Experimental Microbiology*, **56**, 590-594.

Arambel, M.J. and Kent, B. (1990) Effect of yeast culture on nutrient digestibility and milk yield response in early- to midlactation dairy cows. *Journal of Dairy Science*, **73**, 1560-1563.

Arambel, M.J., Weidmeier, R. D. and Walters, J.L. (1987) Influence of donor animal adaptation to added yeast culture and/or *Aspergillus oryzae* fermentation extract on in vitro rumen fermentation. *Nutrition Reports International*, **35**, 433-437.

Aslan, V.S., Thamsborg, M., Jorgensen, R.J. and Basse, A. (1995) Induced acute ruminal acidosis in goats treated with yeast (*Saccharomyces cerevisiae*) and bicarbonate. *Acta Veterinaria Scandinavica*, **36**, 65-68.

Beecham, T.J., Chambers, J.V. and Cunningham, M.D. (1977) Influence of *Lactobacillus acidophilus* on performance of young dairy calves. *Journal of Dairy Science*, **60**, (Supplement 1), 74. (Abstract)

Beharka, A.A., Nagaraja, T.G. and Morrill, J.L. (1991) Performance and ruminal development of young calves fed diets with *Aspergillus oryzae* fermentation extracts. *Journal of Dairy Science*, **74**, 4326-4336.

Bertrand, J.A. and Grimes, L.W. (1997) Influence of tallow and *Aspergillus oryzae* fermentation extract in dairy cattle rations. *Journal of Dairy Science*, **80**, 1179-1184.

Bruce, B.B., Gilliland, S.E., Bush, L.J. and Staley, T.E. (1979) Influence of feeding cells of *Lactobacillus acidophilus* on the fecal flora of young dairy calves. *Oklahoma Animal Science Research Report* **207**, Stillwater, OK.

Chang, J.S., Harper, E.M. and Calza, R.E. (1999) Fermentation extract effects on the morphology and metabolism of the rumen fungus *Neocallimastix frontalis* EB188. *Journal of Applied Microbiology*, **86**, 389-398.

Cerna, B., Cerny, M., Betkova, H., Patricny, P., Soch, M. and Opatrna, I. (1991) Effect of Proma on calves. *Zivocisna Vyroba,* **36**, 381-388.

Chaucheyras, F., Fonty, G., Bertin, G. and Gouet, P. (1995) Effects of live *Saccharomyces cerevisiae* cells on zoospore germination, growth, and cellulolytic activity of the rumen anaerobic fungus, *Neocallimastix frontalis* MCH3. *Current Microbiology*, **31**, 201-205.

Chaucheryras, F., Fonty, G., Bertin, G. and Gouet, P. (1995b) In vitro utilization by a ruminal acetogenic bacterium cultivated alone or in association with an Archea methanogen is stimulated by a probiotic strain of *Saccharomyces cerevisiae*. *Applied Environmental Microbiology*, **61**, 466-3467.

Chaucheyras, F., Fonty, G., Bertin, G., Salmon, J.M. and Gouet, P. (1995c) Effects of a strain of Saccharomyces cerevisiae (Levucell SC), a microbial additive for ruminants, on lactate metabolism in vitro. *Canadian Journal of Microbiology*, **42**, 927-933.

Cruywagen, C.W., Jordaan, A. and Venter, L. (1996) Effect of Lactobacillus acidophilus supplementation of milk replacer on preweaning performance of calves. *Journal of Dairy Science*, **79**, 483-486.

Dann, H.M., Drackely, J.K., McCoy, G.C., Hutjens, M.F. and Garrett, J.E. (2000) Effects of a yeast culture (*Saccharomyces cerevisiae*) on prepartum intake and postpartum intake and milk production of Jersey cows. *Journal of Dairy Science*, **83**, 123-127.

Denigan, M.E., Huber, J.T., Alhadhrami, G. and Al-Deneh, A. (1992) Influence of feeding varying levels of Amaferm on performance of lactating dairy cows. *Journal of Dairy Science*, **75**, 1616-1621

Durand-Chaucheyras, F., Fonty, G., Bertin, G., Theveniot, M. and Gouet, P. (1998) Fate of Levucell SC I-1077 yeast additive during digestive transit in lambs. *Reproduction Nutrition Development*, **38**, 275-280.

Dawson, K.A., Neuman, K.E. and Boling, J.A. (1990) Effects of microbial supplements containing yeast and lactobacilli on roughage-fed ruminal microbial activities. *Journal of Animimal Science*, **68**, 3392-3398.

Dawson, K.A., and Hopkins, D.M. (1991) Differential effects of live yeast on the cellulolytic activities of anaerobic ruminal bacteria. *Journal of Animal Science*, **69**, (Supplement 1) 531. (Abstract)

Erdman, R.A., and Sharma, B.K. (1989) Effects of yeast culture and sodium bicarbonate on milk yield and composition in dairy cows. *Journal of Dairy Science*, **72**, 1929-1932.

Fuller, R. (1989) Probiotics in man and animals. *Journal of Applied Bacteriology*, **66**, 365-378.

Gomez-Alarcon, R.A., Huber, J.T., Higginbotham, G.E., Wiesma, F., Ammon, D. and Taylor, B. (1991) Influence of feeding *Aspergillus oryzae* culture on the milk yields, eating patterns, and body temperatures of lactating cows. *Journal of Animal Science*, **69**, 1733-1740.

Gregg, K., Hamdorf, B., Henderson, K., Kopecny, J. and Wong, C. (1998) Genetically modified ruminal bacteria protect sheep from fluoroacetate poisoning. *Applied Environmental Microbiology*, **64**, 3496-3498.

Harrison, G.A., Hemken, R.W., Dawson, K.A., Harmon, R.J. and Barker, K.B. (1988) Influence of addition of yeast culture to diets of lactating cows on ruminal fermentation and microbial populations. *Journal of Dairy Science*, **71**, 2967-2975.

Huber, J.T., Higginbotham, G., Gomez-Alarcon, R.A., Taylor, R.B., Chen, K.H., Chan, S.C. and Wu, Z. (1994) Heat stress interactions with protein, supplemental fat, and fungal cultures. *Journal of Dairy Science*, **77**, 2080-2090.

Huck, G.L., Kriekemeier, K.K. and Ducharme, G.A. (1999) Effect of feeding *Lactobacillus acidophilus* BG2F04 (Micro cell) and *Propionibacterium freudenrechii* P -63 (MicroCell PB on growth performance of finishing heifers. *Journal of Animal Science*, **77**, (Supplement 1), 264. (Abstract)

Hutchenson, D.P., Cole, N.A., Keaton, W., Graham, G., Dunlap, R. and Pitman, K. (1980) The use of living, nonfreeze-dried *Lactobacillus acidophilus* culture for receiving feedlot calves. *Proceedings Western Section, American Society of Animal Science*, **31**, 213. (Abstract)

Jaquette, R.D., Dennis, R.J., Coalson, J.A., Ware, D.R., Manfredi, E.T. and Read, P.L. (1988) Effect of feeding viable *Lactobacillus acidophilus* (BT1386) on the performance of lactating dairy cows. *Journal of Dairy Science*, **71**, (Supplement 1), 219. (Abstract)

Jeong, H.Y., Kim, J.S., Ahn, B.S., Cho, W.M., Kweon, U.G., Ha, J.K. and Chee, S.H. (1998) Effect of direct-fed microbials (DFM) on milk yield, rumen fermentation and microbial growth in lactating dairy cows. *Korean Journal of Dairy Science*, **20**, 247-252.

Jenny, B.F., Vandijk, H.J. and Collins, J.A. (1991) Performance and fecal flora of calves fed a *Bacillus subtilis* concentrate. *Journal of Dairy Science*, **74**, 1968-1973.

Jones, R.J. and Megarrity, R.G. (1986) Successful transfer of DHP-degrading bacteria from Hawaiian goats to Australian ruminants to overcome the toxicity of Leucaena. *Australian Veterinary Journal*, **63**, 259-262.

Kellems, R.O., Lagerstedt, A. and Wallentine, M.V. (1990) Effect of feeding *Aspergillus oryzae* fermentation extract or *Aspergillus oryzae* plus yeast culture plus mineral and vitamin supplement on performance of Holstein cows during a complete lactation. *Journal of Dairy Science*, **73**, 2922-2928.

Kim, S.W., Standorff, D.G., Roman-Roasario, H., Yokoyama, M.T. and Rust, S.R. (2000) Potential use of *Propionibacterium acidipropionici* DH42 as a direct-fed microbial for cattle. *Journal of Animal Science*, **78**, (Supplement 1), 292. (Abstract)

Kung, L., Jr. and Hession, A.O. (1995) Altering rumen fermentation by microbial inoculation with lactate-utilizing microorganisms. *Journal of Animal Science*, **73**, 250-256.

Kung, L., Jr. Hession, A.O., Tung, R.S. and Maciorowski, K. (1991) Effect of *Propionibacterium shermanii* on ruminal fermentations. *Proceedings 21st Biennial Conference on Rumen Function.* Chicago, IL, p 31. (Abstract)

Kung, L., Jr. Kreck, E.M., Tung, R.S., Hession, A.O., Sheperd, A.C., Cohen, M.A., Swain, H.E. and Leedle, J.A.Z. (1996) Effects of a live yeast culture and enzymes on in vitro ruminal fermentation and milk production of dairy cows. *Journal of Dairy Science*, **80**, 2045-2051.

Kung, L., Jr. and Muck, R.E. (1997) Animal response to silage additives. Proceedings of the conference on Silage: Field to feedbunk. North American Conference Hershey, PA. *Northeast Regional Agricultural Engineering Service*-99, Ithaca, NY.

Lee, R.W., and Botts, R.L. (1988) Evaluation of a single oral dosing and continuous feeding of *Streptococcus faecium* M74 (Syntabac) on the performance of incoming feedlot cattle. *Journal of Animal Science*, **66**, (Supplement 1) 460. (Abstract)

Newbold, C.J., Wallace, R.J. and McIntosh, F.M. (1996) Mode of action of the yeast *Saccharomyces cerevisiae* as a feed additive for ruminants. *British Journal of Nutrition*, **76**, 249.

Newbold, C.J., Brock, R. and Wallace, R.J. (1991) Influence of autoclaved or irradiated *Aspergillus oryzae* fermentation extract on fermentation in the rumen simulation technique (Rusitec). *Journal of Agricultural Science*, Cambridge, **116**, 159-162.

Newbold, C.J., Wallace, R.J., Chen, X.B. and McIntosh, F. (1995) Different strains of *Saccharomyces cerevisiae* differ in their effects on ruminal bacterial numbers in vitro and in sheep. *Journal of Animal Science*, **73**, 1811-1818.

Nisbet, D.J., and Martin, S.A. (1990) Effect of dicarboxylic acids and *Aspergillus oryzae* fermentation extract on lactate uptake by the ruminal bacterium *Selenomonas ruminantium*. *Applied Environmental Microbiology*, **56**, 3515-3518.

Orr, C.L., Ware, D.R., Manfredi, E.T. and Hutchenson, D.P. (1988). The effect of continuous feeding of *Lactobacillus acidophilus* strain BT1386 on gain and feed efficiency of feeder calves. *Journal of Animal Science*, **66**, (Supplement 1) 460. (Abstract)

Piva G., Belladonna, S., Fusconi, G. and Sicbaldi, F. (1993) Effects of yeast on dairy cow performance, ruminal fermentation, blood components, and milk manufacturing properties. *Journal of Dairy Science*, **76**, 2717-2722.

Plata, F.P., Mendoza, G.D., Barcena-Gama, J.R. and Gonzalez, S.M. (1994) Effect of a yeast culture (*Saccharomyces cerevisiae*) on neutral detergent fibre digestion in steers fed oat straw based diets. *Animal Feed Science Technology*, **49**, 203-210.

Robinson, J.A., Smolenski, W.J., Greening, R.C., Ogilvie, R.L., Bell, R.L., Barsuhn, K. and. Peters, J.P. (1992). Prevention of acute acidosis and enhancement of feed intake in the bovine by *Megasphaera elsdenii* 407A. *Journal of Animal Science*, **70**, (Supplement 1), 310. (Abstract)

Robinson, P.H. (1997) Effect of yeast culture (*Saccharomyces cerevisiae*) on adaptation of cows to diets postpartum. *Journal of Dairy Science*, **80**, 1119-1125.

Savage, D.C. (1987) Microorganisms associated with epithelial surfaces and the stability of the indigenous gastrointestinal microflora. *Die Nahrung*, **5-6**, 383-390.

Savoini, G., Mancin, G., Rossi, C.S., Grittini, A., Baldi, A., Dell-Orto, V. (2000) Administration of lactobacilli in transition [peripartum] cows: effects on blood level of glucose, beta-hydroxybutyrate and NEFA and on milk yield. *Obiettivi-e-Documenti-Veterinari*, **21**, 65-70.

Sievert, S. and Shaver, R. D. (1993) Effect of nonfibre carbohydrate level and *Aspergillus oryzae* fermentation extract on intake, digestion, and milk production in lactating dairy cows. *Journal of Animal Science*, **71**, 1032-1040.

Schwartz, D.L., Muller, L.D., Rogers, G.W. and Varga, G.A. (1994) Effect of yeast cultures on performance of lactating dairy cows: a field study. *Journal of Dairy Science*, **77**, 3073-3080.

Soder, K.J., and Holden, L.A. (1999) Dry matter intake and milk yield and composition of cows fed yeast prepartum and postpartum. *Journal of Dairy Science*, **82**, 605-610.

Swinney-Floyd, D, Gardner, B.A., Rehberger, T. and Parrot, T. (1999) Effects of inoculation with either *Propionibacterium* strain P-63 alone or combined with *Lactobacillus acidophilus* strain :LZ 53545 on performance of feedlot cattle. *Journal of Animal Science*, **77**, (Supplement 1) 77. (Abstract)

Vandevoorde, L., Christianens, H. and Verstraete, W. (1991) In vitro appraisal of the probiotic value of intestinal lactobacilli. *World Journal of Microbiology and Biotechnology*, **7**, 587-592.

Varel, V.H., Kreikemeier, K.K., Jung, H.J.G. and Hatfield, R.D. (1993) In vitro stimulation of forage fibre degradation by ruminal microorganisms with *Aspergillus oryzae* fermentation extract. *Applied Environmental Microbiology*, **59**, 3171-3176.

Waldrip, H.M. and Martin, S.A. (1993) Effects of an Aspergillus oryzae fermentation extract and other factors on lactate utilization by the ruminal bacterium *Megasphaera elsdenii*. *Journal of Animal Science*, **71**, 2770-2776.

Ware, D.R., Read, P.L. and Manfredi, E.T. (1988) Lactation performance of two large dairy herds fed *Lactobacillus acidophilus* strain BT 1386. *Journal of Dairy Science*, **71**, (Supplement 1), 219. (Abstract)

Williams, P.E.V., Tait, C.A.G., Innes, G.M. and Newbold, C.J. (1991) Effects of the inclusion of yeast culture (*Saccharomyces cerevisiae* plus growth medium) in the diet of dairy cows on milk yield and forage degradation and fermentation patterns in the rumen of steers. *Journal of Animal Science*, **69**, 3016-3026.

Wohlt, J.E., Finkelstein, A.D. and Chung, C.H. (1991) Yeast culture to improve intake, nutrient digestibility, and performance by dairy cattle during early lactation. *Journal of Dairy Science*, **4**, 1395-1400.

17

THE USE OF ENZYMES IN RUMINANT DIETS

KAREN A. BEAUCHEMIN[1], DIEGO P. MORGAVI[1], TIM A. MCALLISTER[1], WEN Z. YANG[1] and LYLE M. RODE[2]

[1]*Agriculture and Agri-Food Canada, Lethbridge, AB, Canada, T1J 4B1*
[2]*Biovance Technologies Inc., Lethbridge, AB, Canada, T1K 3J5*

Introduction

Enzymes are used extensively by the poultry industry to increase the digestibility and absorption of nutrients, to remove anti-nutritional factors from feeds, and to complement endogenous enzymes (Bedford, 2000). In contrast, the use of feed enzymes in ruminant diets is a technology in development. Research has demonstrated that supplementing diets for dairy cows and feedlot cattle with fibre-degrading enzymes has significant potential to improve feed utilization and animal performance. A limited number of ruminant enzyme products are now commercially available in North America and we expect this list of products to grow. However, research is needed to understand the mode of action of these new products so that on-farm efficacy can be assured.

The use of fibre-degrading enzymes for ruminants was first examined in the 1960s (as reviewed by Beauchemin and Rode, 1996), but many of these early preparations were poorly defined, animal responses were variable, and little or no effort was made to design these products specifically for ruminants. At that time, production of exogenous enzymes was expensive, as fermentation systems were not characterized and the proper isolates and tools of molecular biology were not available to lower the cost of enzyme production. Recent reductions in fermentation costs, together with more active and better-defined enzyme preparations, have prompted a re-examination of the potential use of exogenous enzymes in ruminant production. While much progress has been made in terms of advancing enzyme technology for ruminants, considerable research is still required to reduce the variability in response. This chapter reviews the research on feed enzyme usage in ruminant diets and the possible mechanisms by which these products may improve nutrient utilization.

Feed Enzymes

Feed enzyme products for ruminants are concentrated fermentation products that have specific enzyme activities, although usually only the main activities are controlled. Almost all of the enzyme products for ruminants are from *Trichoderma longibrachiatum*, although other bacterial and fungal species are permitted (Pendleton, 1998). In addition to refined sources of enzymes, many crude fermentation products and direct-fed microbials are marketed, at least partly or implicitly, upon their residual enzymatic content. In contrast to concentrated feed enzyme products, these crude products contain relatively little enzyme activity (< 5%). There is no minimum level of enzyme activity required for products to be registered as feed enzymes in North America, which adds tremendous confusion in the marketplace. It can be difficult to distinguish between "true" concentrated enzyme products and direct-fed microbials with low levels of enzyme activity even though all enzyme products labelled for use in animal feeds must specify enzyme activity. Even though enzyme activities are reported, the conditions of the assays and the enzyme units differ among manufacturers. Furthermore, when enzyme activities are reported they are typically measured at the manufacturers' recommended optima, which is usually a lower pH and a higher temperature than that encountered in the rumen. Thus, the activities quoted are considerably higher than those that would be measured at a pH and temperature similar to that in the rumen. The scope of this discussion is limited to concentrated enzyme products, as the mode of action of direct-fed microbial products is probably very different to that of enzymes.

Enzyme preparations for ruminants are evaluated primarily on the basis of their capacity to degrade plant cell walls. Typically, these enzymes fall into the general classification of cellulases and xylanases. Each of these categories can consist of numerous specific enzymes having activity against cellulose or xylan. For example, enzymes with cellulase activity include numerous types of endoglucanases that cleave internal β-1,4 and β-1,3 linkages, cellobiohydralases (also called exoglucanases) that attack cellulose chains from the non-reducing end and release cellobiose, and β-glucanases that cleave within the main chain of mixed linkage β-glucan. Similarly, numerous other specific enzymes are classified as xylanases. Thus, the specific types of cellulases and xylanases can differ substantially among products, and differences in the relative proportions and activities of these individual enzymes will have an impact on the efficacy of cell wall degradation by these products. In addition to fibre-degrading enzymes, these products also contain secondary enzymes, such as amylases, proteases, and pectinases. The diversity of enzymes within commercially available enzyme preparations is advantageous, because complete digestion of complex ruminant feedstuffs such as hay or grain requires literally dozens of enzymes. Furthermore, a single product can target a wide variety of substrates. However, this diversity presents problems in terms of quality control and comparison of research findings among different preparations.

Production responses to feed enzymes

BEEF CATTLE

Evidence that feed enzymes could improve live-weight gain and feed efficiency in beef cattle was first recorded in a series of feeding trials reported almost forty years ago (Burroughs, Woods, Ewing, Greig, and Theurer, 1960; Nelson and Damon, 1960; Rovics and Ely, 1962), although not all responses to enzyme supplementation were positive (Kercher, 1960; Leatherwood, Mochrie and Thomas, 1960; Perry, Cope and Beeson 1960; Clark, Dyer and Templeton, 1961). Even though these early studies provided valuable information on the potential benefits of enzymes for beef cattle, they did little to address the impact of factors such as the composition of the diet, types and levels of enzymes present, or the method of enzyme application. More recent studies have been designed specifically to address these issues.

In one such study, application of increasing levels (0.25 to 4.0 L/t of dry matter; DM) of an enzyme mixture containing xylanase and cellulase enzymes (Xylanase B, Biovance Technologies Inc., Omaha, NE; and, Spezyme CP®, Genencor, Rochester, NY) increased live-weight gain of steers fed primarily on cubed lucerne hay or timothy hay by 30 and 36%, respectively (Beauchemin, Rode and Sewalt, 1995). However, the effect was non-linear and the optimum dose differed for the two forages. Furthermore, the same enzyme mixture had no effect on live-weight gain when applied to barley silage (Beauchemin *et al.,* 1995). When a similar mixture was applied to a diet containing 95% grain, live-weight was increased by between 6 and 18% and feed efficiency was improved by between 5 and 12%, depending upon level of enzyme supplementation (Table 17.1) (Beauchemin, Jones, Rode and Sewalt, 1997; Iwaasa, Rode, Beauchemin and Eivemark, 1997). Subsequently, this enzyme mixture (Pro-Mote, Biovance Technologies Inc., Omaha, NE; product currently marketed as Promote N.E.T.™ by Agribrands International, St. Louis, MO) was evaluated in a field study at a commercial feedlot using a finishing diet containing a high proportion of grain that was fed to growing heifers (Beauchemin, Rode and Karren, 1999). The diet consisted of rolled barley, supplement and 80 g/kg whole-crop barley silage (DM basis); the cattle were vaccinated, implanted, and melangesterol acetate was provided in the supplement. A dilute enzyme solution was added to the rolled grain as it was augured into the feed wagon. Over the 116-day feeding period, cattle fed enzymes gained 16 kg more and consumed 12.8 kg less feed DM, so feed efficiency increased (Table 17.2).

Table 17.1 Effects of adding a commercial feed enzyme product to high-concentrate diets for feedlot finishing that consist of barley grain, supplement, and barley silage.

	Control	Enzyme level 1 x	Enzyme level 2 x	Change
Study 1 (no ionophore, no implant)[1]				
Initial weight (kg)	407	414	. . .	. . .
Dry matter intake (kg/d)	9.99	9.53	. . .	-5%
Live-weight gain (kg/d)	1.43	1.52	. . .	+6%
Dry matter intake: live-weight gain (kg/kg)	7.11	6.33	. . .	-11%
Study 2 (ionophore and implant)[2]				
Initial weight (kg)	477	477	477	. . .
Dry matter intake (kg/d)	11.1	11.5	11.5	+4%
Live-weight gain (kg/d)	1.70	1.87	2.01	+10 to +18%
Dry matter intake: live-weight gain (kg/kg)	6.50	6.18	5.70	-5 to -12%

[1]From Beauchemin *et al.*, (1997).
[2]From Iwaasa *et al.*, (1997).

Table 17.2 Effects of adding a feed enzyme product to finishing diets in commercial feedlots.[1]

Item	Control	Enzyme	Response
Number of cattle[2]	86	101	
Initial weight (kg)	385	360	-6.5%
Dry matter intake (kg/d)	10.7	10.6	-1%
Live-weight gain (kg)	172	188	+9%
Daily live-weight gain (kg/d)	1.4	1.53	+9%
Dry matter intake: live-weight gain (kg/kg)	7.11	6.33	-11%

[1]From Beauchemin *et al.*, (1999a).
[2] A random sub-sample of cattle were removed from each pen (200 cattle/pen) after 116 days and weighed. Final weight was subtracted from initial weight for each animal to determine live-weight gain.

Another mix of fungal enzyme preparations (Cellulase A and Xylanase B, Finnfeeds International Ltd. Marlborough, UK) was evaluated in a series of experiments with growing cattle. McAllister, Oosting, Popp, Mir, Yanke, Hristov, Treacher and Cheng (1999) treated whole-crop barley silage, which contributed 825 g/kg of the ration DM, with rates up to 5.0 L/t DM. In the first two months, feed intake, live-weight gain and feed efficiency of growing cattle improved as a result of enzyme supplementation, but no effects were observed over the complete 17-week feeding period. Michal, Johnson and Treacher (1996) added this enzyme mixture to a diet consisting of 0.85 lucerne silage and 0.15 rolled barley grain (DM basis). Treatment with enzyme increased feed intake by 9%, but there were only minor effects on live-weight gain. Pritchard, Hunt, Allen and Treacher (1996) also observed increased DM intakes with minor effects on live-weight gain when this enzyme mixture was added to a diet of lucerne silage, chopped grass hay and barley grain (forage to concentrate ratio 70:30; DM basis). The response to using this enzyme mixture in high-grain feedlot finishing diets was also variable. McAllister *et al.*, (1999) reported that treating both the forage (ryegrass silage; 30% of the ration DM) and grain (barley; 70% of the ration DM) portions of the total mixed ration at 3.5 L/t of DM increased live-weight gain by 10%. However, ZoBell, Weidmeier, Olson and Treacher (2000) reported no effect when the enzyme was added to a barley-based feedlot finishing diet containing 17% forage (DM basis).

The registration and commercialisation of feed enzymes for use in beef cattle diets has been fairly slow to date. This reluctance is partially due to the variability in responses obtained in controlled research trials, and partially due to the cost of feed enzymes relative to other technologies such as ionophores, antibiotics, and implants. However, with increased consumer concern over the use of antibiotics in livestock production, the commercial use of enzymes in diets for beef cattle will surely increase, although further research is necessary to ensure consistency of animal response on the farm.

Dairy cattle

There have been a number of recent studies to examine the effects of fibrolytic exogenous enzymes on milk production in dairy cows (Table 17.3). As in studies using beef cattle, production responses by dairy cattle to feed enzymes have been variable. Across 16 studies, the average increase in DM intake was 1.6 ± 4.6 kg/d and the average increase in milk yield was 1.3 ± 1.6 kg/d ($4.2\% \pm 4.9$). Thus, when viewed across a variety of enzyme products and experimental conditions the variability in response is high. While this may be viewed as an indication that feed enzyme supplementation of diets is not a suitable technology for ruminants, we believe that much of the variability can be attributed to factors such as enzyme type, level of supplementation, method of enzyme application, and whether the cows had an energy deficit. It is important to examine individual studies to determine the conditions that are most likely to result in positive responses to enzymes.

Table 17.3 Dry matter intake and milk yield responses to some fibrolytic feed enzyme supplements.

Reference	Enzyme product	Enzyme level[1] (ml or g/kg TMR)	Enzyme application	Diet composition	DMI response (kg/d)	Milk yield response (kg/d)
Beauchemin et al. 1999	Pro-Mote® and pectinase; Biovance Technol, Omaha, NE	2.5	TMR or hulless	barley grain 0.7 barley, BS, LH	-0.7 1.5 (3.8)	0.3 (1%)
Yang et al. 1999	Pro-Mote®; Biovance Technol, Omaha, NE	0.45 1.7 1.7	hay or hay and grain	barley grain, BS, LH	0.3 0.3 0.4	0.9 (4%) 1.9 (8%) 1.6 (7%)
Rode et al. 1999	Pro-Mote®; Biovance Technol, Omaha, NE	1.15	grain	barley grain, LH, CS	0.3	3.6 (10%)
Yang et al. 2000	modified high xylanase, concentrated Pro-Mote®; Biovance Technol, Omaha, NE	0.05 0.05	TMR or grain	barley grain, LH, CS	1.0 0.4	-0.1 (0%) 2.1 (6%)
Beauchemin et al., unpublished	Pro-Mote®; Biovance Technol, Omaha, NE	1	grain	barley grain, LH, CS	1.6 2.0	1.6 (5%) -0.8 (-2%)
Beauchemin et al., unpublished	Pro-Mote® concentrated; Biovance Technol, Omaha, NE	1	grain	maize grain, LH, CS	-0.6 -0.4	1.4 (6%) 3.1 (8%)
Lewis et al. 1999	Cornzyme®; FinnFeeds Int., Marlborough, UK	0.69	forage	barley grain, LS, LH	1.9	1.3 (5%)

Table 17.3 Contd.

Reference	Enzyme product	Enzyme level[1] (ml or g/kg TMR)	Enzyme application	Diet composition	DMI response (kg/d)	Milk yield response (kg/d)
Lewis *et al.* 1999	Cornzyme®; FinnFeeds Int., Marlborough, UK	0.5 1.0 2.0	forage	barley grain, LS, LH	1.8 1.8 2.2	1.2 (3%) 6.3 (16%) 1.6 (4%)
Kung *et al.* 2000	EA; Finn Feeds Int., Marlborough, UK	1 (EA) 2.5 (EA)	forage	maize grain, CS, LH	0.5 -0.2	2.5 (7%) -0.8 (-2%)
Kung *et al.* 2000	EA and EB; FinnFeeds Int., Marlborough, UK	1 (EA) 0.6 (EB)	forage	maizegrain, CS, LH	0.9 0.9	0.7 (2%) 2.5 (8%)
Nussio *et al.* 1997	C/X; Finn Feeds Int., Marlborough, UK	0.3 0.5 0.8	forage	barley grain, LH	1.5 1.9 2.4	-2.1 (-6%) -0.3 (-1%) 0.5 (1%)
Schingo-ethe *et al.* 1999	C/X; Finn Feeds Int., Marlborough, UK	0.39 0.55 0.8	forage	maize grain, CS, LH	0.8 -0.3 1.7	1.1 (4%) 0.9 (4%) 2.7 (11%)
Beauchemin *et al.* 2000	Natugrain, BASF Corp., Ludwigshafen, Germany	1.22 3.67	TMR	barley grain, barley silage, LS	1.5 1.2	-0.5 (-2%) -0.5 (-2%)
Higgin-botham *et al.* 1996	Rumenase, Argri-Science, Liverpool, NY	1.7	TMR	maize and barley grain, CS, LH	ND	≤40 DIM: 0.9 (2%); >40 DIM: 0.2 (0.5%)
Stokes and Zheng 1995	unspecified	0.6	forage	barley grain, LS, LH	2	4.2 (15%)
Zheng and Stokes 1997	unspecified	0.83	forage	unspecified silage, hay, and grain	1	2.1 (7%)

CS = maize silage, DIM = days in milk; LH = lucerne hay, LS = lucerne silage, BS = whole-crop barley silage, ND = not determined, TMR = total mixed ration
[1] Amount of enzyme product added to diet expressed on the basis of the complete TMR, even though the enzyme may have been added to only a component of the diet.

We evaluated the effects of adding an enzyme product (Pro-Mote, Biovance Technologies Inc, Omaha, NE) to diets consisting of either barley or maize grain in a study conducted at two sites (Beauchemin, Rode, Farr, Shelford, Baah and Hartnell, unpublished data). The diet consisted of approximately 0.5 concentrate,0.3 maize silage and 0.20 chopped lucerne hay (DM basis) and was calculated to supply sufficient metabolisable energy and protein for a cow producing 40 kg/d of milk with 3.5% milk fat. When enzymes were added (1 g/kg of the total ration DM) to a maize-based concentrate fed to cows in mid-lactation (Lethbridge site), milk yield increased by 1.4 kg/d (6.3%) and 4% fat-corrected milk increased by 3.1 kg/d (14.5%) (Table 17.4). When the same concentrate was fed to cows in early lactation (Agassiz site), milk yield increased by 3.1 kg/d (7.6%) and 4% fat-corrected milk increased by 2.6 kg/d (8.6%). The increase in milk production was not accompanied by an increase in DM intake at either location.

Table 17.4 Intake, milk production, and digestibility responses to feed enzymes in maize and barley based diets.[1,2]

Item	Lethbridge Site[3]		Agassiz Site[4]	
	Control	Enzyme	Control	Enzyme
Maize Concentrate				
DM intake (kg/d)	22.5	21.9	24.7	24.3
Milk yield (kg/d)	22.1	23.5	40.8	43.9
4% fat-corrected milk (kg/d)	21.4	24.5	30.3	32.9
Milk Composition (g/kg)				
Fat	38.8	43.4	23.3	23.0
Protein	36.1	37.0	33.2	31.6
Barley Concentrate				
DM intake (kg/d)	22.2	23.8	22.4	24.4
Milk yield (kg/d)	29.1	30.7	32.0	31.2
4% fat-corrected milk (kg/d)	27.9	28.8	27.8	27.4
Milk Composition (g/kg)				
Fat	38.9	36.9	32.4	33.0
Protein	34.9	35.3	34.6	35.7

[1] Beauchemin, Rode, Farr, Shelford, Baah and Hartnell, unpublished data.
[2] Enzyme product used was Pro-Mote, Biovance Technologies Inc., Omaha, NE added at 1.0 g/kg dietary DM.
[3] The study was conducted at the Agriculture and Agri-Food Canada Research Centre in Lethbridge, Alberta.
[4] The study was conducted at the University of British Columbia in Agassiz, British Columbia.

However, for diets containing barley grain, the response to enzymes depended upon the site. At the Lethbridge site, milk yield of cows in mid-lactation increased by 1.6 kg/d (5.5%) and fat-corrected milk increased by 0.9 kg/d (3.2%), but there was no response at the Agassiz site. The variability in response to enzymes in this study was probably due to the fact that the energy content of the diet exceeded the requirements for milk production, with the exception of the cows at Agassiz receiving the maize-based diet. Increasing the energy availability of feed using enzymes is most likely to increase milk yield of cows in early lactation due to their negative energy balance.

Several studies have examined the potential benefits of providing feed enzymes to cows in early lactation. In the first study, Rode, Yang and Beauchemin (1999) applied the enzyme (Pro-Mote, Biovance Technologies Inc., Omaha, NE) to the concentrate to provide 1.3 g/kg dietary DM and the diet was formulated to meet the protein and energy requirements of a cow producing 37 kg/d of milk containing 35g/kg fat. The diet contained 0.24 maize silage, 0.15 chopped lucerne hay, and 0.61 concentrate (DM basis). Cows received either the control diet or the enzyme-treated diet for the first 12 weeks of lactation. Cows offered the enzyme-treated diet produced 3.6 kg/d (10%) more milk than cows offered the control diet, yet feed intake was unchanged (Table 17.5). Digestibility of nutrients in the total tract was dramatically increased by enzyme treatment. Concentrations of fat and lactose in milk were significantly lower with enzyme supplementation of the diet, but there was no significant effect on milk protein concentration or yield of milk constituents. Both groups of cows in this study were in negative energy balance. Adding enzymes to the diet increased the digestible energy content of the diet, which lessened the energy deficit and increased milk yield.

Table 17.5 Effects of supplementing diets fed to cows in early lactation with an enzyme mixture.

| Item | Study 1: Rode et al., 1999 | | Study 2: Yang et al., 2000 | | |
	Control	Enzyme in Conc	Control	Enzyme in Conc	Enzyme in TMR
Dry matter intake (kg/d)	18.7	19.0	19.4	19.8	20.4
Milk production (kg/d)	35.9[f]	39.5[g]	35.3[b]	37.4[a]	35.2[b]
Milk composition (g/kg)					
Fat	38.7[a]	33.7[b]	33.4	31.9	31.4
Protein	32.4	30.3	31.8	31.3	31.3
Lactose	47.3[c]	46.2[d]	46.5	46.5	45.6
Live-weight change (kg/d)	-0.63	-0.60	0.15	0.04	0.14
DM digestibility	0.617[a]	0.691[b]	0.639[a]	0.666[b]	0.657[ab]
NDF digestibility	0.425[a]	0.510[b]	0.426	0.443	0.459

[a,b] Means within a study differ ($P < 0.05$),[c,d] Means within a study differ ($P < 0.10$), [f,g] Means within a study differ ($P = 0.11$).

TMR = total mixed ration; Conc = concentrate

In the next study, Yang, Beauchemin and Rode (2000) used a diet similar to that used by Rode *et al.*, (1999), but treated it with a modified version of the enzyme preparation used in the previous lactation study (equivalent to1.5 g/kg DM with respect to xylanase and 0.4 g/kg DM with respect to cellulase). The enzyme was applied either to the concentrate or sprayed daily onto the total mixed ration (TMR). From weeks 3 to 15 of lactation, cows offered the diet with enzyme applied to the concentrate produced 2 kg/d (5.9%) more milk than cows offered the control diet, without a change in the concentrations of milk components (Table 17.5). Total-tract DM digestibility also increased. However, when the enzyme was applied to the TMR, there was no effect on milk production, although total-tract digestibility tended to increase, suggesting that the milk production response to enzymes may depend upon method of application. Similarly, Beauchemin, Yang and Rode (1999) reported that applying enzymes to a TMR prior to feeding did not significantly increase milk production, but increased digestibility of organic matter and fibre in the total tract. In that study, the increase in ADF digestibility over the total tract (60 g/kg) was mainly due to increased post-ruminal digestion.

It is proposed that when an enzyme product is applied to a dry feed a stable enzyme-feed complex is formed which acts as a slow-release mechanism for enzymes in the rumen. Without this stable complex the enzymes dissolve in rumen fluid and flow rapidly from the rumen. In that case, the primary site of action of the enzymes is post-ruminal, which may not result in an improvement in milk production. Rumen fermentation provides the host animal with both energy, in the form of volatile fatty acids (VFA), and amino acids, in the form of protein, whereas post-ruminal digestion can potentially only provide energy. Also, more extensive fermentation in the hindgut may increase the size of the caecum and increased gut size can have a significant impact on energy requirements for maintenance in the ruminant. Consequently, hindgut fermentation may result in no net increase in energy available for productive purposes. Therefore, method of enzyme delivery is crucial for obtaining improvements in ruminal digestibility and milk production.

There are various other studies in the literature that suggest low to moderate levels of some enzyme products can be beneficial in diets for lactating cows, particularly in early lactation (Table 17.1). For example, Schingoethe, Stegeman and Treacher (1999) reported that cows receiving forage (0.6 maize silage and 0.4 alfalfa hay; DM basis) treated with increasing levels (0, 0.39, 0.55, and 0.8 ml/ kg TMR DM) of cellulase and xylanase (FinnFeeds Int., Marlborough, UK) during the first 100 days postpartum produced 9 to 15% more milk and 16 to 23% more energy-corrected milk. However, milk production was not increased for cows that were in mid-lactation when they first received enzyme-treated forage. In other studies (Nussio, Huber, Theurer, Nussio, Santos, Tarazon, Lima-Filho, Riggs, Lamoreaux and Treacher, 1997; Lewis, Sanchez, Hunt, Guy, Pritchard, Swanson and Treacher, 1999; Kung, Treacher, Nauman, Smagala, Endres and Cohen, 2000) in which various fibrolytic enzymes supplied by FinnFeeds Int. (Marlborough, UK) were added to diets for dairy cows, DM intake increased on average by 1.4 kg/d and milk yield increased by 4.6%. Within these studies, Cornzyme® added at 1 mL/kg TMR (DM basis) resulted in the greatest improvement in milk yield (16.3%) (Lewis *et al.*, 1999). With Natugrain® (BASF Corp., Ludwigshafen,Germany), a product marketed for non-ruminants containing high ß-glucanase, cellulase, and xylanase activities, DM intake and digestibility increased, but no response in milk production occurred (Beauchemin, Rode, Maekawa, Morgavi and Kampen, 2000). With Ruminase® (Agri-Science, Liverpool, NY), a product containing cellulase and xylanase, the response was minimal (Higginbotham, dePeters, Berry and Ahmadi, 1996). With other unspecified products, milk production increased by 7 to 15% (Stokes and Zheng, 1995; Zheng and Stokes, 1997).

Learning from animal experiments

When viewed across all enzyme products, responses in animal performance to feed enzymes have been inconsistent. This variability is actually not surprising given the diversity in the formulation of these products. Comparisons among experiments conducted using different enzyme formulations are difficult because enzyme products are poorly defined. Even within particular enzyme products, it is difficult to assess the variability in response because the formulation of individual products often changed over time as new knowledge became available. In addition, some of the variability within enzyme products is due to supplementation with insufficient or excessive enzyme activity. *In vivo* responses to enzyme supplementation are typically non-linear (Beauchemin *et al.*, 1995; Kung *et al.*, 2000), and it is possible to over-supplement. For example, in the study reported in Table 17.4, we also used higher levels of enzyme product (2, 5, and 10 g/kg dietary DM) in addition to the 1.0 g/kg reported (Beauchemin, Rode, Farr, Shelford, Baah and Hartnell, unpublished data). While the responses to enzymes at a low level were generally positive, the response was variable at 2 g/kg, and nonexistent at the higher levels. Similarly, Kung *et al.*, (2000) offered forage (0.6 maize silage and 0.4 lucerne hay; DM basis) treated with two different enzyme products (FinnFeeds Int., Marlborough, UK) to cows, and responses depended upon the enzyme formulation. However, in that study, a high level of enzyme supplementation was actually less effective than a low level.

Method of application also has a major impact on production responses. Enzymes are less effective if infused directly into the rumen than when they are applied directly to feed (Lewis, Hunt, Sanchez, Treacher, Pritchard and Feng,1996; McAllister *et al.*, 1999; Wang, McAllister, Rode, Beauchemin, Morgavi, Nsereko, Iwaasa and Yang, 2001). Furthermore, the response to enzyme supplementation is greater when the enzyme preparation is applied to dry feed (concentrate or hay) rather than fresh forage or silage (Beauchemin *et al.*, 1995; Feng, Hunt, Pritchard, and Julien 1996; Yang *et al.*, 2000). Feng *et al.*, (1996) applied an enzyme solution directly to grass, and observed no effect when added to fresh or wilted grass, but when applied to dried grass, enzymes increased DM and fibre digestion. When we applied a low level of a fibrolytic enzyme preparation onto alfalfa silage prior to feeding, no effects on DM digestibility were observed (Beauchemin and Rode, unpublished data). However, when the enzyme was added to the silage after it had been dehydrated, DM digestibility increased by 2.9%. Silage pH values are generally around the optimum for most fungal enzymes (4 - 5.5) and these amenable pH conditions may be expected to enhance the efficacy of feed enzymes. However, Nsereko, Morgavi, Beauchemin and Rode (2000) showed that silages contained factors that inhibited xylanase activities of feed enzyme additives.

Applying the enzymes to dry feed creates a stable enzyme-feed complex that increases enzyme effectiveness. This stable complex occurs quickly (within hours) and once stabilized onto dry feed, the enzymes are stable and effective for at least several months. Enzymes that are applied to silage or TMR immediately prior to feeding may be released into the rumen and leave quickly before they can be effective. This explains the effects in the large intestine when enzymes were applied daily to a TMR (Beauchemin *et al.*, 1999b) rather than to forage or concentrate (Yang, Beauchemin and Rode, 1999).

Production responses to enzymes can be attributed mainly to improvements in the proportion of fibre digested in the rumen (Yang *et al.*, 1999; Table 17.6) as a result of increased rate of digestion (Feng *et al.*, 1996). This increase in digestibility provides more digestible energy to the animal. Thus, animal responses will be greatest in situations in which fibre digestion is impaired, and when energy is the first-limiting nutrient. In high producing ruminants, such as dairy cows or feedlot cattle, fibre digestion is often impaired, due to low ruminal pH and rapid transit time through the rumen. In fact, the National Research Council (1989) assumes a 4% reduction in digestibility for each multiple increase in intake over maintenance intake (Figure 17.1). It is in this situation that feed enzymes can be beneficial.

This concept is illustrated in a study in which dairy cows and sheep were fed a TMR with and without supplemental enzymes added to the concentrate (Yang *et al.*, 2000). For dairy cows fed the control diet, total-tract digestion was 0.639 for DM and 0.318 for acid detergent fibre. Digestibility was higher for sheep; total tract digestion was 0.771 for DM and 0.498 for acid detergent fibre. Supplementing the diet with an enzyme product improved digestion by dairy cows, but not sheep. This study indicates that feed enzymes improve feed digestion only when the potential digestion of the diet is not attained because digestion is impaired. It is this "loss" in digestible energy that potentially becomes digested with the use of feed enzymes. Thus, existing enzyme technology is not likely to benefit cattle fed at maintenance; the greatest responses will be for cattle fed for maximum productivity.

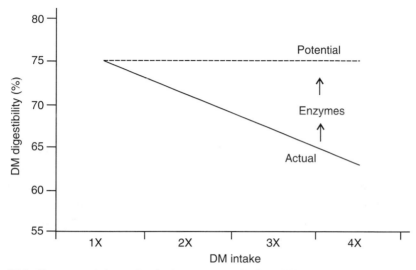

Figure 17.1 The expected depression in dry matter (DM) digestibility with increased intake of cattle above maintenance energy requirements. Use of fibrolytic enzymes recaptures some or all of this loss in digestible dry matter.

Table 17.6 Effects of supplementing a dairy cow diet with low or medium levels of a fibre-degrading enzyme mixture applied to lucerne hay cubes or cubes and concentrate (Yang et al. 1999).

| Item | Control cubes | Enzyme Added to cubes | | Added to cubes and concentrate | SE |
		Low	Medium	Medium	
Dry matter intake (kg/d)	20.4	20.7	20.7	20.8	0.7
Milk production (kg/d)	23.7[b]	24.6[ab]	25.6[a]	25.3[ab]	0.6
Digestibility					
Total tract OM	0.644[b]	0.659[ab]	0.670[a]	0.665[a]	0.007
Ruminal OM (true)	0.541	0.543	0.584	0.572	0.030
Total tract NDF	0.388[b]	0.412[ab]	0.436[a]	0.424[ab]	0.013
Ruminal NDF	0.307	0.349	0.369	0.356	0.048

[a,b] Means in the same row with different superscripts differ ($P < 0.05$).
NDF = neutral detergent fibre; OM = organic matter

It is obvious that many factors influence enzyme efficacy in ruminants. Therefore, an understanding of the mechanisms by which enzymes improve nutrient utilization in ruminants is crucial to obtaining consistent positive responses to enzyme additives over a broad range of diets and animal types.

Modes of action

For non-ruminant animals, alteration of intestinal viscosity is the primary mode by which feed enzymes function to improve animal performance (Bedford, 1995). In ruminants, the mode of action appears to be more complex. Various modes of action for ruminants have been documented. However, not all of them act concurrently and their relative importance will depend on the enzyme product, substrate, method of application, and physiological status of the animal. When enzymes are applied to dry feeds, the primary site of action is the rumen. Enzyme supplementation increases the hydrolytic capacity of the rumen as a result of synergistic effects with the rumen microbial population. The net effect is that the overall ability of the rumen to digest feed is enhanced. Pre-ingestive effects also occur when enzymes are applied to feed, but these effects are relatively minor. Intestinal effects of enzymes have also been reported in ruminants, particularly when enzymes are applied to moist feeds (Beauchemin *et al.*, 1999b; Yang *et al.* 2000) or provided as a liquid supplement (Hristov, McAllister, Treacher and Cheng, 1998). The mode of action of exogenous enzymes in ruminants is complex and continues to be a major focus of the research presently being conducted with these additives.

PRE-INGESTIVE EFFECTS

When fibrolytic enzymes are applied in a liquid form to feeds prior to consumption, they can have a positive effect on animal performance. Therefore, it is not unreasonable to assume that the mode of action is through some form of pre-ingestive attack of the enzymes upon the plant fibre. This would suggest that exogenous enzymes should be more efficacious when applied to high moisture feeds such as silages compared with dry feeds. However, there is evidence that exogenous enzymes are more effective when applied in a liquid form to dry forage as opposed to wet forage (Beauchemin *et al.*, 1995; Feng *et al.*, 1996; Yang *et al.*, 2000). At first this seems improbable, given that the role of water in the hydrolysis of soluble sugars from complex polymers is a fundamental biochemical principle (Lehninger, 1982). However, feed offered to ruminants is seldom absolutely dry; even feeds that nutritionists would describe as "dry" (e.g., grain, hay) contain 60 to 120 g/kg moisture. Release of soluble sugars from these "dry" feeds suggests that their free water content is sufficient to enable hydrolysis.

Release of sugars from feeds arises at least partially from the solubilisation of NDF and ADF (Hristov, Rode, Beauchemin and Weurfel, 1996; Gwayumba and Christensen, 1997). This is consistent with observed increases in the soluble fraction and rate of *in situ* digestion (Feng *et al.*, 1996; Hristov *et al.*, 1996; Yang *et al.*, 1999). However, most studies have not found exogenous enzymes to improve the extent of *in situ* or *in vitro* DM digestion (Feng *et al.*, 1996; Hristov *et al.*, 1996). These results suggest that enzyme additives only degrade substrates that would be naturally digested by the endogenous enzymes of the rumen microflora. Although exogenous enzymes do affect release of soluble carbohydrates, the amount liberated represents only a minute portion of the total carbohydrate present in the diet, and probably has little effect in the rumen.

Applying fibrolytic enzymes to feed prior to feeding enhances ruminal fibre digestion by altering the structure of the feed thereby making it more susceptible to degradation. This was demonstrated in a study by Nsereko, Morgavi, Rode, Beauchemin and McAllister (2000) in which enzyme preparations were applied to lucerne hay which was then autoclaved to denature enzyme activities and washed to remove any products of hydrolysis. *In vitro* NDF digestion was higher at 12 and 48 h for treated hay than non-treated hay and generally this effect was enhanced by a longer pre-incubation time with enzymes. Endo-1,4-ß-xylanase and acetyl xylan esterase activities were positively correlated with NDF digestion, and it was suggested that these activities caused structural changes to the forages which improved digestion.

A pre-ingestive effect does not account for why exogenous enzymes can be effective in improving overall fibre digestion when they are only applied to the concentrate (low fibre) portion of the diet (Yang *et al.*, 1999, 2000). The most important reason for applying enzymes to feed is to provide a slow-release mechanism for enzymes in the rumen. Thus, the greater the proportion of the diet treated with enzymes, the greater the chances that enzymes endure in the rumen. Without this stable feed-enzyme complex, the enzymes are soluble in rumen fluid and flow rapidly from the rumen. The major portion of the positive production responses observed with the use of enzyme supplements is apparently due to post-ingestive effects.

RUMINAL EFFECTS

Direct hydrolysis

Until recently, it was assumed that the proteolytic activity in the rumen ecosystem would rapidly inactivate unprotected enzyme feed additives (Chesson, 1994). This was substantiated by Kopecny, Marounek and Holub (1987) who reported rapid inactivation of a *Trichoderma reesei* cellulase preparation by rumen bacterial proteases. However, more recent studies have shown that feed enzyme additives are generally more stable than previously thought, although there are significant differences depending on the type of enzyme preparation and activity measured (Hristov, McAllister and Cheng, 1998; Morgavi, Beauchemin, Nsereko, Rode, McAllister, Iwaasa, Wang and Yang, 2001). Xylanases tend to be resistant to microbial degradation (Fontes, Hall, Hirst, Hazlewood and Gilbert, 1995), possibly due to their high degree of glycosylation (Gorbacheva and Rodionova, 1977). Glycosidases tend to be less resistant (Morgavi *et al.*, 2001). In addition, the survival of enzymes within the rumen varies with different ruminal conditions in the host animal (Morgavi *et al.*, 2001). However, in practice, feed enzyme additives are administered with the feed and the presence of the feed substrate is a known factor increasing enzyme resistance to proteolytic inactivation (Fontes *et al.*, 1995; Morgavi *et al.*, 2000a).

The fact that exogenous enzymes remain active in the rumen raises the possibility that they improve digestion through the direct hydrolysis of ingested feed within the rumen environment. Researchers have shown that exogenous enzymes can enhance fibre degradation by ruminal microorganisms *in vitro* (Forwood, Sleper and Henning, 1990; Varel, Kreikemeier, Jung and Hatfield, 1993; Hristov *et al.*, 1996; Feng *et al.*, 1996), *in situ* (Feng *et al.,* 1996; Lewis *et al.*, 1996), and *in vivo* (Yang *et al.*, 1999). However, the contribution of added exogenous enzymes to total ruminal activity is small. Based on the average fibrolytic activity normally present in the rumen, exogenous enzymes probably increase cellulase activity by less than 15% (Beauchemin and Rode, 1996), and even this may be an inflated amount depending upon how enzyme activity is measured. Furthermore, increasing the level of enzyme supplementation is usually not beneficial. Given that exogenous enzymes represent only a small fraction of the ruminal enzyme activity, and that the ruminal microbiota is inherently capable of readily digesting fibre (McAllister, Bae, Jones and Cheng, 1994), it is difficult to envision how exogenous enzymes would enhance ruminal fibre digestion through direct hydrolysis. This, of course, is assuming that measured activities, such as xylanase or cellulase, are causing the observed improvement in animal performance. However, if the causative activity is one of the minor activities contained in the enzyme products, it is then likely that enzyme supplementation can significantly increase the overall level in the rumen.

Synergism with ruminal microorganisms

Increased enzyme activity

Morgavi, Beauchemin, Nsereko, Rode, Iwaasa, Yang, McAllister and Wang (2000) demonstrated synergism between exogenous enzymes and ruminal enzymes such that the net combined hydrolytic effect in the rumen was much greater than estimated from the individual activities. By combining exogenous enzymes with ruminal enzymes, hydrolysis increased by up to 35, 40, and 100% in the case of soluble cellulose, corn silage and xylan, respectively. Furthermore, the synergistic effect was greatest at a pH range of 5.0 to 6.0, pH values that are not uncommon in high producing dairy cows or feedlot cattle.

Stimulation of rumen microbial populations

Relatively low levels of enzyme supplementation increase numbers of non-fibrolytic as well as fibrolytic bacteria in the rumen (Nsereko, Rode, Beauchemin, McAllister, Morgavi, Furtado, Wang, Iwaasa and Yang, 1999; Wang et al., 2001). The increase in non-fibrolytic bacterial numbers may be due to an enzyme-induced release of polysaccharides that are utilized by these bacteria. Interestingly, the increase in bacterial numbers disappeared when higher levels of enzyme supplementation were used (Nsereko et al., 1999) which may explain the non-linear dose responses observed in vivo (Beauchemin et al., 1995; Kung et al., 2000). Stimulation of rumen microbial numbers due to the use of enzymes results in greater microbial biomass, which provides more total polysaccharidase activity to digest feedstuffs and greater flow of microbial nitrogen to the small intestine. Although concentration of cellulolytic bacteria in ruminal fluid is not considered to be a limiting factor for animals given high-forage diets (Dehority and Tirabasso, 1998), this may not be the case for cattle consuming large quantities of high quality forage and grain.

Bacterial attachment

Close association between microorganisms and fibre is essential for the digestion of feedstuffs in the rumen and the rate of degradation of fibre *in vivo* depends on the rate at which the adherent cellulolytic microbial population develops (Silva, Wallace and Ørskov, 1987). Exogenous enzymes stimulate the attachment of rumen microbes to plant fibre, which may explain how small quantities of enzymes can have such significant effects on fibre degradation *in vivo*. Treating lucerne hay with exogenous enzymes, prior to ingestion, increased bacterial colonization and *in situ* DM disappearance of forage between 3 h and 24 h of ruminal incubation (Yang *et al.*, 1999). These responses were supported by concurrent increases in fibre digestibility in the rumen and total gastrointestinal tract. Similarly, Wang *et al.*, (2001) reported that supplemental exogenous enzymes increased microbial attachment of ruminal microbes to feed and increased activity of enzymes associated with feed particles. Applying enzymes to feed causes the release of soluble sugars from fibre, thereby increasing the chemotactic attraction and eventual attachment of fibrolytic rumen bacteria to the plant surface. Furthermore, application of enzymes onto feed exposes additional adhesion sites on the feed surface making it more suitable for microbial colonization (Morgavi, Nsereko, Rode, Beauchemin, McAllister and Wang, 2000).

Supplying critical enzyme activities

Limitations to digestion of plant cell walls in the rumen can result from insufficient quantities or types of enzyme production by the ruminal microbes, from an inability of degrading enzyme(s) to interact with the target substrates, or from conditions in the rumen not being optimal for enzyme activity (e.g., low ruminal pH). More than 20 different enzymatic activities have been identified as being involved in the hydrolysis of the structural polysaccharides of the plant cell wall, all of which are produced by a normally functioning ruminal microflora (White, Mackie and Doerner, 1993). Exogenous enzyme preparations may contain enzymatic activities that would normally limit digestion of plant cell walls by ruminal microorganisms. The extent of cross-linking by *p*-coumaryl and feruloyl groups to arabinoxylans has been identified as one factor that limits the digestion of plant cell walls (Hatfield, 1993). *Aspergillus oryzae* has been shown to produce an esterase capable of breaking the ester bridges that cause these cross-linkages (Tenkanen, Schuseil, Puls and Poutanen, 1991). However, the fact that exogenous enzymes usually only increase the rate and not the extent of digestion (Varel *et al.*, 1993; Feng *et al.*, 1996; Hristov *et al.*, 1996) suggests that the activities contributed by exogenous enzymes are not novel to the ruminal environment.

Ruminal pH

Considering the low fibre content of high-concentrate diets, it is surprising that fibrolytic enzymes have improved feed digestion (Krause, Beauchemin, Rode, Farr and Nørgaard, 1998) and performance of cattle given high-cereal diets (Beauchemin *et al.*, 1997; Iwassa *et al.*, 1997). An explanation of this phenomenon may come from comparing the pH optima of the fibrolytic enzymes produced by ruminal microorganisms with the pH optima of commercial fibrolytic enzymes (Figure 17.2). It is well documented that growth of fibrolytic bacteria is inhibited (Russell and Dombrowski, 1980), and that fibre digestion is severely impaired when pH falls below 6.2 (Hoover, Kincaid, Varga, Thayne and Junkins, 1984). Isolated and purified fibrolytic enzymes from pure cultures of rumen bacteria have pH optima above 6.2 (Greve, Labavitch, and Hungate, 1984; Matte and Forsberg, 1992) in contrast to the pH optima of 4.5 to 5.5 for many commercial fibrolytic enzymes (Gashe, 1992).

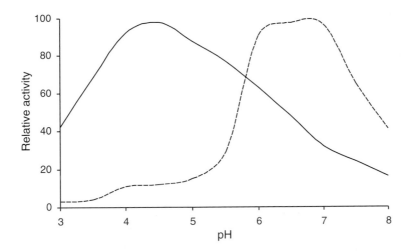

Figure 17.2 Effect of pH on relative xylanase activity from a feed enzyme preparation (———) and rumen fluid (- - - -) (Nsereko, Rode, Beauchemin and Morgavi; unpublished data).

The response to enzymes is highest when pH is low. For example, the extent to which *T. longibrachiatum* enzymes enhanced gas production increased as the pH declined from 6.5 to 5.5 (Morgavi, Beauchemin and Rode, unpublished data). Although a decline in pH from 6.5 to 5.5 reduced DM disappearance from maize silage in mixed ruminal cultures supplemented with *T. longibrachiatum* enzymes, the negative effect of low pH on DM disappearance was more pronounced in the absence of added enzyme (Morgavi, Nsereko, Rode, Beauchemin, McAllister, Wang, Iwaasa and Yang, 2000). Ruminal pH can be below 6.0 for a significant portion of the day in dairy cattle (Beauchemin, 2000) and feedlot cattle (Krause *et al.*, 1998). Under these conditions, exogenous enzymes could make a significant contribution to ruminal fibre digestion.

Post-ruminal effects

Because viscosity of duodenal digesta increases with increasing levels of grain in the diet (Mir, Mears, Mir and Morgan Jones, 1998), enzyme-mediated reductions in viscosity could improve nutrient absorption in the small intestine of cattle fed on grain diets. Reduced intestinal viscosity was associated with 1.2% and 1.5% increases in total tract digestibility of DM when enzymes were applied to the feed or infused into the abomasum, respectively (Hristov *et al.*, 1998a). However, intestinal viscosity in cattle is only between 1 and 2 cPoise (Mir *et al.*, 1998) whereas intestinal viscosity in poultry may exceed 400 cPoise (Bedford, 1995).

Improved growth performance in poultry supplemented with enzymes is often associated with 10-fold reductions in intestinal viscosity (Bedford, 1995). Consequently, it is likely that the relatively modest declines in intestinal viscosity observed in ruminants fed on diets supplemented with enzymes are associated with small or negligible improvements in nutrient absorption in the small intestine.

Conclusions

Over the past decade, there has been a flurry of interest in the potential of using fibre-degrading enzymes in ruminant diets to improve animal performance. Results have been inconsistent, particularly when viewed across all products. However, for certain products, there is sufficient evidence that animal responses are actually quite consistent and predictable. Furthermore, some of the variation can be attributed to under or over-supplementation of enzymes and inappropriate method of application. Most enzyme preparations are currently being used with no attempt to define the types or activities of the enzymes they contain. Such random employment of enzymes on feeds, without consideration for specific substrate targets, will only discourage or delay adoption of exogenous enzymes for ruminants.

Production responses to enzymes can be attributed mainly to improvements in fibre digestion, and ultimately an increase in digestible energy intake. Thus, animal responses will be greatest in situations in which fibre digestion is impaired, and when energy is the first limiting nutrient. There is evidence that exogenous enzymes initiate digestion of feeds prior to consumption and that they improve feed digestion in the rumen and lower digestive tract. The opportunity to improve performance of ruminants using feed enzymes lies in complementing a biological system that already has an active fibrolytic system. In modern feeding scenarios, the environmental conditions within the rumen (e.g. pH, retention time, osmolality) often fall outside the range for maximum feed digestion. This means that new feed enzymes will need to be synergistic with the endogenous rumen system to provide the greatest opportunity for improvement. However, this can be difficult to achieve as it requires an understanding of both the substrate and the underlying endogenous enzyme system itself.

With increasing consumer concern about the use of growth promoters and antibiotics in ruminant production, and the magnitude of increased animal performance obtainable using feed enzymes, there is no doubt that these products will play an increasingly important role in future ruminant production. A more complete understanding of the mode of action of these products will allow development of low-cost enzyme products designed specifically to improve feed digestion by ruminants.

References

Beauchemin, K.A., Rode, L.M. and Sewalt, V.J.H. (1995) Fibrolytic enzymes increase fibre digestibility and growth rate of steers fed dry forages. *Canadian Journal of Animal Science,* **75**, 641-644.

Beauchemin, K.A. and Rode, L.M. (1996) Use of feed enzymes in ruminant nutrition. In *Animal Science Research and Development - Meeting Future Challenges,* pp. 103-131. Edited by L.M. Rode. Minister of Supply and Services Canada, Ottawa, ON.

Beauchemin, K.A., Jones, S.D.M., Rode, L.M. and Sewalt, V.J.H. (1997) Effects of fibrolytic enzyme in corn or barley diets on performance and carcass characteristics of feedlot cattle. *Canadian Journal of Animal Science,* **77**, 645-653.

Beauchemin, K.A., Rode, L.M. and Karren, D. (1999a) Use of feed enzymes in feedlot finishing diets. *Canadian Journal of Animal Science,* **79**, 243-246.

Beauchemin, K.A., Yang, W.Z. and Rode, L.M. (1999b) Effects of grain source and enzyme additive on site and extent of nutrient digestion in dairy cows. *Journal of Dairy Science,* **82**, 378-390.

Beauchemin, K.A. (2000) Managing rumen fermentation in barley-based diets: balance between high production and acidosis. *In Advances in Dairy Technology - Volume 12,* pp109 - 125. Edited by J. Kennelly. University of Alberta, Edmonton, AB.

Beauchemin, K.A., Rode, L.M., Maekawa, M., Morgavi D. and Kampen R. (2000) Evaluation of a non-starch polysaccharidase feed enzyme in dairy cow diets. *Journal of Dairy Science,* **83**, 543-553.

Bedford, M.R. (1995) Mechanisms of action and potential environmental benefits from the use of feed enzymes. *Animal Feed Science and Technology,* **53**, 145-155.

Bedford, M.R. (2000) Exogenous enzymes in monogastric nutrition - their current value and future benefits. *Animal Feed Science and Technology,* **86**, 1-13.

Burroughs, W., Woods, W., Ewing, S.A., Greig, J. and Theurer, B. (1960) Enzyme additions to fattening cattle rations. *Journal of Animal Science,* **19**, 458-464.

Chesson, A. (1994) Manipulation of fibre degradation: an old theme revisited. In *Biotechnology in the feed industry, Proceedings of Alltech's Tenth Annual Symposium,* pp 83-98. Edited by T.P. Lyons and K.A. Jacques, Nottingham University Press, Loughborough, UK.

Clark, J.D., Dyer, I.A. and Templeton, J.A. (1961) Some nutritional and physiological effects of enzymes for fattening cattle, *Journal of Animal Science,* **20**, 928 (Abstract).

Dehority B.A. and Tirabasso, P.A. (1998) Effect of ruminal cellulolytic bacterial concentrations on in situ digestion of forage cellulose. *Journal of Animal Science,* **76**, 2905-2911.

Feng, P., Hunt, C.W., Pritchard, G.T. and Julien, W.E. (1996) Effect of enzyme preparations on in situ and in vitro degradation and in vivo digestive characteristics of mature cool-season grass forage in beef steers. *Journal of Animal Science,* **74**, 1349-1357.

Fontes, C.M.G.A., Hall, J., Hirst, B.H., Hazlewood, G.P. and Gilbert, H.J. (1995) The resistance of cellulases and xylanases to proteolytic inactivation. *Applied Microbial Biotechnology,* **43**, 52-57.

Forwood, J.R., Sleper, D.A. and Henning, J.A. (1990) Topical cellulase application effects on tall fescue digestibility. *Agronomy Journal,* **82**, 900-913.

Gashe, B.A. (1992) Cellulase production and activity by *Trichoderma* sp. A-001. *Journal of Applied Bacteriology,* **73**, 79-82.

Gorbacheva, I. and Rodionova, N.A. (1977) Studies on xylan degrading enzymes. I. Purification and characterization of endo-1,4-b-xylanase from *Aspergillus niger* str. 14. *Biochemistry, Biophysica Acta,* **484**, 79-93.

Greve, L.C., Labavitch, J.M. and Hungate, R.E. (1984) α-L-Arabinofuranosidase from *Ruminococcus albus* 8: Purification and possible role in the hydrolysis of alfalfa cell wall. *Applied and Environmental Microbiology,* **47**, 1135-1140.

Gwayumba, W. and Christensen, D.A. (1997) The effect of fibrolytic enzymes on protein and carbohydrate degradation fractions in forages. *Canadian Journal of Animal Science,* **77**, 541-542.

Hatfield, R.D. (1993) Cell wall polysaccharide interactions and degradability. Phenolic constituents of plant cell walls and wall biodegradability. In *Forage Cell Wall Structure and Digestibility*, pp. 285-313. Edited by H.G. Jung, D.R. Buxton, R.D. Hatfield and J. Ralph. American Society of Agronomy, Crop Science Society of America, Soil Science Society of America, Madison, Wisconsin.

Higginbotham, G.E., dePetters, E.J., Berry S.L. and Ahmadi, A. (1996) Effect of adding a cell wall degrading enzyme to a total mixed ration for lactating dairy cows. *The Professional Animal Scientist,* **12**, 81-85.

Hoover, W.H., Kincaid, C.R., Varga, G.A., Thayne, W.V. and Junkins, L.L. Jr. (1984) Effects of solids and liquid flows of fermentation in continuous cultures. IV. pH and dilution rates. *Journal of Animal Science,* **58**, 692-699.

Hristov, A.N., Rode, L.M., Beauchemin, K.A. and Wuerfel, R.L. (1996) Effect of a commercial enzyme preparation on barley silage in vitro and in sacco dry matter degradability. *Proceedings, Western Section, American Society of Animal Science*, Rapid City, South Dakota, **47**, 282-284.

Hristov, A.N., McAllister, T.A. and Cheng, K.-J. (1998a) Stability of exogenous polysaccharide-degrading enzyme in the rumen. *Animal Feed Science and Technology*, **76**, 161-168.

Hristov, A.N., McAllister, T.A., Treacher, R.J. and Cheng, K.-J. (1998b) Effect of dietary or abomasal supplementation of exogenous polysaccharide-degrading enzyme supplementation on rumen fermentation and nutrient digestibility. *Journal of Animal Science*, **76**, 3146-3156.

Iwaasa, A.D., Rode, L.M., Beauchemin, K.A. and Eivemark, S. (1997) Effect of fibrolytic enzymes in barley-based diets on performance of feedlot cattle and in vitro gas production. *Evolution of the Rumen Microbial Ecosystem, Joint Rowett Research Institute - Institute National de Recherche Agronomique Rumen Microbiology Symposium*, Aberdeen, Scotland, March 20 & 21 (Poster 39).

Kercher, J. (1960) Value of diallylstilbesterol and Zymo-pabst enzyme preparation for fattening yearling steers. *Journal of Animal Science*, **19**, 966 (Abstract).

Kopecny, J., Marounek, M. and Holub, K. (1987) Testing the suitability of the addition of *Trichoderma viride* cellulases to feed rations for ruminants. *Zivocisna Vyroba*, **32**, 587-592.

Krause, M., Beauchemin, K.A., Rode, L.M., Farr, B.I. and Nørgaard, P. (1998) Fibrolytic enzyme treatment of barley grain and source of forage in high grain diets fed to growing cattle. *Journal of Animal Science*, **96**, 1010-1015.

Kung, L. Jr., Treacher, R.J., Nauman, G.A, Smagala, A.M., Endres K.M. and Cohen, M.A. (2000) The effect of treating forages with fibrolytic enzymes on its nutritive value and lactation performance of dairy cows. *Journal of Dairy Science*, **83**,115-122.

Leatherwood, J.M., Mochrie, R.D. and Thomas, W.E. (1960) Some effects of a supplementary cellulase preparation on feed utilization by ruminants. *Journal of Dairy Science*, **43**, 1460-1464.

Lehninger, A.L. (1982) *Principles of Biochemistry*, Worth Publishers, Inc., New York, NY.

Lewis, G.E., Sanchez, W.K., Hunt, C.W., Guy M.A., Pritchard, G.T., Swanson, B.I. and Treacher, R.J. (1999) Effect of direct-fed fibrolytic enzymes on the lactational performance of dairy cows. *Journal of Dairy Science*, **82**, 611-617.

Lewis, G.E., Hunt, C.W., Sanchez, W.K., Treacher, R., Pritchard, G.T. and Feng, P. (1996) Effect of direct-fed fibrolytic enzymes on the digestive characteristics of a forage-based diet fed to beef steers. *Journal of Animal Science,* **74**, 3020-3028.

Matte, A. and Forsberg, C.W. (1992) Purification, characterization, and mode of action of endoxylanases 1 and 2 from *Fibrobacter succinogenes* S85. *Applied and Environmental Microbiology,* **58**, 157-168.

McAllister, T.A., Bae, H.D., Jones, G.A. and Cheng, K.-J. (1994) Microbial attachment and feed digestion in the rumen. *Journal of Animal Science,* **72**, 3004-3018.

McAllister, T.A., Oosting, S.J., Popp, J.D., Mir, Z., Yanke, L.J., Hristov, A.N., Treacher, R.J. and Cheng, K.-J. (1999) Effect of exogenous enzymes on digestibility of barley silage and growth performance of feedlot cattle. *Canadian Journal of Animal Science,* **79**, 353-360.

Michal, J.J, Johnson, K.A. and Treacher, R.J. (1996) The impact of direct fed fibrolytic enzymes on the growth rate and feed efficiency of growing beef steers and heifers. *Journal of Animal Science,* **74**, 296 (Abstract).

Mir, P.S., Mears, G.J., Mir, Z. and Morgan Jones, S.D. (1998) Effects of increasing dietary grain on viscosity of duodenal digesta and plasma hormone and glucose concentrations in steers. *Journal of Animal Science,* **76**, Suppl. 1, 247 (Abstract).

Morgavi, D.P., Beauchemin, K.A., Nsereko, V.L., Rode, L.M., Iwaasa, A.D., Yang, W.Z., McAllister, T.A and Wang, Y. (2000a) Synergy between the ruminal fibrolytic enzymes and enzymes from *Trichoderma longibrachiatum. Journal of Dairy Science,* **83**, 1310-1321.

Morgavi, D.P., Nsereko, V.L., Rode, L.M., Beauchemin, K.A., McAllister,T.A. and Wang Y. (2000c) A *trichoderma* feed enzyme preparation enhances adhesion of *Fibrobacter succinogenes* to complex substrates but not to pure cellulose. Proceedings of the *XV Conference on Rumen Function,* Chicago, IL. Page 31 (Abstract.).

Morgavi, D.P., Nsereko,V.L., Rode, L.M, Beauchemin, K.A., McAllister, T.A., Wang,Y. Iwaasa, A.D. and Yang, W.Z. (2000c) Feed enzymes increase the potential of rumen microorganisms to degrade fibrous feeds at low pH. *3rd European Symposium on Feed Enzymes,* Netherlands.

Morgavi, D. P., Beauchemin, K. A., Nsereko, V. L., Rode, L. M., McAllister, T. A, Iwaasa, A. D., Wang, Y. and Yang, W. Z. (20001) Resistance of feed enzymes to proteolytic inactivation by rumen microorganisms and gastrointestinal proteases. *Journal of Animal Science.*

National Research Council (1989) *Nutrient Requirements of Dairy Cattle, 6th revised edition.* National Academy of Science, Washington, DC.

Nelson, F. and Damon, V. (1960) Comparison of different supplemental enzymes with and without diethylstilbestrol for fattening steers. *Journal of Animal Science,* **19**, 1279 (Abstract).

Nsereko, V.L., Rode, L.M., Beauchemin, K.A., McAllister, T.A., Morgavi, D.P., Furtado, A., Wang, Y., Iwaasa, A.D. and Yang, W.Z. (1999) Effects of feeding a fungal enzyme preparation from *Trichoderma longibrachiatum* on the ruminal microbial population of dairy cows. *IX International Symposium on Rumen Physiology*, Pretoria, October 17-22.

Nsereko, V.L., Morgavi, D.P., Beauchemin, K.A. and Rode L.M. 2000. Inhibition of ruminant feed enzyme polysaccharidase activities by extracts from silages. *Canadian Journal of Animal Science*, **80**, 523-526.

Nsereko, V.L., Morgavi, D.P., Rode, L.M., Beauchemin, K.A. and McAllister, T.A.. (2000b) Effects of fungal enzyme preparations on hydrolysis and subsequent degradation of alfalfa hay fibre by mixed rumen microorganisms in vitro. *Animal Feed Science and Technology*, **88**, 153-170.

Nussio, L.G., Huber, J.T., Theurer, C.B., Nussio, C.B., Santos, J., Tarazon, M., Lima-Filho, R.O., Riggs, B., Lamoreaux, M. and Treacher, R.J. (1997) Influence of a cellulase/xylanase complex (C/X) on lactational performance of dairy cows fed alfalfa hay based diets. *Journal of Dairy Science,* 80, Suppl. 1, 220 (Abstract).

Pendleton, B. (1998) The regulatory environment. In *1998-1999 Direct-Fed Microbial, Enzyme and Forage Additive Compendium, Volume 4*, pp. 47-52. Edited by S. Muirhead. The Miller Publishing Company, Minnetonka, Minnesota,

Perry, T.W., Cope, D.D. and Beeson, W.M. (1960) Low vs high moisture shelled corn with and without enzymes and stilbestrol for fattening steers. *Journal of Animal Science*, **19**, 1284 (Abstract).

Pritchard, G., Hunt, C., Allen, A. and Treacher, R. (1996) Effect of direct-fed fibrolytic enzymes on digestion and growth performance in beef cattle. *Journal of Animal Science,* **74**, Suppl. 1, 296 (Abstract).

Rode, L.M., Yang, W.Z. and Beauchemin, K.A. (1999) Fibrolytic enzyme supplements for dairy cows in early lactation. *Journal of Dairy Science,* 82: 2121-2126.

Rovics, J.J. and Ely, C.M. (1962) Response of beef cattle to enzyme supplement, *Journal of Animal Science*, **21**, 1012 (Abstract).

Russell, J.B. and Dombrowski, D.B. (1980) Effect of pH on the efficiency of growth by pure cultures of rumen bacteria in continuous culture. *Applied and Environmental Microbiology*, **39**, 606-610.

Schingoethe, D.J., Stegeman, G.A. and Treacher R.J. (1999) Response of lactating dairy cows to a cellulase and xylanase enzyme mixture applied to forages at the time of feeding. *Journal of Dairy Science*, **82**, 996-1003.

Silva, A.T., Wallace, R.J. and Ørskov, E.R. (1987) Use of particle bound microbial enzyme activity to predict the rate and extent of fibre degradation in the rumen. *British Journal of Nutrition*, **47**, 407-415.

Stokes, M.R. and Zheng, S. (1995) The use of carbohydrase enzymes as feed additives for early lactation cows. *Proceedings of the 23rd Conference on Rumen Function*, Chicago, Illinois, p. 34 (Abstract).

Tenkanen, M., Schuseil, J., Puls, J. and Poutanen, K. (1991) Production, purification, and characterization of an esterase liberating phenolic acids from lignocellulosics. *Journal of Biotechnology*, **18**, 69-84.

Varel, V.H., Kreikemeier, K.K., Jung, H.G. and Hatfield, R.D. (1993) In vitro stimulation of forage fibre degradation by ruminal microorganisms with *Aspergillus oryzae* fermentation extract. *Applied and Environmental Microbiology*, **59**, 3171-3176.

Wang, Y., McAllister, T.A., Rode, L.M., Beauchemin, K.A, Morgavi, D.P., Nsereko, V.L., Iwaasa, A.D. and Yang, W. (2001) Effects of an exogenous enzyme preparation on microbial protein synthesis, enzyme activity and attachment to feed in the Rumen Simulation Technique (Rusitec). *British Journal of Nutrition*, **85**, 325-332.

White, B.A., Mackie, R.I. and Doerner, K.C. (1993) Enzymatic hydrolysis of forage cell walls. In *Forage Cell Wall Structure and Digestibility*, pp. 455-484. Edited by H. G. Jung, D. R. Buxton, R. D. Hatfield and J. Ralph. American Society of Agronomy, Crop Science Society of America, Soil Science Society of America, Madison, Wisconsin.

Yang, W.Z., Beauchemin, K.A, and Rode, L.M. (1999) Effects of enzyme feed additives on extent of digestion and milk production of lactating dairy cows. *Journal of Dairy Science*, **82**: 391-403.

Yang, W.Z., Beauchemin, K.A. and Rode, L.M. (2000) A comparison of methods of adding fibrolytic enzymes to lactating cow diets. *Journal of Dairy Science*, **83**, 2512-2520.

Zheng, S. and Stokes, M.R. (1997) Effects of fibrolytic enzymes on feed stability anf performance of lactating cows. *Journal of Dairy Science*, Suppl. 1, 278 (Abstract).

ZoBell, D.R, Weidmeier, R.D., Olson, K.C., and Treacher, R. (2000) The effect of an exogenous enzyme treatment on production and carcass characteristics of growing and finishing steers. *Animal Feed Science and Technology*, **87**, 279-285.

LIST OF PARTICIPANTS

The thirty-fifth University of Nottingham Feed Conference was organised by the following committee:

Mr N.J. Chandler *(National Renderers Association)*
Mr R. Duran *(Nutreco)*
Mr D. Filmer *(FLOCKMAN)*
Mr P.W. Garland *(BOCM PAULS Ltd.)*
Mr M. Lister *(Computer Applications Ltd)*
Dr S.A. Papasolomontos *(Kego S.A.)*
Mr M. Partridge *(Pen Mill Feeds)*
Mr J.R. Pickford
Dr I.H. Pike *(IFOMA)*
Mr J. Twigge *(Trouw Nutrition)*
Dr P. Wilcock *(ABN Ltd)*
Dr K.N. Boorman
Prof P.J. Buttery
Dr J.M. Dawson
Dr P.C. Garnsworthy *(Secretary)* } *University of Nottingham*
Prof G.E. Lamming
Dr A.M. Salter
Dr J. Wiseman *(Chairman)*

The conference was held at the University of Nottingham Sutton Bonington Campus, 3-5 January 2001. The following persons registered for the meeting.

Adams, Dr C A	Kemin Europa N.V., Industriezone Wolfstee, 2200 Herentals, Belgium
Allder, Mr M J	Eurotec Nutrition Ltd, Unit 24 Lancaster Way Business Park, Ely, Cambridge CB6 3NW, UK
Allegaert, Miss L	INVE Technologies NV, Oeverstraat 7, 9200 Baasrode , Belgium
Atherton, Dr D	Thompson & Joseph Ltd, 119 Plumstead Rd, Norwich , UK
Ball, Mr J A	Roche Products Ltd, Heanorgate, Heanor, Derbys DE75 75G, UK
Bargeman, Mr G	Trouw Nutrition UK, Wincham, Northwich, Cheshire CW9 6DF
Bartram, Dr C	ABN Ltd, ABN House, Oundle Road Peterborough, UK
Bauman, Prof D	Cornell University, Dept of Animal Science, 262 Morrison Hall Ithaca, NY 14853, USA
Baxendale, Miss L	Sun Valley Foods Ltd, Grandstand Road, Hereford HR4 9PB, UK
Baynes, Dr P	Nutec, Eastern Ave, Lichfield, Staffs WS13 7SE, UK
Beardsworth, Dr P M	Roche Products Ltd, Heanor Gate, Heanor, Derbys, UK

323

Beauchemin, Dr K	Agriculture and Agri-Food Canada, Research Centre, 5403 1st Ave South, P O Box 3000, Lethbridge Alberta, T1K 4B1, Canada
Beaumont, Mr D	Pancosma (UK) Ltd, Crompton Rd Industrial Est, Ilkeston , UK
Beckerton, Miss C	University of Nottingham, Sutton Bonington Campus, Loughborough, Leics LE12 5RD, UK
Bedford, Dr M R	Finnfeeds, 10 Box 777, Marlborough, Wilts SN8 1XN, UK
Beer, Dr J	Gamebird Consultancy, 127 Church Rd, Salisbury SP1 1RB, UK
Beerlandt, Miss G	INVE Technologies NV, Oeverstraat 7, 9200 Baasrode , Belgium
Berni, Mr N	Heygate & Sons Ltd, Bugbrooke Mills, Northampton NN7 3QH, UK
Best, Mr P	Feed International, 18 Chapel St, Petersfield, Hampshire GU32 3DZ, UK
Birchall, Mr T	Trevor Birchall Agriculture, The Old Yarn Mills, Sherborne, Dorset, UK
Bishop, Ms R	Spillers, Old Wolverton Road, Old Wolverton, Milton Keynes MK12 5PZ, UK
Blake, Dr J	Consultant, Highfield, Little London, Andover, Hants, UK
Bone, Mr P	Thompson & Joseph Ltd, 119 Plumstead Rd, Norwich , UK
Boorman, Dr K N	University of Nottingham, Sutton Bonington Campus, Loughborough, Leics LE12 5RD, UK
Boswerger, Mr B	ABCTA u.a., Postbus gl, 7240AB Lochen , The Netherlands
Boyd, Dr J	BOCM Pauls, P O Box 39, 47 Key St, Ipswich IP4 1BX, UK
Brackenbury, Miss J	Nutec, Eastern Ave, Lichfield, Staffs WS13 7SE, UK
Brealey, Mr R	Intervet UK Ltd, Walton Manor, Walton, Milton Keynes MK7 7AJ, UK
Brooks, Prof P	Seale-Hayne Faculty, University of Plymouth, Newton Abbot TQ12 6NQ, UK
Broom, Mr L	University of Leeds, School of Biology, Leeds LS2 9JT, UK
Brown, Mr G	Roche Products Ltd, Heanorgate, Heanor, Derbys DE75 7SG, UK
Bruce, Dr D W	John Thompson & Sons Ltd, 35-39 York Road, Belfast BT15 3GW, Northern Ireland
Buckle, Mr M	BMS Computer Solutions Ltd, Sproughton House, Sproughton, Ipswich IP8 3AW, UK
Bulbrook, Mrs E	F H Nash Ltd, Four Elms Mills, Bardfield, Saling, Braintree, Essex CN7 5ES, UK

Burt, Dr A W A	23 Stow Road, Kimbolton Huntingdon, PE28 OHU
Buttery, Prof P J	University of Nottingham, Sutton Bonington Campus, Loughborough, Leics LE12 5RD, UK
Campani, Dr I	F.lli Martini & C. S.p.A., Via Emilia 2614, Longiano, Forli, Italy
Carmichael, Mr D	Elanco Animal Health, Kingsclere Rd, Basingstoke, Hants RG21 6XA, UK
Chandler, Mr N J	National Renderers Assoc Inc, 52 Packhorse Road, Gerrards Cross, Bucks SL9 8EF, UK
Chapman, Mr N	United Fish Industries, Gilbey Rd, Grimsby, N E Lincs DN31 2SL, UK
Chappell, Mr T	Computer Applications Ltd, Rivington House, Drumhead Road, Chorley Lancs PR6 7BX, UK
Charles, Dr D R	DC R&D Ltd, 62 Main Street, Willoughby, Loughborough, UK
Chaudhry, Dr A S	University of Newcastle, Dept of Agriculture, KGV1 Building, Newcastle upon Tyne NE1 7RU, UK
Cheetham, Mr J	Nottinghamshire County Council, Trading Standards Service, 4th Floor, County Hall, West Bridgford Nottingham, NG2 7QP, UK
Clarke, Mr N	Britphos Ltd, Rawdon House, Green Lane, Yeadon, Leeds LS19 7BY, UK
Close, Dr W	Close Consultancy, 129 Barkham Rd, Wokingham, Berkshire RG41 2RD, UK
Coates, Mr G	Bugico SA, 16 rue de Bugnons, CH - 1217 Meyrin , Switzerland
Cocker, Miss M J	Nutec, Eastern Ave, Lichfield, Staffs WS13 7SE, UK
Cole, Dr D J A	Nottingham Nutrition International, 14 Potters Lane, East Leake, Loughborough Leics LE12 6NQ, UK
Cole, Mr J	International Additives, Old Gunary Lane, Wallasey, Merseyside CH44 4AH, UK
Collyer, Mr M F	Kemin (UK) Ltd, Becor House, Green Lane, Lincoln LN6 7DL, UK
Cooke, Dr B	1 Jenkins Orchard, Wick St Lawrence, Weston-Super-Mare, N Somerset BS22 7YP, UK
Coope, Mr R	Pen Mill Feeds Ltd, Babylon View, Pen Mill Trading Estate, Yeovil, Somerset BA21 5HR, UK
Cox, Mr N	SCA Nutrition Ltd, Maple Mill, Dalton Airfield Ind Est, Thirsk N Yorks, YO7 3HE, UK
Creasey, Ms A	BASF, Blenheim House, Blenheim Rd, Ashbourne, Derbyshire DE6 1HA, UK

Cullin, Mr A W

Forum Products, 41-51 Brighton Road, Redhill, Surrey RH1 6YS, UK

Curtis, Miss M

Farmers Weekly, Quadrant House, The Quadrant, Sutton Surrey SM2 5AS, UK

Daniel, Miss Z

University of Nottingham, Sutton Bonington Campus, Loughborough, Leics LE12 5RD, UK

Davies, Ms C

Computer Applications, Rivington House, Drumhead Road, Chorley Lancs PR6 7BX, UK

Davis, Dr R

Bernard Matthews, Great Witchingham Hall, Norfolk NR9 5QD, UK

Dean, Mr J

Agil Ltd, Hercules 2, Calleva Park, Aldermaston RG7 8DN, UK

Dickerson, Mr C A

Eurotec Nutrition Ltd, Unit 24 Lancaster Way Business Park, Ely, Cambridge CB6 3NW, UK

Dixon, Mr D

Brown & Gillmer Ltd, P O Box 3154, Florence Lodge, 199 Strand Rd, Merrion Dublin 4, Ireland

Donovan, Mr E

Connally's Red Mills, Goresbridge, Co Kilkenny , Ireland

Doorenbos, Mr J

Hendrix Ltd, Postbus 1, 5830 M A Boxmeer , The Netherlands

Doran, Mr B

Trouw Nutrition, Wincham, Northwich, Cheshire CW9 6DF, UK

Drakley, Miss C

University of Nottingham, Sutton Bonington Campus, Loughborough, Leics LE12 5RD, UK

Dugan, Miss K

University of Nottingham, Sutton Bonington Campus, Loughborough, Leics LE12 5RD, UK

Duran, Mr R

Trouw Nutrition Espana, Ronda de Poniente 9, 28760 Tres Cantos, Madrid, Spain

Ellis, Mr S

Queens University of Belfast, Agricultural & Environmental Science, Newforge Lane, Belfast BT9 5PX, Northern Ireland

Eskinazi, Mr S

International Additives, Old Gurney Lane, Wallasey, Merseyside CH44 4AH, UK

Evans, Dr J

Trouw Nutrition, Wincham, Northwich, Cheshire CW9 6DF, UK

Ewing, Dr W

Context, 53 Mill St, Packington, Leics LE65 1WN, UK

Fahey, Miss A

University of Nottingham, Sutton Bonington Campus, Loughborough, Leics LE12 5RD, UK

Farley, Mr R

Trouw Nutrition, Wincham, Northwich, Cheshire CW9 6DF, UK

Feedt, Mr R

Felleskjopet Rogaland Agder, P O Box 208, N-4001 Stavanger, Norway

Filmer, Mr D

FLOCKMAN, Wascelyn, Brent Street, Brent Knoll, Somerset

Fitt, Dr T	Roche Products Ltd, Heanorgate, Heanor, Derbys DE75 7SG, UK
Fjermedal, Mr A	Fiska Molle AS, Sjolystvn 8, 4610 KR Sands , Norway
Flint, Prof A P F	University of Nottingham, Sutton Bonington Campus, Loughborough, Leics, LE12 5RD
Fullarton, Mr P J	Forum Products Ltd, 41-51 Brighton Road, Redhill, Surrey RH1 6YS, UK
Gair, Mrs D	Forum Products Ltd, 41-51 Brighton Road, Redhill, Surrey RH1 6YS, UK
Garland, Mr P	BOCM PAULS Ltd, 47 Key Street, Ipswich IP4 1BX, UK
Garnsworthy, Dr P C	University of Nottingham, Sutton Bonington Campus, Loughborough, Leics LE12 5RD, UK
Garwes, Dr D	MAFF, 704a, 1a Page St, London SW1P 4PQ, UK
Gibson, Mr J E	Parnutt Foods Ltd, Hadley Road, Woodbridge Ind Est, Sleaford, Lincs NG34 7EG, UK
Gillespie, Miss F	United Molasses, Gibraltar House, Crown Square, 1st Avenue, Burton on Trent, Staffs DE14 2WE, UK
Golds, Mrs S P	University of Nottingham, Sutton Bonington Campus, Loughborough Leics LE12 5RD, UK
Gould, Mrs M	Volac International Ltd, Volac House, Orwell, Royston Herts, UK
Graham, Dr H	Finfeeds International, P O Box 777, Marlborough, Wilts SN8 1XN, UK
Graham, Mr M	Premier Nutrition Products Ltd, The Levels, Rugeley Staffs, WS15 1RD, UK
Gray, Mr W	Kemira Chemicals UK Ltd, Orm House, 2 Hookstone Park, Harrogate HG2 7DB, UK
Green, Mrs K A	Fishmeal Information Network, 4 The Forum, Minerva Business Park, Peterborough, Cambs PE2 6FT, UK
Hall, Miss C L	Countrywide Farmers Ltd, Bradford Road, Melksham, Wiltshire SN12 8LQ, UK
Harland, Dr J	Harlandhall, Bridge Cottage, Castle Easton, Swindon SN6 6JZ, UK
Hazzledine, Mr M	Premier Nutrition Products Ltd, The Levels, Rugeley, Staffs WS15 1RD, UK
Henderson, Dr A	BOCM PAULS Ltd, P O Box 39, 47 Key St, Ipswich IP4 1BX, UK
Higginbotham, Dr J D	United Molasses, Gibraltar House, Crown Square, 1st Avenue, Burton on Trent Staffs DE14 2WE, UK
Hillman, Dr K	SAC, Ferguson Building, Craibstone, Aberdeen AB21 9YA, UK

Holder, Dr P

SVG Intermol, Alexandra House, Regent Rd, Bootle, Liverpool, UK

Hook, Mr S

NMR Software Solutions, Pen Hill House, Pen Hill, Chippenham, Wiltshire SN15 1DN, UK

Hughes, Mr G D

Dynamic Nutrition Services Ltd, Hollyeat Farm, South Brentnor, Tavistock, Devon PL19 ONW, UK

Ilsley, Miss S

University of Leeds, School of Biology, Leeds LS2 9JT, UK

Ingham, Mr R W

The Homestead, Troutbeck Bridge, Windermere LA23 1HF, UK

Jackson, Mr J

Nutec, Eastern Ave, Lichfoeld, Staffs WS13 7SE, UK

Jagger, Dr S

ABN Limited, ABN House, P O Box 250, Oundle Rd Peterborough, PE2 9QF, UK

Johnson-Ord, Mrs S

Kemira Chemicals UK Ltd, Orm House, 2 Hookstone Park, Harrogate, N Yorks HG2 7DB, UK

Johnston, Mr R D

BOCM Pauls Ltd, 47 Key Street, Ipswich IP4 1BX, UK

Jones, Mr H

Heygate & Sons Ltd, Bugbrooke Mills, Northampton NN7 3QU, UK

Jordan, Mr G

Regal Processors, 2 Silverwood Ind Est, Craigavon, Co Armagh, Northern Ireland

Kamel, Dr C

AXISS France S.A.S, Site d'Archamps, Immeuble ABC10Allee A 74160 Archamps, France

Kemp, Ms C

Ross Breeders Limited, Newbridge, Midlothian EH28 8SZ, UK

Kenyon, Mr B

Cherry Valley Farms Ltd, Divisional Offices, Caistor Rd, North Kelsey Moor, Market Rasen, Lincoln LN7 6HH, UK

Kenyon, Mr P

Harbro Farm Sales Ltd, Markethill, Turriff, Aberdeenshire AB53 4PA, UK

Kitchen, Dr D I

Kemin UK Ltd, Becor House, Green Lane Lincoln LN6 7DL, UK

Klasing, Dr K

University of California, Dept of Animal Science, Davis CA 95616, USA

Knock, Mr W

Food Standards Agency, Animal Feed Division, Ergon House, P O Box 31037 17 Smith Square, London SW1P 3WG, UK

Knowles, Mr S

Kemira Chemicals (UK) Ltd, 2 Hookstone Park, Harrogate HG2 7DB, UK

Korsager, Mr H

United Fish Industries, Greenwell Place, Aberdeen AB12 3AY, UK

Krober, Dr K

Bausenbergweg 7, D-56651 Niederzissen , Germany

Kung, Dr L

University of Delaware, Dept of Animal & Food Sciences, 531 South College Ave - TNS 33 Newark DE19717-13-3, USA

Lamming, Prof G E	University of Nottingham, Sutton Bonington Campus, Loughborough, Leics LE12 5RD, UK
Lampkin, Dr N	Organic Centre Wales, Institute of Rural Studies, University of Wales Aberystwyth SY23 3AL, UK
Le Floc'h, Dr N	UMRVP - INRA, Domaine de la Prise, 35590 Saint Gilles , France
Lee, Mrs G	University of Nottingham, Sutton Bonington Campus, Loughborough, Leics LE12 5RD, UK
Lima, Mr S	Felleskjopet Rogaland Agder, Postboks 208, N-4001 Stavanger , Norway
Lock, Mr A	University of Nottingham, Sutton Bonington Campus, Loughborough, Leics LE12 5RD, UK
Lowe, Dr J	Tuttons Hill Nutrition, Tuttons Hill Cottage, Southdown, Charle, I.O.W PO38 2LJ, UK
Lugsden, Miss K	F H Nash Ltd, Four Elms Mills, Bardfield, Saling, Braintree, Essex CN7 5ES, UK
Macdonald, Mr P	Intervet UK Ltd, Walton Manor, Walton, Milton Keynes, Bucks MK7 7AJ, UK
Mackinson, Mr I	Premier Nutrition Products Ltd, The Levels, Rugeley, Staffs WS15 1RD, UK
MacLeod, Miss C A	Feedmark, St Cross, Harleston, Norfolk IP20 ONY, UK
Mafo, Mr A	Hi Peak Feeds, Proctors (Bakewell) Ltd, High Peak Feeds Mill, Sheffield Rd Killamarsh, Derbys, SZ1 1ED, UK
Malandra, Dr F	Agribrands Europe, Italia S. oa, Via Pavia 4, 27010 Spessa Pavia, Italy
Mann, Mrs S	W C P N, Freeby Lane, Waltham-on-the-Wolds, Melton Mowbray, Leics LE14 4RT, UK
Marsden, Dr M	ABN Ltd, ABN House, P O Box 250, Oundle Rd, Peterborough, UK
Marsh, Mr S P	Harper Adams University College, Newport, Shropshire TF10 8NB, UK
McAdoo, Mr S	Regal Processors, 2 Silverwood Ind Est, Craigavon Co Armagh, Northern Ireland
McCarthy, Mr P	Volac Feeds, Killeshandra, Co Cavan , Ireland
McCracken, Dr K J	The Queens University of Belfast, Agricultural & Environmental Science, Dept of Agric. & Rural Development, Newforge Lane, Belfast BT9 5PX, Northern Ireland
McGee, Dr M	IAWS Group plc, 151 Thomas St, Dublin 8, Ireland
McGrane, Mr M	Trouw Nutrition, 36 Ship Street, Belfast BT15 1JL, Northern Ireland

McTiffin, Mr P J	McT Nutrition, Yew Tree Cottage, off The Green, Badgeworth Cheltenham GL51 4UL, UK
Meakin, Mr S	Elanco Animal Health, Kingsclere Road, Basingstoke, Hants RG21 6XA, UK
Meijer, Dr J	Nutreco, Swine Research Centre, P O Box 240, 5830AE Boxmeer, Holland
Mellor, Ms S	Feed Mix/Feed Tech, Elsevier International, Hanzestraat 1 7006 RH Doetinchem, The Netherlands
Metcalf, Dr J A	Borregaard UK Ltd, Clayton Road, Risley, Warrington WA3 6QQ, UK
Metz, Dr S	Provimi R & T Centre, Lenneke Marelaan 2, B 1932 Sint-Stevens-Wolowe, Belgium
Miller, Dr H	University of Leeds, School of Biology, Leeds LS2 9JT, UK
Mills, Mr C	University of Nottingham, Sutton Bonington Campus, Loughborough, Leics LE12 5RD, UK
Millward, Mr J	Royal Pharmaceutical Society, Animal Medicines Inspectorate, NAC, Stoneleigh Park, Warks CV8 2LZ, UK
Mounsey, Mr A	HGM Publications, HGM House, Nether End, Baslow, Bakewell, Derbyshire DE45 1SR, UK
Mounsey, Mr S	HGM Publications, HGM House, Nether End, Baslow, Bakewell, Derbyshire DE45 1SR, UK
Mulder, Mr K	Nutreco RRC, P O Box 220, 5830 AE Boxmeer , The Netherlands
Newbold, Dr C J	Rowett Research Institute, Bucksburn, Aberdeen, UK
Newbold, Dr J R	Provimi Research and Tech. Centre, Lenneke Marelaan 2, B 1932 Sint-Stevens-Woluwe , Belgium
Newcombe, Mrs J	University of Nottingham, Sutton Bonington Campus, Loughborough, Leics LE12 5RD, UK
Nixey, Dr C	British United Turkeys Ltd, Hockenhull Hall, Tarvin, Chester CH3 8LE, UK
Nottage, Mr C	Fishmeal Information Network, 4 The Forum, Minerva Business Park, Peterborough, Cambs PE2 6FT, UK
Offer, Dr N	SAC Auchincruive, Ayr, KA6 5HW, UK
Overend, Dr M	Nutec Ltd, Eastern Ave, Lichfield, Staffs WS13 7SE, UK
Owers, Dr M	BOCM Pauls Ltd, P O Box 39, 47 Key St, Ipswich IP4 1BX, UK
Papasolomontos, Dr S A	Kego S A, 1st km Artaki-Psahna Rd, GR - 34600, N Artaki , Greece
Partridge, Mr M	Pen Mill Feeds Ltd, Babylon View, Pen Trading Estate, Yeovil, Somerset BA21 5HR, UK

Pass, Mr R T	U.D.V, Spirit Supply Centre, Carsebridge Road, Alloa, Clackmannanshire FK10 3BB, UK
Peyraud, Dr J-L	Dairy Production Research Unit, UMR INRA-ENSAR, 35590 St Gilles, France
Phillips, Mr G	Silo Guard Europe, Greenway Farm, Charlton Kings, Cheltenham GGL52 6PL, UK
Pickard, Miss J	University of Nottingham, Sutton Bonington Campus, Loughborough, Leics LE12 5 RD, UK
Pickford, Mr J R	Bocking Hall, Bocking Church St, Braintree, Essex CM7 5JY, UK
Pike, Dr I H	IFOMA, 2 College Yard, St Albans AL3 4PA, UK
Piva, Prof G	ISAN - Facolta di Agraria, Via E Parmense, 84, 29100 Piacenza , Italy
Powell, Mr P	Agil Ltd, Hercules 2, Caleva Park, Aldermaston RG7 8DN, UK
Pritchard, Mr S	Premier Nutrition Products Ltd, The Levels, Rugeley, Staffs WS15 1RD, UK
Probert, Miss L D	Finfeeds International, P O Box 777, Marlborough, Wilts SN8 1XN, UK
Putnam, Mr M	Consultant, 61 Hempstead Lane, Potten End, Birkhamsted, Herts HP4 2RZ, UK
Ratcliffe, Mrs J	Frank Wright, Blenheim House, Blenheim Rd, Ashbourne Derby,
Reece, Mr L	Braes Feed Research, Old Gorsey Lane, Wallasey CH44 4AH, UK
Reeve, Dr A	C & H Nutrition, Alexander House, Crown Gate, Runcorn WA7 2UP, UK
Richards, Dr S	Nutec Ltd, Eastern Ave, Lichfield, Staffs WS13 7SE, UK
Robinson, Mr S	Nottingham University Press, Manor Farm, Thrumpton NG11 0AX, UK
Rodiek, Dr A	California State University, 2415 E San Ramon Ave, M/S AS75 Fresno, CA 93740-8033, USA
Roele, Mr R	The Homestead, Troutbeck Bridge, Windermere LA23 1HF, UK
Rogers, Mr M	Volac International, Volac House, Orwell, Royston Herts, UK
Rosen, Dr G D	Consultant, 66 Bath Gate Road, London SW19 5PH, UK
Rowney, Mr S	Sun Valley Foods Ltd, Tram Inn, Allensmore, Hereford HR2 9AW, UK
Russell, Ms S	The Cottage, 9 New Road, Heage, Derbys DE56 2BA, UK
Rutjes, Mr G	ORFFA Nederland Feed BV, Burgstraat 12, 4283 GG Giessen , The Netherlands

Rymer, Dr C   ADAS Nutritional Sciences R.U., Alcester Rd, Stratford-on-Avon CV37 9RQ, UK

Salter, Dr A M   University of Nottingham, Sutton Bonington Campus, Loughborough, Leics LE12 5RD, UK

Sazili, Mr A   University of Nottingham, Sutton Bonington Campus, Loughborough, Leics LE12 5RD, UK

Schons, Mr D   Format International Ltd, Format House, Poole Road, Woking, Surrey GU21 104, UK

Scragg, Mr R   Optivite Ltd, Main St, Laneham, Retford Notts   DN22 ONA, UK

Shepperson, Dr N   Nutec Ltd, Eastern Ave, Lichfield, Staffs WS13 7SE, UK

Shorrock, Dr C   FSL Bells, Hartham, Corsham, Wilts SN13 OQB, UK

Short, Dr F J   ADAS, Meden Vale, Mansfield, Nottinghamshire NG20 9PF, UK

Silley, Dr P   Don Whitley Scientific Ltd, 14 Otley Road, Shipley, West Yorkshire BD17 7SE, UK

Simmins, Dr P H   Finfeeds International Ltd, P O Box 777, Marlborough SN8 1XN, UK

Slade, Mr R   University of Leeds, School of Biology, Leeds LS2 9JT, UK

Slee, Mr A   University of Nottingham, Sutton Bonington Campus, Loughborough, Leics LE12 5RD, UK

Slevin, Mrs J   University of Nottingham, Sutton Bonington Campus, Loughborough, Leics LE12 5RD, UK

Smiley, Mr J   North Eastern Farmers, Rose Hall, Turriff, Aberdeenshire AB53 4PT, UK

Smith, Dr A   Nutec, Eastern Ave, Lichfield, Staffs WS13 7SE, UK

Steinbock, Mr M   Forum Products Ltd, 41-51 Brighton Rd, Redhill RH1 6YS, UK

Sylvester, Mr D   Alpharma, 1 Bishop Street, Mansfield, Notts NG18 1HJ, UK

Tan, Dr H M   Kemin Industries (Asia) Pte Ltd, No 12 Senoko Drive, Singapore 758200 , Asia

Taylor, Dr A J   Roche Products Ltd, Heanor Gate, Heanor, Derbys , UK

Taylor, Mr D   Pancosma (UK) Ltd, Crompton Road Ind Estate, Ilkeston , UK

Taylor, Dr S J   Wine Tavern, Stratford on Scaney Co Wicklow, Ireland

Teo, Dr A   Kemin Industries (Asia) Pte Ltd, No 12 Senoko Drive, Singapore 758200 , Asia

Thomas, Prof P   Artilus Ltd, 33 Cherry Tree Park, Balerno, Midlothian EH14 5AJ,

Thompson, Miss J	University of Nottingham, Sutton Bonington Campus, Loughborough, Leics LE12 5RD, UK
Toplis, Mr P	Primary Diets Ltd, Melmerby Ind Estate, Melmerby, Ripon HG4 5HP, UK
Torrance, Mrs L	BMS Computer Solutions Ltd, Sproughton House, Sproughton IP8 3AW, UK
Tucker, Dr L	Finnfeeds Int Ltd, P O Box 777, Marlborough SN8, UK
Twigge, Dr J	Trouw Nutrition, Wincham, Northwich, Cheshire CW9 6DF, UK
Uprichard, Mr J	Trouw Nutrition, 36 Ship Street, Belfast BT15 1YL, Northern Ireland
Van Cauwenberge, Mrs S	Ajinomoto Eurolysine, 153 rue de Courcelles, 75817 Paris Cedex 17 , France
Van der Aar, Mr P J	De Schothorst, P O Box 533, 8200AM Lelystad , The Netherlands
Van der Ploeg, Mr H	"de Molenaar", Stationsweg 4, Maarsson, The Netherlands
Van Hoecke, Mr P	EBS R & D Centre, Havenstaaat 84, B 1800 Vilvoorde , Belgium
Varley, Dr M	SCA Nutrition Ltd, Maple Mill, Dalton Airfield Ind Est, Thirsk, N Yorkshire YO7 3HE, UK
Vernon, Dr B	BOCM Pauls Ltd, P O Box 39, 47 Key St, Ipswich IP4 1BX, UK
Vik, Mr K-J	Fiska Molle AS, Damsoardsveien 106, 5058 Bergen, Norway
Webb, Prof R	University of Nottingham, Sutton Bonington Campus, Loughborough, Leics LE12 5RD, UK
Weber, Dr G	F Hoffman-la Roche Ltd, Vitamins Division, Industry Unit Feed, CH 4070 Basel, Switzerland,
Whalley, Mrs L	Trouw UK, Wincham, Northwich, Cheshire CW9 6DF, UK
White, Miss A	Nutec, Eastern Ave, Lichfield, Staffs WS15 7SE, UK
Wilcock, Dr P	Trouw Nutrition, Wincham, Northwich, Cheshire CW9 6DF, UK
Williams, Mr P G	Azko Nobel Surface Chemicals Ltd, St Albans, Herts , UK
Williams, Mr S	Computer Applications, Rivington House, Drumhead Rd, Chorley, Lancs PR6 7BX, UK
Wilson, Dr S	BOCM Pauls Ltd, P O Box 39, 47 Key St, Ipswich IP4 1BX, UK
Wiseman, Dr J	University of Nottingham, Sutton Bonington Campus, Loughborough, Leics LE12 5RD, UK
Wynn, Mrs F	Trouw Nutrition, Wincham, Northwich, Cheshire CW9 6DF, UK
Wynn, Mr R	University of Nottingham, Sutton Bonington Campus, Loughborough, Leics LE12 5RD, UK

Yeo, Dr G   Premier Nutrition Products Ltd, The Levels, Rugeley, Staffs WS15
                                    1RD, UK

Zwart, Mr S   Tessenderlo Chemie, P O Box 133, NL-3130 AC Vlaardingen,
                                    The Netherlands

INDEX

POSTERS

The following posters were presented during the meeting. For details, contact the individual authors.

Conjugated Linoleic Acids (CLA): 1 Background Information
A.L. Lock and P.C. Garnsworthy
University of Nottingham, Division of Agriculture and Horticulture, Sutton Bonington Campus, Loughborough, Leics. LE12 5RD

Conjugated Linoleic Acids (CLA): 2 Seasonal changes in the CLA content of milk in the UK
A.L. Lock and P.C. Garnsworthy
University of Nottingham, Division of Agriculture and Horticulture, Sutton Bonington Campus, Loughborough, Leics. LE12 5RD

Conjugated Linoleic Acids (CLA): 3 Individual effects of linoleic and linolenic acids on CLA production in dairy cows
A.L. Lock and P.C. Garnsworthy
University of Nottingham, Division of Agriculture and Horticulture, Sutton Bonington Campus, Loughborough, Leics. LE12 5RD

Diet composition affects apparent energy value of wheat and the response to xylanase in broiler diets
K.J. McCracken [1] and M.R. Bedford [2]
[1] Agricultural and Environmental Science, Department of Agriculture and Rural Development, N. Ireland and The Queen's University of Belfast, Newforge Lane, Belfast,BT9 5PX, [2] Finnfeeds Ltd., Marlborough, Wiltshire SN8 1XN.

Interactions between copper sulphate and in-feed xylanase in diets for broilers.
K.J. McCracken[1], M.R. Bedford[2] and L. Marron[1].
[1] Agricultural and Environmental Science Division, Department of Agriculture and Rural Development and The Queen's University of Belfast, Newforge Lane, Belfast BT9 5PX [2] Finnfeeds Ltd, Marlborough, Wiltshire. U.K.

Nutritional value to broilers of wheat of low specific weight
K.J.McCracken[1] and J.M.McNab[2]
[1] Agricultural and Environmental Science Division, Department of Agriculture for Northern Ireland and The Queen's University of Belfast, Newforge Lane, Belfast BT9 5PX, N. Ireland [2] Roslin Institute (Edinburgh), Roslin, Midlothian EH25 9PS, Scotland

Variety differences and impact of IBIR rye translocation on ileal digestibility of nutrients in broiler diets containing high levels of wheat
K.J.McCracken
Agricultural and Environmental Science Division, Department of Agriculture and Rural Development and The Queen's University of Belfast, Newforge Lane, Belfast BT9 5PX, N. Ireland

Postmortem proteolysis in muscle & meat quality - variable activity of the calpain proteolytic system
P.L. Sensky, T. Parr, R.G. Bardsley & P.J. Buttery
University of Nottingham, Division of Nutritional Biochemistry, Sutton Bonington Campus, Loughborough, Leics. LE12 5RD

Dietary tannins acting as anthelmintic agents?
N.L. Butter, J.M. Dawson, D. Wakelin & P.J. Buttery
University of Nottingham, Division of Nutritional Biochemistry, Sutton Bonington Campus, Loughborough, Leics. LE12 5RD